Geophysical Monograph Series

American Geophysical Union

Geophysical Monograph Series

A. F. SPILHAUS, JR., *managing editor*

Flow and Fracture of Rocks

series:
American Geophysical Union,
geophysical monograph 16

Flow and Fracture of Rocks

II

H. C. HEARD
I. Y. BORG
N. L. CARTER
C. B. RALEIGH
editors

American Geophysical Union
Washington, D. C.
1972

*Published with the aid of a grant
from the Charles F. Kettering Foundation*

International Standard Book Number 87590-016-X

Copyright © 1972 by the American Geophysical Union
1707 L Street, N. W.
Washington, D. C.   20036

Library of Congress Catalog Card Number 72-91609

WILLIAM BYRD PRESS, RICHMOND, VIRGINIA

Dedication

With great appreciation, this volume is dedicated to Professor David T. Griggs on the occasion of his 60th birthday by his colleagues and former students.

David Tressel Griggs was born October 6, 1911, in Columbus, Ohio. His parents were Robert Fiske and Laura Amelia Tressel Griggs. His father was a widely known professor of botany and a leading ecologist and environmental conservationist at a time when these viewpoints were less familiar than they are today. David was an undergraduate at Ohio State University in 1930 when he participated in a National Geographic Society expedition, led by his father, to the Valley of Ten Thousand Smokes. This Alaskan experience and the encouragement he received from a gifted and enthusiastic teacher, Professor Edmund M. Spieker, led him to choose for his life work the application of physics to the problems of the earth. After a year of graduate studies in geology at Ohio State, David moved on in 1933 to Harvard, where a new program of high-pressure studies devoted to geophysical problems had just been initiated under the inspired guidance of Percy W. Bridgman, pre-eminent leader in the experimental exploration of the physics of very high pressures and in the philosophical analysis of the logical processes of scientific thinking.

Young Griggs shared something of Bridgman's flair for the invention and the manipulation of laboratory apparatus, and with this 'gift for gadgets,' as he calls it, he was soon successfully applying Bridgman's techniques and designing new ones for investigating the physical properties of rocks and minerals. He felt strongly that it was essential as background for his laboratory studies that he maintain close contact with field problems of structural geology, and consequently he spent part of nearly every summer in geologic field work in Montana, Utah, and Nevada. David Griggs was a member of the Society of Fellows at Harvard from 1934 until he left there in 1941. During this period he published eleven papers that are all still rated as important contributions. With the appearance of these papers the experimental deformation of rocks was established as an important new subdiscipline of earth science. His presentation of laboratory results was accompanied by studies of the theory of rock behavior under extreme conditions. One of these papers, 'A Theory of Mountain-Building,' published in 1939, presented a graphic model of how convection currents in the earth's mantle could conceivably account for major deformation of the earth's crust and the building of mountain ranges at the borders between continents and oceans. It matters not that the exact form of this model can now be seen as an oversimplification and that the processes illustrated by it can at best be only part of the total story; the fact remains that Griggs's model of convection currents and the origin of mountains has been in the minds of earth scientists and has influenced geological thinking for more than thirty years. It stands today as a remarkably prescient if rough approximation of what probably is one of the important processes of major deformation of the earth.

As it did to the careers of many others, World War II drastically interrupted the career of David Griggs. Before Pearl Harbor he realized something of what lay ahead for this country, and he left Harvard in 1941 to become a research associate at the Radiation Laboratory at the Massachusetts Institute of Technology, where the exciting new tool of radar was under active development. This assignment led quickly to his appointment soon after the United States entered the war as Special Assistant for Scientific Matters to the Secretary of War. In this capacity he played a foremost part in the application of new scientific and technological principles to military problems and in liaison between scientists and military men, as is described in some detail by Dr. Ivan A. Getting in the accompanying comments on Griggs's contributions to military science and technology during and after the war. Following the war Griggs had some of the same difficulties of readjustment suffered by many others. He became a section chief in the 'Rand' project at the Douglas Aircraft Company and took a leading part in establishing the independent successor organization, The RAND Corporation. All his classified work done during and after the war remains unpublished.

In this difficult period of post-war readjustment, Dr. Eleanora Bliss Knopf played a most important part in helping him to decide to return to the field of high-pressure research, in which he had been so productive before the war.

In 1948 Griggs joined the Institute of Geophysics at the University of California at Los Angeles as Professor of Geophysics. There he re-established his now much improved high-pressure apparatus and resumed his studies of the behavior of rocks under high pressures. Since then a steady flow of scientific papers, many of them authored jointly with one or more of his students and former students and with a UCLA colleague, Professor John M. Christie, have come from his laboratory. With Francis J. Turner and graduate students at UCLA and Berkeley he undertook an exhaustive and rewarding study of the mechanism of deformation of the Yule marble and other carbonate rocks. His investigations were soon extended to quartz and other common rock-making minerals, and he became interested in the aspects of seismology that involve the strength and failure of rocks. In 1956 Griggs and George C. Kennedy developed an ingenious 'simple squeezer' device that readily attains high pressures in the laboratory; this device soon became standard equipment in high-pressure laboratories everywhere. Griggs was a leader in the development of the concept that very small amounts of water as impurities in quartz and other silicates play an amazingly major role in the type of deformation undergone by these materials.

Professor Griggs's achievements have brought him many honors. He is a member of the National Academy of Sciences and the American Academy of Arts and Sciences. He has been president of the Section of Tectonophysics of the American Geophysical Union and a member of the Council of the Geological Society of America. He was one of a group of four who in 1969 received the Interdisciplinary Award of the Intersociety Committee on Rock Mechanics 'in recognition of an outstanding interdisciplinary paper in rock mechanics.' This paper was 'The Denver Earthquakes,' published in *Science* in 1968. In 1970 he was a recipient of the Centennial Achievement Award granted to alumni of the Ohio State University who have made outstanding contributions to the advancement of their professions and society. In the same year he also received the Walter H. Bucher Medal of the American Geophysical Union 'for original contributions to the basic knowledge of the earth's crust.' More important, perhaps, and more enduring than any of these honors is the fact that a whole school of enthusiastic young scientists trained by Griggs or at one time closely associated with him now occupy many of the leading academic and governmental positions in this country in the field of experimental deformation of rocks.

W. W. RUBEY

Dave Griggs is an activist. Gifted with an imaginative mind and a superb memory, believing in the fundamental goodness of his country, and living in an age of technological growth and international tension and strife, he reacted creatively to make great contributions to military science and technology. Oblivious of personal danger and clear in his objectives, he had a singleness of purpose that helped to achieve the necessary ends.

Dave loves the mountains, as a scientist and as a sportsman. During one of his geology trips an unrelated automobile accident left Dave facing the amputation of his legs. With characteristic determination he prevailed against odds and saved his legs. This incident was important in that it affected Dave's entry to military science and technology. Since he was denied skiing and mountain climbing, Dave purchased a small plane and learned to fly over his mountains. And so, when in 1940 the Radiation Laboratory was established at MIT, in the absence of availability of military aircraft Dave and his plane were hired at $10 an hour as a target in the development of the first automatic tracking radar, which was intended for ground antiaircraft fire control. Dave recognized the applicability of these principles to air-to-air combat and soon became the program manager of the prototype of all interceptor radars, the AGL-1. Dave's desire to 'get out where the action is' was soon satisfied when, as Special Assistant to the Secretary of War, he went to join General James Doolittle, the Commander of the 8th Air Force. The 8th Air Force had just received microwave radar equipment for blind bombing. The introduction of this equipment was at first unsuccessful. Radar was a mystery to operational pilots. It was technically complex, and it appeared to pilots flying combat missions over Germany as another complication in an already dangerous mission. In Jimmy Doolittle's words, 'Dave did the only thing that could have straightened out the equipment. He became one of the boys; he flew on combat missions and demonstrated the use of the equipment under combat conditions. Dave took some flak in the leg, and I had to take him off of flying. Dave had an unusual capability of dealing in an understandable way with operational people and when he left the operators carried on in the image established by Dave.'

Dave then went to the 15th Air Force under General Nate Twining, where, as the General stated, 'he did the same thing, changing from no success to full success.' While he was still in the European Theatre, Dave established the concept of critical courier in which he provided a personal communication link between the theatre commanders, such as Doolittle and Twining, and the leaders back home, Secretary Stimpson and General Arnold. As the war moved to the Pacific, he applied the same concept of trouble shooter and direct tie-line communicator as a member of General MacArthur's staff.

Dave's post-war contributions are fully known only to himself. Although his principal interests have been in support of the Air Force, he has contributed also to the work of the Atomic Energy Commission and to the Army. In 1947 he was instrumental in setting up the RAND Corporation and became the first head of its Physics Department. As a member of the Air Force Scientific Board and as Chief Scientist of the Air Force (1952), he labored for a better understanding of the effects of nuclear weapons, the development of the hydrogen bomb, and the establishment of underground testing. The early fifties were characterized by the great debate as to whether the United States should develop thermonuclear weapons. The details of this debate are voluminous, but it is clear that Dave Griggs as Chief Scientist of the Air Force was able to project the official Air Force position in support of thermonuclear weapon development and the need for establishing a second AEC weapons laboratory, the Lawrence Radiation Laboratory at Livermore, California. In the words of Edward Teller, who too was an advocate for the development of thermonuclear weapons and the establishment of a second lab, 'Why, Dave even convinced me that I should leave Chicago and go to Livermore.'

Dave's scientific work in geology brought him to considerations of large earth motions that might result from underground nuclear explosions. Although the so-called Plowshare underground nuclear explosion experiment was principally for scientific and economic studies, Dave wrote with Edward Teller the classic paper advocating underground testing of military nuclear weapons. Fortunately, the basic work was finished in time to meet the ban on atmospheric testing.

Dave Griggs with his colleagues at the Rand Corporation led in the understanding of the effects of nuclear weapons. In the words of Bill McMillan, 'When it became apparent in 1961 that the Soviet Union was about to abrogate the nuclear test moratorium that had existed since 1958, Professor Griggs helped organize the Defense Atomic Support Agency Scientific Advisory Group on Effects. Over the subsequent half-dozen years he contributed immeasurably to guiding the national nuclear test program along lines best suited to provide essential information for strengthening the U.S. deterrent.'

Dave commenced his contributions to military science and technology in 1940. A quarter century later he donned fatigues to help the fighting soldier in the national commitment in Vietnam. While a strong policy supporter of scientific effort in Southeast Asia, Dave himself made three extended trips to the theatre. He helped design an organizational structure to provide competent scientific advice to the component commands, with an information focal point at MACV Headquarters. Again he was in the forefront of military technology: during the 1968 Tet offensive he and an associate developed on-the-spot software that markedly improved the performance of the newly introduced sensor equipment system in the interdiction of enemy supply lines.

As many as have been Dave's personal technical contributions to military science and technology, perhaps his greatest contribution came from recognizing critical problem areas and calling on his prodigious acquaintance with other scientists whom he marshalled in support. Throughout he has been motivated by a strong feeling of patriotism and devotion to his country. Perhaps sometimes he has been abrasive—but so are diamonds.

I. A. GETTING

Contents

Experimental Folding of Rocks under Confining Pressure: Buckling of Single-Layer Rock Beams

JOHN HANDIN, MELVIN FRIEDMAN, JOHN M. LOGAN, LINDA J. PATTISON, AND HENRI S. SWOLFS

Center for Tectonophysics, Texas A&M University
College Station, Texas 77843

Experimental folding of single-layer beams of limestone, sandstone, and brass (for comparison) has been achieved under moderate confining pressures (0.3–2.5 kb) in a screw-driven rock deformation apparatus that accomodates the long specimens (20 cm) needed for the buckling instability. The deflection of 3.6 cm, the greatest allowed by the free space in the triaxial pressure cell, was reached in a beam of limestone 0.6 × 3.2 × 20 cm under 1-kb confining pressure. The associated values are 7.6% for axial shortening, 28° for limb dip, and 7% for maximum compressive fiber strain. The half wavelength is about 15 cm. For all the materials, thin beams with aspect ratios of about ≥30 are clearly geometrically unstable. Their axial stress-strain curves show sharp peaks that signal elastic buckling, and their critical buckling stresses are well predicted by the Euler formula for an unsupported beam fixed at both ends. Thick beams of sandstone do not fold; they fail first by shear fracturing near the ends. Thick beams of limestone with aspect ratios as low as 4.4 do fold, but their stress-strain curves rise monotonically and show no evidence of a geometric instability. Once elastic buckling has occurred and the breaking strength of the brittle sandstone or the yield stresses of the ductile limestone and the brass have been attained, physical instabilities that largely control the ultimate shapes of the folds arise. Plastic flow in the brass and cataclastic flow in the sandstone are localized at the hinges, the limbs rotate rigidly, and chevron folds develop. The folds in ductile limestone, however, are very different. A central anticline and two adjacent synclines form, and their profiles are closely approximated by cosine curves. The petrofabric data on surface strains and macrofracturing, as well as calcite deformation in the limestone, are qualitatively consistent with the superposition of bending stresses on the axial compressive prestress. However, the secondary principal strain axes and the compression-extension axes derived from analyses of calcite twin lamellas are both definitely skewed to the dip direction on the limbs of the fold. There are three arrays of fractures that have commonly observed natural counterparts.

The problem of the folding of ordinarily brittle rocks has intrigued and challenged structural geologists since the beginning of our science, and the phenomenon of folding is of utmost practical importance to the mineral industries. Yet, despite the extensive literature on mathematical and physical model studies of the folding of layered media [*Biot*, 1961, 1964, 1965a, d; *Biot et al.*, 1961; *Chapple*, 1968a, b, 1969, 1970a, b; *Spang and Chapple*, 1970; *Sherwin and Chapple*, 1968; *Currie et al.*, 1962; *Bell and Currie*, 1964; *Dieterich*, 1969; *Dieterich and Carter*, 1969; *Dieterich and Onat*, 1969; *Ghosh*, 1968; *Johnson*, 1969; *Price*, 1967; *Ramberg*, 1961a, b, 1963a, b, 1964, 1967; *Weiss*, 1969] and on the kinematics of natural folds, as summarized, e.g., by *de Sitter* [1964], *Goguel* [1962], *Whitten* [1966], and *Ramsay* [1967], the mechanics of folding is still not understood well enough.

Our purpose is to acquire a better understanding of what *Donath and Parker* [1964] have called 'flexural slip' and 'flexural flow' folding, which are characteristic of the shallow crust where confining pressures and temperatures are relatively low and for which a pre-existing planar anisotropy plays a significant role. This first paper describes the results of preliminary experiments on single-layer rock beams. Experiments are now being done on more geologically meaningful multilayered specimens,

1

and all data are being integrated with the results of concurrent theoretical studies by George M. Sowers and field studies by David W. Stearns.

Kinking of laminated rocks has been obtained in the laboratory repeatedly [*Borg and Handin*, 1966; *Donath*, 1968; *Paterson and Weiss*, 1966, 1968; *Williams and Means*, 1971]. However, we are unaware of any previous work on single-layer rock beams that fold without kinking [*Handin and Logan*, 1970].

APPARATUS

Press. The rock deformation apparatus consists essentially of a screw-driven press that axially loads a specimen confined in a triaxial compression cell (Figure 1).

The 100-ton Saginaw ball-bearing screw is driven through a gear train by a 0.5-kw 10:1 variable-speed direct-current gear motor. At the maximum motor speed of 1750 rpm the rate of translation of the screw can be set at 2×10^{-2}, 2×10^{-4}, and 2×10^{-6} cm/sec by changing the positions of a single idler gear. At the minimum motor speed these rates are reduced by a factor of 10, so that in a specimen 20 cm in length the axial strain rates can be varied from 10^{-3} to 10^{-8} sec^{-1}.

To eliminate the need for a thrust bearing,

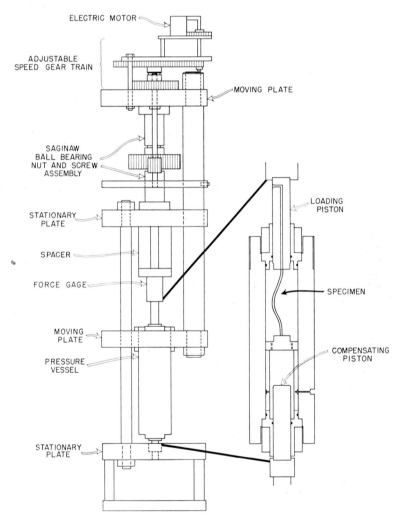

Fig. 1.   Schematic diagram of the rock deformation apparatus. For scale, the outside diameter of the pressure cell is 20 cm.

one half of the screw is right-hand threaded, and the other half is left-hand threaded. Since this arrangement doubles the translation per revolution, it reduces the mismatch between the main spur gear on the screw shaft and the two pinions that drive it. The maximum travel length is 10 cm.

The diameter of the loading piston in the pressure cell is 5.7 cm, and the maximum confining pressure is 3 kb, so that the force due to the confining pressure alone can reach 80 tons. To reduce the size and cost of the Saginaw screw, and incidentally to maintain constant pressure, this force is isolated from the screw by yoking the upper (loading) piston to the lower (compensating) piston of the same diameter, much as *Griggs* [1936] did many years ago. The pistons are supported by the upper and lower fixed plates that are joined by four tie rods. The pressure cell is carried by the lower moving plate that is tied to the upper moving plate by two rods. The screw is located between upper fixed and moving plates. As it opens, the pressure cell is drawn upward, pressing the upper piston against the force gage and superimposing an axial compression on the specimen (compression test). As it closes, the cell is pushed downward to relieve the axial pressure (extension test). The spacer between the force gage and the upper fixed plate is a slotted tube, the purpose of which is to allow insertion and extraction of the very long piston specimen assembly without removing the massive pressure cell.

Pressure cell. Being of conventional design, the cell is 54 cm long, has inside and outside diameters of 10 and 20 cm, respectively, is made of Omega steel with a Rockwell hardness of C50, and is threaded to a collar bolted to the lower moving plate. Closures and pistons are sealed by O rings.

To induce folding, the rock specimens must be geometrically unstable; i.e., the length of a rock beam must be large in relation to the cross section. The maximum diameter and length of the cylindrical columns are 5 and 20 cm, respectively. Square columns can be 20 cm long and $5/(2)^{1/2}$ cm on a side. The most satisfactory room temperature jacketing material for non-cylindrical specimens has been polyolefin shrink-fit tubing, which conforms nicely to any shaped specimen under confining pressure.

For pore pressure tests, interstitial water can be pumped into the specimen through the hollow loading piston. For high-temperature tests (to 400°C) an internal furnace surrounds the specimen, and power and thermocouple leads are taken out through the piston.

Instrumentation. The axial differential force, which is transmitted to the specimen through the loading piston, is determined by the external force gage, located between the piston and the spacer below the upper fixed plate (Figure 1). This gage is a water-cooled heat-treated steel tube to which four foil resistance strain gages are bonded. Two of these gages measure axial strain, whereas the other pair records transverse strain. All together the four gages form a Wheatstone bridge arranged so that thermal expansions cancel, whereas the voltage output associated with axial strain is $2(1 + \nu)$ times that of a bridge with a single active gage (where ν is Poisson's ratio, about 0.3 for steel). The output of the bridge is fed to the vertical channel of an X-Y recorder 30 × 45 cm. The gage is calibrated against a proving ring.

The shortening or elongation of a specimen is measured by an external transducer placed between the lower fixed and moving plates. The output of this transducer is fed to the horizontal channel of the X-Y recorder. Calibration is accomplished with a dial gage, and recorded displacements are corrected for the elastic distortion of the apparatus, about 3×10^{11} dynes/cm.

Confining pressures of oil-kerosene to 3 kb are generated by a duplex plunger pump, measured by a SR-4 pressure transducer, and indicated by a strip-chart recorder. The accuracies of the force, displacement, and confining-pressure records are of the order of 1%.

EXPERIMENTAL RESULTS

Sample Preparation and Starting Material

The earliest exploratory experiments were done on cylindrical specimens 5 × 20 cm that were easily prepared by coring with a diamond drill bit and trueing up the ends in the lathe. However, stress and strain distributions in cylindrical beams are difficult to determine, and the symmetry is hardly typical of natural folds. Most tests, therefore, have been made on rectangular rock beams, the preparation of which is rather time consuming. With a diamond saw the prisms are cut to roughly the desired size and shape from a large oriented block of starting material. All surfaces are then flattened, and pairs of faces made parallel to each other and orthogonal to both other pairs with a surface grinder. A rectangular grid is then stamped on the top, the bottom, and both sides of the beam to measure the surface strains after deformation. Merely to preserve them after large deformations that are often accompanied by fracturing, most beams have been sandwiched between thin sheets of lead.

Because the specimens are 20 cm long, large blocks of starting material that are as statistically homogeneous and isotropic as practical are needed. Another important requirement is that the materials pass through the macroscopic brittle-ductile transition within the available range of confining pressure, 0–3 kb, at room temperature, and at an average rate of shortening of 10^{-4} sec^{-1}. We chose two limestones and one sandstone for our first experiments.

Coconino sandstone (Permian, Arizona). A pink, fine-grained, well-sorted, and silica-cemented rock, Coconino sandstone is composed of 90% quartz grains, 9% rock fragments, and about 1% clay. Quartz overgrowths are commonly euhedral. The quartz grains are essentially unfractured, about 22% of them show undulatory extinction, and only 1% contain deformation lamellas. The blocks are indistinctly bedded; bedding is manifest only by the dimensional alignment of detrital grains and by concentrations of clay and heavy minerals in places. Porosity is 16%.

Lueders limestone (Permian, north-central Texas). A fine-grained, grayish-tan, and macroscopically homogeneous and isotropic rock, Lueders limestone is composed almost exclusively of calcium carbonate in the form of micrite, diagenetically altered organic particles, echinoid fragments, sparry calcite cement, and fossil fillings. Even in thin sections the trace of bedding is obscure. Only the sparry calcite is sufficiently coarse grained for optical study. The starting material contains essentially no twin lamellas, and the crystals are randomly oriented dimensionally and crystallographically. Porosity is about 17%.

Indiana limestone (Mississippian, Bedford, Indiana). A medium-grained, grayish, and macroscopically homogeneous rock, Indiana lime-

stone, like the Lueders, consists chiefly of calcium carbonate in the form of oolites, diagenetically altered organic particles, echinoid, crinoid, and other fossil fragments, abundant sparry calcite cement, and fossil fillings. Bedding is indistinct. The amount of micrite is smaller and the size of the sparry crystals is larger than that in the Lueders limestone. The calcite crystals in the starting block are essentially untwinned, and they are randomly oriented crystallographically. Young's modulus is about 0.4 mb at atmospheric pressure [*Birch*, 1966, p. 165]. Porosity is about 10%.

Definitions. The initial dimensions of the specimens are designated as follows: L is the length and D the diameter of a cylindrical specimen, and b is the width and h the thickness (or height) of a prismatic specimen (Figure 2). The aspect ratios are defined as L/D and L/h for cylindrical and prismatic beams, respectively.

The final permanent length, exclusive of recoverable elastic deformation, is Lp. In folded specimens the arc length L_a is measured between inflection points on the limbs, and L_a is one half of the wavelength λ. The amplitude of the fold a is a measure in the y direction in the xy plane of the distance between the intersections of a curved surface with the inflection points on a limb and the anticlinal hinge. It is one half of the deflection d.

In our experimental folds the hinge lines are always parallel to the fold axes, and the hinge surfaces are plane and normal to the direction of axial shortening and coincide with the axial planes.

Procedures. All experiments have been made on jacketed room-dry specimens loaded in compression parallel to bedding at room temperature and, with a single exception noted later, at an axial strain rate of 10^{-4} sec^{-1}.

To determine the important mechanical properties of the rocks, conventional triaxial compression tests have been made on both cylinders and prisms with an aspect ratio of about 3. All these specimens are 10 cm long; the cylinders are 3.5 cm in diameter, and the prisms are 3.2 cm square. True stress-strain curves are derived from the axial force-displacement records by dividing the differential force by the cross-sectional area of the specimen (corrected for shortening) and the displacement (corrected for

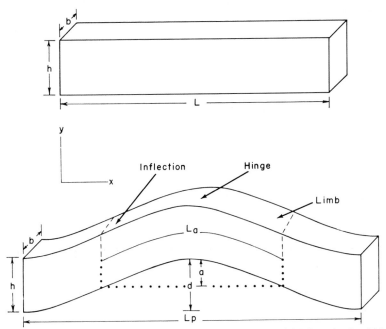

Fig. 2. Definitions of the dimensions of the specimens: L, initial length; h, thickness; b, width; L/h, aspect ratio; L_p, final permanent length; L_a, arc length; d, deflection; a, amplitude.

apparatus distortion) or the change in length by the initial length.

To study folding processes, tests have been carried out on specimens with aspect ratios of 4.4 or more. To compare the behavior of speci-

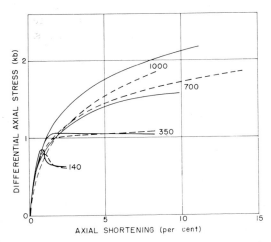

Fig. 3. Stress-strain curves for cylinders and prisms of Lueders limestone with an aspect ratio of 3 compressed under different confining pressures in bars. Solid line, the square cross section; dashed line, the circular cross section.

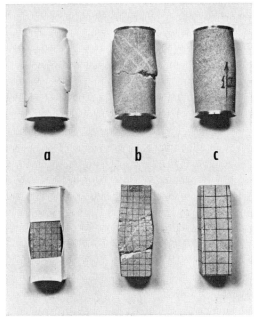

Fig. 4. Brittle-ductile transition illustrated in Lueders limestone for circular and square specimens. This transition is unaffected by cross-sectional shape, as is shown in Figure 3. Confining pressures were (a) 140 bars, (b) 350 bars, and (c) 700 bars.

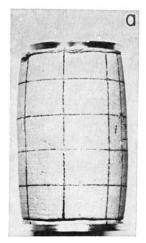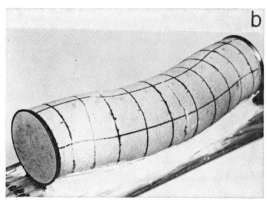

Fig. 5. Cylinders of Lueders limestone shortened under 1.4-kb confining pressure. Aspect ratios are (a) 3 and (b) 4.4.

mens with different cross-sectional areas and aspect ratios, the axial force-displacement records have been normalized by dividing the differential force by the initial cross-sectional area of the specimen and the displacement or change in length by the initial length. The permanent axial shortening ($L - Lp/L$, Figure 2) includes both the inelastic axial strain and the shortening due to folding. The procedure yields an apparent stress-strain curve that is not, of course, a measure of the fiber stress in a folded beam.

As work progressed, early ideas were discarded, and the course of the research changed accordingly. It thus seems best to report our results chronologically.

LUEDERS LIMESTONE

Conventional tests. Since we wished ultimately to deform prismatic rather than cylindrical specimens, we have investigated the influence of shape on the stress-strain curves that have been determined under otherwise conventional conditions. Let us compare the curves for cylinders with those for square prisms (Figure 3). Each curve represents the average of two or more identical tests, and the reproducibility is better than 5%. Thus at pressures of 140, 350, and 700 bars any differences due solely to shape lie within the experimental error, and hence they are indistinguishable. The higher strength of prisms at 1000 bars may be significant, but the difference is not large. All the curves reflect the usual increases in ultimate strength and ductility

with increasing confining pressure. The brittle-ductile transition occurs between 350 and 700 bars, and the effects of the confining pressure on the mode of deformation are the same for both shapes (Figure 4).

Folding experiments. Having determined that the effect of cross-sectional shape alone was unimportant, we considered the effect of aspect ratio. Folding was achieved for the first time in a cylinder 4.5 × 20 cm of this rock shortened under 1.4-kb confining pressure. In Figure 5 this specimen can be compared with a conventional one deformed under the same confining pressure. The stress-strain curves (Figure 6) reflect the difference in behavior. At small strains the two curves are essentially similar, but at large strains the short specimen 'work

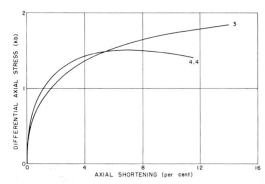

Fig. 6. Stress-strain curves for cylinders of Lueders limestone with different aspect ratios shortened under 700-bars confining pressure.

TABLE 1. Data for Experimentally Folded Lueders Limestone

Specimen	Confining Pressure, kb	Permanent Shortening, %	Fold Deflection, cm	Arc Length, cm	Maximum Limb Dip, deg	Average Strains Normal to Major Fold Axis at Maximum Deflection*			Strains Parallel to Major Fold Axis		Comments†
						Uppermost Fiber, %	Middle Fiber, %	Lowermost Fiber, %	Uppermost Fiber, %	Lowermost Fiber, %	
					Cylindrical Specimens, 4.6 × 20.3 cm						
2	0.7	7.4	1.4	13.5	11.0						Fractures A and B and fault zone parallel to C; ∠ B shear fractures = 50°–55°; ∠ C shear fractures = 60°–66°
3	0.7	7.7	1.6	11.0	11.0	−15.4‡		23.0			Fractures A and B and fault zone parallel to C; ∠ B shear fractures = 50° ∠ C shear fractures = 66°
4	1.4	10.0	0.5	9.7	4.0	<3.0		18.5			Unfractured
					Prismatic Specimens, 3.2 × 3.2 × 20.3 cm						
13	0.3	1.2	0.3			−1.0		5.0	−10.0	−2.0	Fractures B and C and fault zone parallel to C
12	0.7	1.4	0.7	11.7	5.5	−8.2‡		8.0‡	−18.0	1.0	Fractures A and B; ∠ B shear fractures = 45°
43	1.0	1.5	<0.1		0.5	0.0		0.5	−0.3	−0.3	Unfractured
44	1.0	2.8	<0.1		1.0	1.1		1.9	−1.0	−0.3	Unfractured
42	1.0	4.0	0.4	14.6	3.0	0.8	3.7	5.5	−2.8	−0.6	Unfractured
40	1.0	6.6	0.8	14.8	6.0	0.0	4.0	10.0	−6.2	−0.4	Fracture A
14	1.0	9.2	0.5	14.3	5.0	−2.5‡	5.0‡	6.5‡	−12.0	2.0	Fracture A
15	1.4	9.6	0.6	11.0	5.0	1.0	4.6	15.0			Fracture A
47	2.0	4.1	<0.1			3.1		3.6	−0.7	−0.5	Unfractured
55	2.0	5.8	0.3	15.2	2.0	3.5	5.0	7.0	−1.6	−0.1	Unfractured
53	2.0	6.3	0.6	15.2	4.5	2.9	5.1	7.3	−2.3	−0.1	Unfractured
54	2.0	6.5	0.5	14.6	4.0	3.9	6.3	8.3	−2.9	−0.2	Fracture A
57	2.4	8.3	0.6	15.4	5.0	2.5	4.9	8.2	−3.6	−0.3	Fracture A

* Strains are calculated from distortion of a rectangular grid placed on the specimens before folding; shortenings are counted as positive. Angles are averages of at least 10 measurements.

† Type and orientation of fractures and faults in relation to the folded specimen are shown in Figures 13–15.

‡ Strain measurements are only estimates because of separation along fractures.

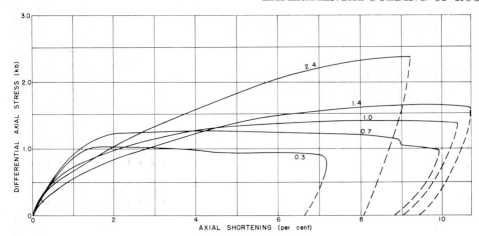

Fig. 7. Stress-strain curves for thick square beams of Lueders limestone shortened under different confining pressures in kilobars.

hardens' in bulk, whereas the long cylinder is unstable and 'work softens.'

A series of tests was made on thick beams with $L = 20$ cm and $h = b = 3.2$ cm under different confining pressures (Table 1). For shortening of less than about 0.5 cm, the axial differential stress tends to decrease as confining pressure increases over the range 0.3–2.4 kb for reasons that are not understood (Figure 7). For shortenings of 5% or more the reverse is true: work hardening increases with confining pressure.

The following maximum values were achieved in this test series: permanent axial shortening of 9.6%, deflection of 0.8 cm, and limb dip of 6°. The half wavelength is about 15 cm, which is longer than that in cylindrical beams, and it appears to be independent of confining pressure.

To get a better idea of the sequence of events, a series of five tests (14, 40, 42, 43, and 44) was made under identical conditions, but each test was terminated after a different permanent shortening over the range 1.5–9.2% (Table 1). Reproducibility over the same range of shortening is better than 3%. Folding evidently begins very early in the axial stress-strain history, since specimen 43, shortened only 1.5%, is already folded and, at the lower confining pressure of 700 bars, specimen 12, shortened 1.4%, has a limb dip of about 5°. The lower the confining pressure and/or the greater the axial shortening, the more important the role of fracturing in the folding process.

INDIANA LIMESTONE

Conventional tests. During the course of our work we shifted from Lueders to Indiana limestone, because Indiana limestone has an even lower brittle-ductile transition at about 200-bars confining pressure. Again, the cross-sectional shapes of specimens with the conventional aspect ratio of 3 have little or no effect on the stress-strain curves (Figure 8). The de-

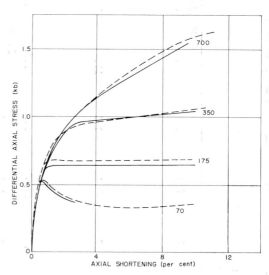

Fig. 8. Stress-strain curves for cylinders and prisms of Indiana limestone with an aspect ratio of 3 compressed under different confining pressures in bars. Solid line, the square cross section; dashed line, the circular cross section.

TABLE 2. Data for Experimentally Folded Prismatic Specimens of Indiana Limestone

Specimen	Confining Pressure, kb	Permanent Shortening, %	Fold Deflection, cm	Arc Length, cm	Maximum Limb Dip, deg	Average Strains Normal to Major Fold Axis at Maximum Deflection*		Strains Parallel to Major Fold Axis†		Aspect Ratio	Comments‡
						Uppermost Fiber, %	Lowermost Fiber, %	Uppermost Fiber, %	Lowermost Fiber, %		
Specimen Thickness of ~0.630 cm											
131	1.0	0.3	0.5		3.5	0.0	1.7		−0.2	33	Fractures B and D
130	1.0	0.6	0.6	17.2	4.5	0.0	2.0		−0.2	33	Fractures B and D
58	1.0		1.5							33	
84	1.0	4.3	2.2	15.2	17.5	−1.6	4.2		−0.5	36	Fractures A and B
83	1.0	5.1	3.3	15.5	25.5	−3.1	5.6		−0.5	32	Fractures A, B, and D
111	1.0	7.6	3.6	15.3	27.5	−3.7	6.5		−1.0	33	Fractures A, B, and D
Specimen Thickness of ~1.06 cm											
129	1.0	0.7	0.2		0.5	+0.7	1.5		0.0	19	Unfractured
128	1.0	1.3	0.5	14.5	3.5	0.0	2.5		−0.1	19	Fractures B and C
76	1.0	1.8	0.6	15.1	5.5				−1.0	19	Fracture A
77	1.0	2.8	1.5	16.1	11.3	−1.9	3.9	+1.1	−0.6	19	Fractures B and D
79	1.0	3.8	1.9	14.1	13.0	−2.1	5.6	+2.0	−1.3	19	Fractures A and D
82	1.0	5.9	2.5	14.9	15.0	−3.0	7.6	+2.1	−4.4	19	Fractures A and D
Specimen Thickness of ~1.59 cm											
108	1.0	0.6	<0.1	17.8	0.0	+1.1	1.4	+0.3	−0.1	13	Unfractured
109	1.0	1.3	0.2	16.0	1.5	0.0	2.1	+0.1	−0.1	13	Unfractured
106	1.0	2.0	0.4	15.7	3.0	−0.6	2.6	+0.2	−0.7	13	Unfractured
110	1.0	2.6	0.9	15.2	7.5	−1.0	2.9	0.0	−1.0	13	Fracture B
86	1.0	3.3	0.9	17.0	12.5	−1.5	3.4	+1.9	−0.4	13	Fracture D
103	1.0	4.9	1.9	16.2	14.5	−3.1	7.8	+2.3	−2.9	13	Fractures B and D
90	1.0		2.6	15.1	19.5	−5.4	9.7	+4.6	−3.8	13	Fractures B and D
100	1.0	7.4	2.6		21.0	−6.5	8.1	+1.9	−6.4	13	Fractures B and D
Specimen Thickness of ~3.18 cm											
48	1.0	3.4	0.2		0.5	−0.8	3.7	−0.9	−1.1	7	Unfractured

* Strains are calculated from distortion of a rectangular grid placed on the specimens before folding; shortenings are counted as positive.
† The first eight percentages listed are mean values.
‡ Type and orientation of fractures and faults in relation to the folded specimen are shown in Figures 13–15. Angles are averages of at least 10 measurements.

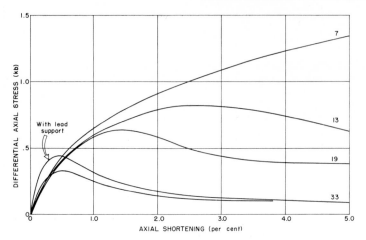

Fig. 9. Stress-strain curves for 20-cm beams of Indiana limestone with different aspect ratios shortened under 1-kb confining pressure.

formed specimens closely resemble those of Lueders limestone shown in Figure 4.

Folding experiments. About 21 prisms of this rock, all 20 cm long and 3.5 cm wide, were shortened under a confining pressure of 1 kb (Table 2). The effect of the different aspect ratios of about 6, 13, 19, or 33 on the normalized force-displacement curves is dramatic (Figure 9). The thick beams with an aspect ratio of 6.3 behave much as those composed of Lueders limestone behave, and the stress-strain curves rise monotonically as the beam work hardens. For higher aspect ratios the curves reach a peak stress and then fall off as the beam work softens in bulk. The higher the aspect ratio, the sharper

the peak and the lower the axial strain at which the peak occurs. Sandwiching of thin beams between lead sheets increases the peak stress somewhat but does not affect the shape of the stress-strain curve.

Again, to understand better the sequence of events in the history of folding, a series of tests was made for each aspect ratio (except the smallest). The permanent axial shortening was gradually increased from <1% to nearly 8%. The reproducibility of these tests over the same range of strain is better than 3%. As an example, Figure 10 shows the stress-strain curves for the series with the aspect ratio of 13. The unloading curves reflect the large elastic recoveries of these

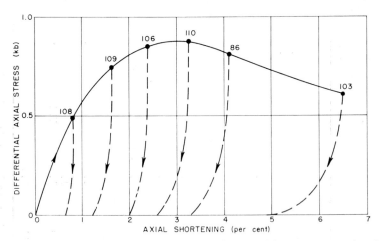

Fig. 10. Stress-strain curves for Indiana limestone beams with an aspect ratio of 13 compressed under 1-kb confining pressure. Note the large elastic recovery.

bent beams. The greater the total shortening, the larger the elastic 'unbending.'

As we had observed in Leuders specimens, the folding of Indiana limestone beams also must have commenced after a very small total shortening. Specimens 129 and 131, for example, are gently folded, although their permanent shortenings are only 0.7 and 0.3%, respectively. The thicker the beam for a given shortening and the greater the shortening for a given aspect ratio, the more prevalent fracturing tends to be.

The following maximum values were achieved in tests on Indiana beams: permanent axial shortening of 7.6%, deflection of 3.6 cm, and limb dip of 28°. This deflection is the maximum that can be accommodated by the 10-cm bore of the pressure cell. The half wavelengths are of the order of 15 cm, and there is no apparent dependence on aspect ratio.

COCONINO SANDSTONE

Conventional tests. Figure 11, the stress-strain curves of cylinders and prisms of this sandstone with an aspect ratio of 3, shows that the effect of shape is not negligible, as it is for limestone. The prismatic specimens are of the order of 30% or stronger on the average. The sandstone is essentially brittle at all confining pressures in the range 0.7–2.5 kb, although the transition to macroscopic ductility is nearly attained at the highest pressure. Failure results from shear fracturing or from faulting without total loss of cohesion at the higher pressures.

Folding experiments. About a dozen beams of this rock, all 3.2 cm wide and 20 cm long with aspect ratios of 7, 13, 19, 35, or 45, were shortened under confining pressures of 1 and 2 kb (Table 3). Under our experimental conditions the sandstone is essentially brittle. Thick beams with aspect ratios of 7 or 13 do not fold; they fail first by shear fracturing near their ends. However, appreciable folding can be achieved in thin beams, because the rock in bulk can deform by macroscopic cataclastic flow.

The axial stress-strain curves for thick beams with aspect ratios of 7 or 13 reflect the breaking strength of the material, whereas for higher ratios the sharp peaks signal the onset of folding (Figure 12).

The following maximum values were achieved in these experiments on sandstone: axial shortening of 3%, deflection of 2.7 cm, limb dip of

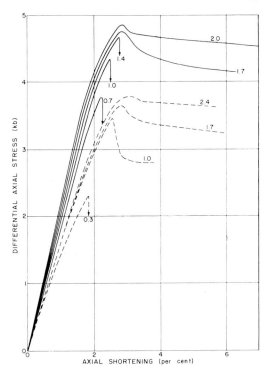

Fig. 11. Stress-strain curves for cylinders (dashed line) and prisms (solid line) of Coconino sandstone with an aspect ratio of 3 compressed under different confining pressures in kilobars.

16°, and compressive fiber strain of 3%. The half wavelengths are 17–19 cm, i.e., nearly the same as the original length of the beams.

BRASS

We wished to compare our data on rocks with those on a nearly perfectly elasticoplastic metal, specifically on rectangular beams of brass with shapes and end loadings identical to those of our rock beams. A series of tests was carried out at atmospheric pressure on about 10 beams with $L = 20$, $b = 3.5$, and $h = 0.6$ cm (Figure 25). The axial stress-strain data will be compared with those of similar rock beams in the discussion that follows. A typical example is shown in Figure 23. Several beams were deformed under 1-kb confining pressure, and no significant pressure effects were observed.

PETROFABRIC RESULTS

Procedure. Macroscopic petrofabric studies of all the folded specimens include (1) measurements of arc length, deflection, and limb dips,

TABLE 3. Data for Experimentally Folded Prismatic Specimens of Coconino Sandstone

Specimen	Confining Pressure, kb	Permanent Shortening, %	Fold Deflection, cm	Arc Length, cm	Average Strains Normal to Major Fold Axis at Maximum Deflection*			Strains Parallel to Major Fold Axis, %	Aspect Ratio	Comments†
					Maximum Limb Dip, deg	Uppermost Fiber, %	Lowermost Fiber, %			
Specimen Thickness of ~0.045 cm										
78	1.0		1.6	16.9	11.5	0.2	2.6	−0.2	44	Fractures A, B, and D
85	1.0		1.6	16.9	10.5	−1.2	1.5	−0.2	44	Fractures B and D
135	2.0	1.5	1.6	17.1	11.0	−1.0	−1.0	−0.2	45	Fractures A, B, and D
134	2.0	3.0	2.7	16.5	16.0	−1.4	1.7	−0.3	45	Fractures A and B
133	2.0	7.6	‡	‡	‡	‡	‡	‡	45	Fractures A and B
Specimen Thickness of ~0.63 cm										
72	1.0	1.1	1.4	19.1	9.5	−2.0	3.3	−0.1	35	Fractures A, B, and D
73	1.0	0.3	1.4	17.0	13.5			0.9	32	Fractures A and D
120	2.0	0.4	1.4	19.2	9.0	−1.8	2.1	−0.2	35	Fractures B and D
Specimen Thickness of ~1.06 cm										
75	1.0	0.1						0.4	19	Unfractured
74	1.0	<0.1	1.4		<0.5				19	Fracture A
Specimen Thickness of ~1.59 cm										
37	2.0	0.4	0.2						13	Fractures A and B
38	2.0	0.6							13	Fractures A and B
Specimen Thickness of ~3.18 cm										
	2.0	0.5							7	Faulted

* Strains are calculated from distortion of a rectangular grid placed on the specimens before folding; shortenings are counted as positive.
† Type and orientation of fractures and faults in relation to the folded specimen are shown in Figures 13–15. Angles are averages of at least 10 measurements.
‡ The fold is asymmetrical.

(2) calculations of surface strains from distortions of the initially rectangular grid pattern, and (3) observations of the development and orientations of fractures (Tables 1–3). Microscope studies were made of only Lueders limestone specimens 3 and 14. The trends of calcite twin-lamella indices were determined, and twin-lamella compression and extension axes were derived. The thin-section work was done with a Leitz research microscope and a universal stage; the data were plotted on the lower hemisphere of the equal-area projection.

Macrofractures. Three systems of macrofractures occur in the folded thick beams of Lueders limestone (Table 1). Fracture type A (Figure 13) usually transects the specimen, is oriented at high angles to the long dimension, and occurs in the regions around fold crests and troughs. In nearly all specimens the fracture is wider on the compressive side of the fold, and it tapers toward the extensile side (Figure 13a, b). In at least one specimen the fracture dies out before it reaches the surface on the extensile side. Type A fractures originate on the compressive side of the fold, where a crush zone also occurs, and they then propagate into the extensile region of the fold. Some of them make a marked change in direction. When the specimen is viewed as a whole, these jogs are symmetrical about the hinge of the central anticline. The location of the jog either coincides with the central fiber of the fold (specimen 12, Figure 13a) or is displaced somewhat toward the extensile side (specimen 14, Figure 17). It is demonstrated later that this jog coincides with the neutral fiber as defined by the calcite fabrics. Type A fractures occur late in the folding history.

The type B fracture assemblage (Figure 13a, c) occurs only on the extensile side of the anticlines and the synclines in the hinge regions (Table 1). The geometry of the array suggests a conjugate system of shear and extension fractures that relate to a state of stress with the maximum principal compressive stress σ_1 parallel to the fold axis, the stress σ_2 normal to the surface, and the stress σ_3 normal to the fold axis and parallel to the surface. The angle between the two shear fractures is 45°–55°, i.e., less by about 10°–15° than that for the other set of shear fractures (type C) in the same specimen. In the third dimension, type B fractures die out either at or short of the crossover into the

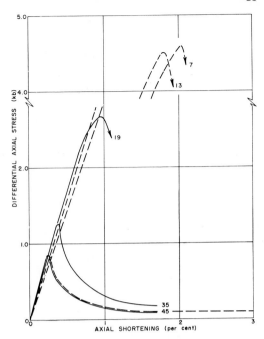

Fig. 12. Stress-strain curves for beams of Coconino sandstone with different aspect ratios compressed under 1-kb (solid line) or 2-kb (dashed line) confining pressure.

compressive side of the beam (Figure 13b). Displacements along the fractures are both strike-slip and dip-slip (Figure 13c). These fractures occur only in those specimens in which the strain parallel to the surface and perpendicular to the fold axis is extensile, not in those specimens in which that strain is still compressive.

Type C fractures are parallel to or conjugate to faults that develop in a few specimens (Table 1). The fault zones occur along the flanks of the folds in the vicinity of the inflection between the anticline and the syncline (Figure 13d). The strikes of the faults and the conjugate shear fractures are parallel to the fold axis. The angle between the conjugate shear fractures is 60°–66°, about equal to that observed in conventional cylindrical specimens deformed in the brittle regime. In specimen 3 (Figure 13d), type B shear fractures also occur, and the corresponding angle is only 50°.

In thin beams of Indiana limestone and Coconino sandstone, macrofractures of types A and B occur with those of a type D not recognized in the thick beams. Type A fractures

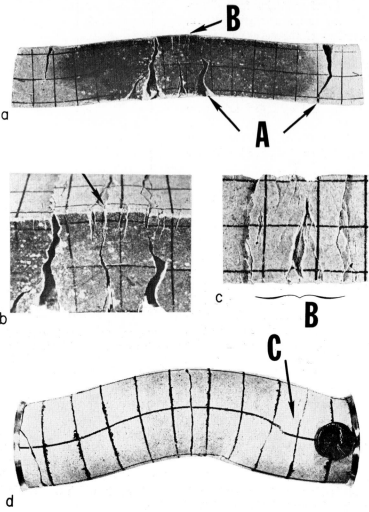

Fig. 13. Photographs showing fracture types in thick folded beams of Lueders limestone. (a) Fractures of type A (specimen 12) are throughgoing, or nearly so, and are oriented subperpendicular to the axial load. They probably originate in a crush zone (lost material) on the compressive side of the fold and propagate toward the extensile side [cf. *Price*, 1967, Figure 6]. There is a marked change in trajectory at what is probably the neutral surface. Grid spacing is 1.2 cm. (b) Fractures of type B (specimen 12) occur only on the extensile side of the fold, only where the strain parallel to the load axis is actually an extension, and only when the rock is in the brittle regime. They die out before reaching the apparent neutral surface. The angle between conjugate shear fractures is less by 10°–15° than that observed for shear fractures associated with fractures of type C. Grid spacing is 1.2 cm. (c) Top surface view of fractures of type B. (d) Fractures of type C (specimen 3) are 'thrust' faults and related conjugate shear fractures formed across the flank of the fold. The strike of the fault parallels the fold hinge. Grid spacing is 1.9 cm.

Fig. 14. (Opposite) Photographs showing fracture arrays developed in the hinge regions of thin folded beams of Indiana limestone. (a) Fractures of type A (specimen 76). Note sudden changes in fracture trajectories. (b) Fractures of type B (specimen 90). Fold axis is parallel to the paper and strikes north-south. View is of the extended side of the beam. (c) Fractures of type D (specimen 77). These fractures occur only on the compressive side of the beam. (d) Fractures of type D (specimen 79). Grid spacing on all specimens is 0.6 cm.

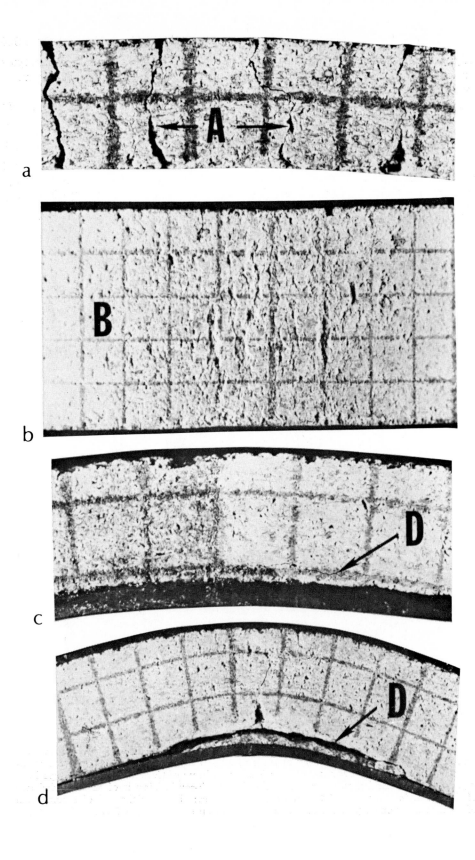

a

b

c

d

typically exhibit crushed zones at both the upper and the lower surface of the specimen, and they also tend to take pronounced jogs (Figure 14a). Type B fractures are similar to those in the thick beams except that strike-slip and dip-slip displacements along the surfaces are not as great (cf. Figures 13, 14, and 15). Type D fractures occur on the compressive side of the thin beam and lie either parallel to or inclined at angles of up to 30° to the lower surface (Fig-

ures 14c, d and 15c). In some specimens types B and D merge. The strikes of type D fractures are parallel to the fold axis.

Surface strains. Strains at the surfaces of the specimens were calculated from measurements of the distortions of an originally square grid marked on the specimens before deformation. Two types of strain data were obtained: (1) percentage of elongation or shortening perpendicular and parallel to the hinge in the region

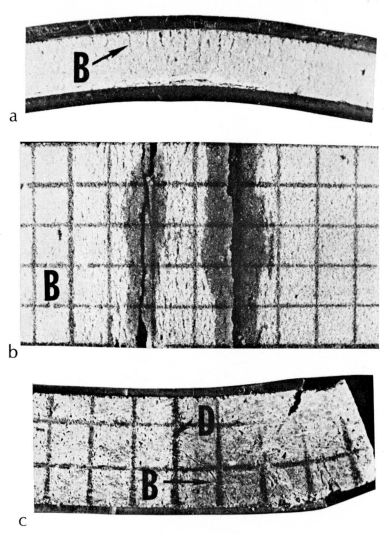

Fig. 15. Photographs showing fracture arrays developed in folded beams of Indiana limestone and Coconino sandstone. (a) Fractures of type B in a thick beam of Coconino sandstone (specimen 85). (b) Fractures of type B in a thin beam of Coconino sandstone (specimen 78). Fold axis is parallel to the paper and strikes north-south. View is of the extended side of the beam. (c) Fractures of types B and D in the syncline near the end of a thin beam of Indiana limestone (specimen 90). Grid spacing in all specimens is 0.6 cm.

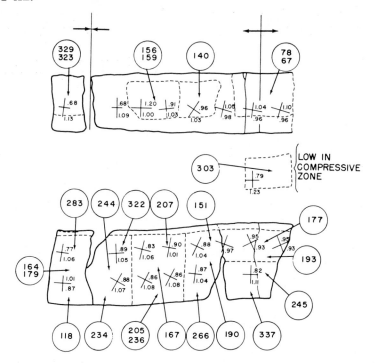

Fig. 16. Secondary principal quadric strains with calcite twin-lamella spacing indices at different locations in a folded thick beam of Lueders limestone (specimen 14). The double values of an index represent two independent measurements to test reproducibility. (Top) Top view showing the uppermost fiber. (Bottom) View perpendicular to the fold axes.

of maximum amplitude of the central anticline (Tables 1–3) and (2) secondary principal quadric elongations $(1 \pm e)^2$ [*Ramsay*, 1967, p. 52] for each grid cell on several specimens (Figure 16). Each cell was treated as a 45° rosette; the means of the two parallel sides and one diagonal formed the three-gage configuration. The secondary principal quadric elongations were calculated from the major and minor axes of the ellipsoid obtained from these three strains. Calculations are based on the not entirely realistic assumptions that the state of strain within each cell is homogeneous, the original dimensions of all the cells are identical, and no displacements are due to fracturing. However, the computed directions and magnitudes of strain are qualitatively significant.

Strains perpendicular to the fold axis vary with position in relation to the uppermost, central, and lowermost fibers of the central anticline (Tables 1–3). In only four specimens is the net strain in the uppermost fiber an extension; in most of the folds the permanent axial strain exceeds the extension due to bending. All lowermost fibers show compressive strains with a maximum of 23% in specimen 3 (Figure 13d). All strains along the central fiber are also compressive, and they are intermediate in magnitude between those along the uppermost and those along the lowermost fiber.

Strains parallel to the fold axis also vary between the uppermost and lowermost positions (Table 1–3). In most specimens these strains are all extensions. Maximum elongations occur at the lowermost fiber; the maximum is 18% in specimen 12. These strains can readily be visualized by bending a rectangular rubber eraser.

A map of the secondary principal quadric strains along two faces of half of specimen 14 is shown in Figure 16. The strains are symmetrically disposed on the other half of the specimen. Along the uppermost fiber the principal compressive (<1.0) and extensile (>1.0) strain axes at the left end of the specimen (adjacent to the piston) are perpendicular and parallel, respectively, to the synclinal fold axis.

These axes progressively rotate toward the right until at the anticlinal crest they are reversed; i.e., the maximum shortening is subparallel to the fold axis. Along the opposite parallel surface at the lowermost fiber the strains are the reverse of those along the uppermost fiber, as is indicated at one locality in Figure 16. Along the top face the axes of quadric shortening, especially in the inflection region, are systematically inclined at low angles to the axis of shortening (Figure 17).

Calcite twin-lamella index. The abundance of calcite twin lamellas has been expressed as a spacing index: the mean number of e_1 (best developed) lamellas per millimeter as viewed on edge [*Turner and Ch'ih*, 1951]. The mean index of grains in the aggregate is known to increase with shortening or elongation up to about 15% [*Turner and Ch'ih*, 1951; *Friedman*, 1964]. This trend is meaningful provided that the thicknesses of lamellas are about the same in all specimens under comparison and the data are from the best-developed set in each grain. Here the mean indices from 60 grains at each of 21 localities in specimen 14 were determined to detect any trends in relation to the position within the folded specimen (Figure 16). Reproducibility was tested by duplicate sampling at five places and is of the order of 10–15%.

Several trends are evident from the data: (1) along the uppermost fiber the index decreases from near the trough of the syncline to the crest of the anticline, (2) in the plane perpendicular to the fold axis the index decreases from the lowermost to the uppermost fiber in the vicinity of the hinge of the anticline, (3) in the region of the synclinal hinge the index decreases toward the lowermost fiber, and (4) in the inflection region the index is lowest near the central fiber of the fold.

The indices can be compared with quadric strains only where both parameters were measured at nearly the same location or where extrapolation can be parallel to a fiber of the fold. In general, the indices increase with local strain and axial shortening, i.e., with decreasing quadric shortening (Figure 18). These trends are only qualitative, because the quadric strains are only approximations and the strains in the third dimension are neglected.

Calcite compression and extension axes. Compression and extension axes derived from calcite e_1 lamellas after the method of *Turner* [1953] were determined at 16 localities in specimen 14 (Figures 19 and 20) and 2 localities in specimen 3 (Figures 13d and 21). Several sets of three perpendicular thin sections were prepared from one half of symmetrical specimen 14. They are oriented perpendicular to the hinge, parallel to the hinge and the uppermost and lowermost surfaces, and perpendicular to the direction of axial shortening. Data are combined for structurally comparable domains within the fold to assess the over-all pattern of the compression

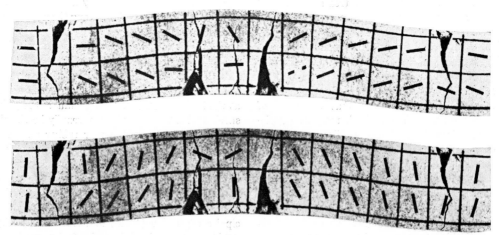

Fig. 17. Secondary principal quadric strain axes calculated from distortion of grid (originally 1.25-cm square) on a folded thick beam of Leuders limestone (specimen 14). Strain values in left half are shown in Figure 16. (Top) Compressions. (Bottom) Elongations.

and extension axes. It is preferable, however, to illustrate the data from single thin sections cut perpendicular and parallel to the hinge (Figures 19 and 20). The patterns in the partial diagrams are the same as those found in corresponding composite diagrams of two or three mutually perpendicular sections.

For the surface perpendicular to the hinge (Figure 19) several repeated trends are apparent.

1. At localities a, b, and c the compression axes are concentrated subparallel to the direction of axial shortening, which is east-west on the diagram.

2. At d the compression axes are noticeably skewed to the horizontal axis. The angle of inclination is essentially the same as that of the secondary principal quadric shortening at the same locality (Figures 16 and 17). A similar correlation is also apparent at localities e, f, and h.

3. At locality i in the inflection region the compression axes are more widely scattered than they are elsewhere.

4. At locality g high in the extension zone of the central anticline the compression axes

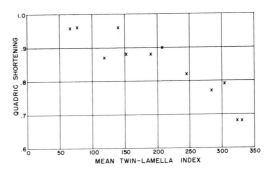

Fig. 18. Mean calcite twin-lamella spacing index as a function of quadric shortening in a folded thick beam of Lueders limestone (specimen 14).

are rotated 90°, and they are grouped perpendicular to the hinge and the uppermost surface of the fold, i.e., north-south on the diagram.

5. Comparison of the patterns for localities f and g shows that in f there is a minor north-south concentration, whereas in g there is a minor east-west concentration. Clearly, these patterns define the crossover between the compressive and extensile regions in the beam. Note that the apparent neutral fiber, defined by the twin-lamella analysis, coincides with the marked

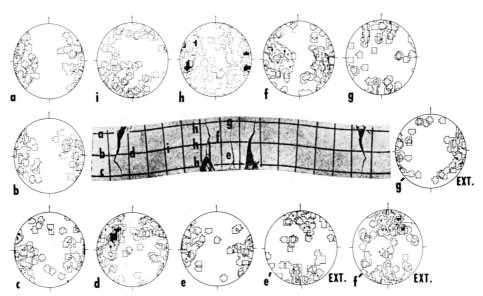

Fig. 19. Calcite compression and extension axes at nine locations (a–g) on specimen 14. The plane of each diagram is parallel to the surface of the specimen and perpendicular to the fold axes. Diagrams a–i are of 50, 50, 50, 100, 48, 86, 48, 142, and 62 compression axes, respectively; and diagrams e'–g' are of 48, 86, and 50 extension axes, respectively. Contours are at 1, 2, 4, and 6 points per 1% area.

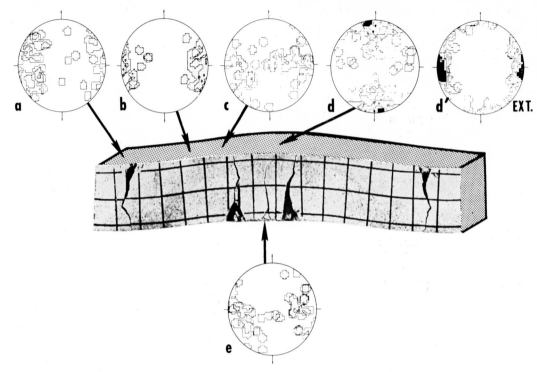

Fig. 20. Calcite compression and extension axes at five locations (*a–e*) on specimen 14. Plane of each diagram is parallel to the top surface of the fold; the fold axes are indicated by the fine arrows (north-south). Diagrams *a–e* are of 50, 36, 49, 88, and 49 compression axes, respectively; and diagram *d'* contains 88 extension axes. Contours are at 1, 2, 4, and 6 points per 1% area.

change in orientations of nearby type A macro-fractures.

Extension axes are illustrated at only three localities (Figure 19), because these patterns are typical of those found throughout. These axes fall into partial or complete girdles oriented 90° to the corresponding patterns of the compression axes. Concentrations within the girdles do not necessarily parallel the directions of the greatest elongations at given localities. For example, at locality *e* the extension axes are grouped only roughly north-south, whereas the secondary principal quadric elongation (1.11 or 5.5%) is located exactly north-south (Figure 16). The true greatest elongation is here parallel to the fold axis and is probably >8.0%, since the maximum elongation at the lowermost fiber is an extension of 12% parallel to the fold axis and locality *e* is about half way between the central and lowermost fibers (Table 1, specimen

14). Clearly, a concentration of extension axes parallel to the hinge (center of the diagram) is missing in diagram *e'* (Figure 19), and its absence is not the result of a 'blind spot' sampling problem.

For the surfaces parallel to the direction of axial shortening and the fold axes (Figure 20) the patterns of the compression axes show the following trends.

1. In the trough of the syncline (locations *a* and *b*) and at the lowermost fiber of the anticline (location *e*) the axes are grouped parallel to the direction of axial shortening.

2. In the inflection region (location *c*) the axes form a girdle around the fold axis.

3. At the crest of the anticline (location *d*) at the uppermost fiber the axes are strongly oriented parallel to the fold axis. A composite diagram at *d* and *g* (from Figures 19 and 20) yields a girdle parallel to the axial plane of the

anticline. Minor concentrations in both d and g are oriented east-west, parallel to the direction of axial shortening.

The extension axes at d show a strong orientation parallel to the direction of axial shortening. The strain in this direction (-8.5%) is the maximum elongation at that location on the fold, where the strain parallel to the fold axis is -3% (Table 1). The pattern at d agrees with the extension axes at nearby location g (Figure 19). Extension axes at location e (not illustrated) form a uniform girdle parallel to the axial plane of the anticline, even though the strain parallel to the fold axis is -12%.

Compression and extension axes also were obtained for two thin sections from specimen 3 cut parallel to the plane defined by the direction of axial shortening and the fold axes. These sections were taken from the crest of the central anticline, where type B shear fractures are well developed. The section from which the compression axes (Figure 21a) were determined is located just 3 mm from the surface of the sample; the second section lies 10 mm directly below the first. The compression axes from the top section are grouped in a concentration subparallel to the fold axis, in agreement with inference from the adjacent type B shear fractures. The compression axes in the lower section do not show this concentration (Figure 21b). Instead, they are more randomly distributed in a broad peripheral girdle. The type B shear fractures have all but vanished at this depth (13 mm).

DISCUSSION

For the first time, as far as we are aware, folding of single-layer beams of real rocks has been achieved in the laboratory. Two significant questions naturally arise: Are these experimental folds stable bends (forced folds), unstable buckles, or some combination of stable bends and unstable buckles? Do they have any natural counterparts?

Instabilities. Since folding does not occur for aspect ratios of less than about 4, it seems clear that in the history of folding a geometric instability first leads to buckling probably in the elastic range of axial shortening. To test this idea, we have compared our data with predictions from the classical buckling theory of Euler [*Chapple*, 1968b; *Shames*, 1964], having first investigated the applicability of the theory to the folding of nearly perfectly elasticoplastic brass beams with dimensions and loadings similar to those of the rock beams.

Under our experimental conditions the flat ends of the beams are in contact with, but not attached to, the steel loading pistons. Whether the beam should be treated as if it were fixed or hinged was not clear. When both ends are hinged, the Euler buckling formula is

$$\sigma_{cr} = \pi^2 E/(L/r)^2 \tag{1}$$

where σ_{cr} is the critical axial stress for elastic buckling, E is Young's modulus (about 0.6 mb for brass), L/r is the slenderness ratio, L is the unsupported length, and r is the radius of gyration. For elastic buckling the half wave-

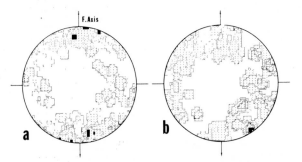

Fig. 21. Petrofabric diagrams of 100 compression axes derived from calcite e_1 twin lamellas in each of two thin sections from Lueders limestone (specimen 3). Planes of diagrams are parallel to the fold axis (north-south) and the axis of shortening (east-west). (*a*) Diagram from section located within 3 mm of crest of central anticline where type B fracturing is conspicuous (Figure 13). (*b*) Diagram from 10 mm directly below (*a*) where type B fractures have virtually vanished. Contours are at 1, 2, 4, and 6 points per 1% area.

length is initially equal to L, as it is in brass beams, and, for small deflections, $L \approx L_a$, the arc length.

When both ends of the beam are fixed, the quantity σ_{cr} increases by a factor of 4. In Figure 22 we plot critical stress as a function of the slenderness ratio for both cases. The data points taken from the normalized force-displacement curves for nine beams fall very close to the Euler curve for a beam that is fixed at both ends.

Since the axial stress-strain curves for thin sandstone beams are similar to those for brass (Figure 23), the sandstone results have also been compared with the Euler predictions (Figure 24). Young's modulus of this rock was taken to be 0.3 mb, the best average value derived from the slopes of the linearly elastic portions of the axial stress-strain curves of about a dozen beams. Again the data points taken from the normalized force-displacement records fall close to the Euler curve for slenderness ratios of about

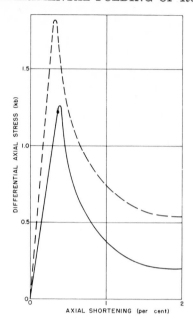

Fig. 23. Typical axial stress-strain curves for beams of brass (dashed line) and Coconino sandstone (specimen 73) (solid line) of similar dimensions, shortened under 1-kb confining pressure. The peak stresses are equivalent to the critical buckling stresses of Euler.

70 and greater. Thick beams of sandstone do not fold; they fracture.

Once buckling has occurred, a bending moment is superimposed on the axial compressive loading, and the net fiber stress in an elastic beam is the sum of the triaxial prestress and the bending stresses that are related to the curvature of the fold [*Price*, 1967, p. 182]. At the crest of the anticline the outermost fiber stresses σ_f are

$$\sigma_f = \Delta\sigma_x + p_c \pm (Eh/2R) \qquad (2)$$

where $\Delta\sigma_x$ is the differential axial compressive prestress, p_c is the confining pressure, and R is the radius of curvature. The bending stress is tensile (negative) in the uppermost fibers and compressive in the lowermost ones.

When a fiber stress reaches the breaking strength or the yield stress of the material, further deformation of the fold is inelastic, and a physical instability arises, the nature of which depends on the rheology of the material. This rheology in turn is intimately related to the mechanisms of inelastic deformation. The physi-

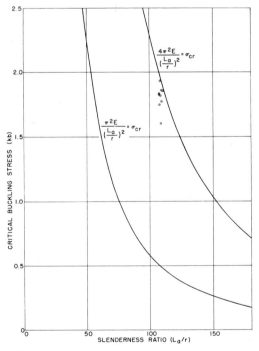

Fig. 22. Critical buckling stress as a function of the slenderness ratio of brass beams, hinged at both ends or fixed at both ends. The data points taken from the normalized force-displacement records fall very close to the theoretical curve for the latter condition.

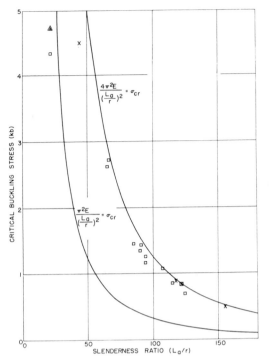

$$\frac{4\pi^2 E}{\left(\frac{L_a}{r}\right)^2} = \sigma_{cr}$$

$$\frac{\pi^2 E}{\left(\frac{L_a}{r}\right)^2} = \sigma_{cr}$$

CRITICAL BUCKLING STRESS (kb)

SLENDERNESS RATIO (L_a/r)

Fig. 24. Critical buckling stress as a function of the slenderness ratio of sandstone beams, hinged at both ends or fixed at both ends. The data points taken from the normalized force-displacement records fall close to the theoretical curve for the latter condition at slenderness ratios of about 70 and greater. The confining pressures are 2100 bars (crosses), 1400 bars (triangles), 1000 bars (squares), and 700 bars (circles).

a

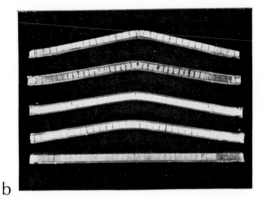

b

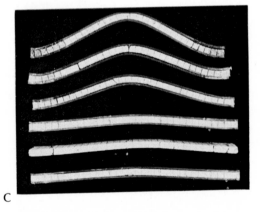

c

cal instability largely determines the ultimate shape of the fold, at least until the deflection becomes very large.

Deformation mechanisms and fold shape. Brass is a nearly perfectly elasticoplastic material. When the fiber stresses (2) in the crestal region of the elastic buckle reach the yield stress, failure occurs by intracrystalline flow. As shortening continues, the limbs rotate essentially rigidly, so that the fold becomes V shaped (Figure 25; cf. *Chapple* [1968a, Figure 8] and *Price* [1967, Figure 6]). Confining pressure has little or no effect on the stress-strain relations and fold shape. The critical buckling stress appears as a sharp peak in the axial stress-strain curve.

The sandstone is elastic-triboplastic, so that confining pressure does tend to increase ultimate strength and ductility, and it affects fold shape perhaps to a small degree. Although the sand-

Fig. 25. Photographs of beams experimentally folded under 1-kb confining pressure. All are 0.6 cm thick, 3.2 cm wide, and 20 cm long. (a) Brass beams permanently shortened (from bottom to top) 0.1, 0.5, 0.7, 1.3, 3.0, and 6.4%. (b) Sandstone beams permanently shortened (from bottom to top) <0.1, 0.4 (2 kb), 1.1, 0.3 (?), and 1.5%. (c) Indiana limestone beams permanently shortened (from bottom to top) 0.3, 0.9, 0.6, 4.3, 5.1, and 7.6%.

stone flows cataclastically and hence has macroscopic ductility, it is locally brittle under our test conditions, and failure first occurs in the crestal region by fracturing. Again, as folding progresses, the limbs rotate rigidly while the inelastic deformation spreads outward from the anticlinal hinge region. The final form is again a chevron fold that closely resembles that developed in the brass (Figure 25). The axial stress-strain curve has a sharp peak that signals buckling (Figure 12).

Although some fracturing does eventually occur in the limestone beams, especially at low confining pressures and large deflections, most specimens are in the ductile regime, and their inelastic deformation is clearly due to intracrystalline twin and translation gliding in the calcite grains. The axial stress-strain relations and the ultimate shapes of the folds are, however, very different from those of brass and sandstone. There is no sharply defined critical buckling stress, the limbs of the folds are not straight, the half wavelength is not equal to the original length, and a central anticline and two adjacent synclines are always developed (Figure 25).

Let us consider the axial stress-strain curves of Indiana limestone folds (Figure 9). The curve for the thin beam (aspect ratio of 33) reaches a maximum and then drops off. Although the peak is not sharp, the form of the curve is not unlike that of the brass and the sandstone (Figure 23),

and the maximum stress is equivalent to the critical buckling stress. As the thickness of the limestone beams increases, this peak disappears. No instability is evident in the thickest beams (aspect ratio of 6.3), yet they are certainly folded (Table 2). The beams are evidently work hardening in bulk, and, in the sense that the axial load must be increased to continue the deformation, the folds would be forced.

The maximums in the curves for the limestone beams of intermediate thickness (and for Lueders limestone beams under low confining pressures, Figure 7) cannot be associated with elastic buckling, since they occur at axial shortenings well into the inelastic range. Instead, they seem to be related to the physical instability associated with the onset of fracturing or faulting.

Another clue to the process of folding is the rate of change of deflection with permanent shortening (Figure 26). The curves for thin beams of brass and Indiana limestone are similar. Both curves are smooth and reflect the rate of growth increasing with shortening. The curves for limestone beams of intermediate thicknesses show two sudden changes in slope. Knick points A, which occur at very small axial strains, may well signal a geometric buckling instability. This instability probably cannot be the ideal elastic instability of Euler (1), but it may be some kind of plastic buckling instability that becomes possible only when the axial differential prestress exceeds the yield stress of the

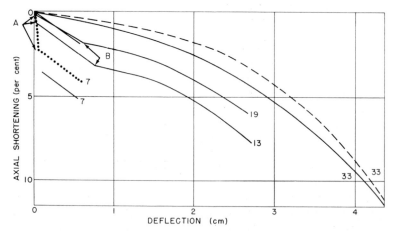

Fig. 26. Permanent axial shortening as a function of deflection in beams of brass and of Indiana and Lueders limestones with different aspect ratios. Knick points A and B signal instabilities where rates of fold growth change suddenly. Dashed line, brass; dotted line, Lueders limestone; solid line, Indiana limestone.

rock. On the other hand, the petrofabric data, as discussed below, show that little if any flow occurs before bending.

Knick points B coincide with the maximums of the axial strain curves; hence they are probably related to the physical instability associated with fracturing. The curves for thick beams do not reflect any physical instability, but knick point A is prominent. Again, it probably represents a geometric instability of some sort.

Although the inelastic deformation of the limestone is due to intracrystalline slip, as it is in the brass, the behavior of the limestone cannot be simply elasticoplastic. The shapes of the folds are closely approximated by cosine curves (Figure 27), and they resemble the viscoplastic shape of *Chapple* [1969, Figure 10], who points out that the relative contributions of viscous and plastic deformations to the growth of the fold will depend on the loading history. All our experiments have been done at a constant rate of axial shortening of 10^{-4} sec^{-1}. The single experiment done at 10^{-6} sec^{-1} on specimen 40, a thick beam of Lueders limestone, reveals no effect of the hundredfold variation in the rate of shortening.

Our data are insufficient to establish the proper rheologic law for the deformation of limestone at low pressure and temperature. It is significant, however, that the petrofabric data are consistent with the idealized stress pattern predicted by (2) for the bending of a simple beam.

Stresses and Strains in Folds

Petrofabric data on macrofractures, surface strains, and deformed calcite in folds of Lueders limestone bear on four important problems: (1) superposition of axial and bending stresses, (2) initiation of folding, (3) development of macrofractures and comparison with natural counterparts, and (4) internal stress distribution as revealed by distributions of calcite compression and extension axes.

Superposition of stresses. In an experiment the rock beam is initially shortened axially, parallel to its length, and then folded. The bending moments superpose tensile and compressive stresses on the axial compressive prestress. It can be shown analytically [e.g., *Friedman and Sowers*, 1970, Figure 14a] that the 'neutral fiber' at the axial planes of the anticlines and the synclines is displaced away from the central fiber to enlarge and reinforce the regions in compression at the expense of those in extension.

Exactly this situation is recorded by the fabric elements. Compression axes derived from e_1 twin lamellas show the crossover high in the fold from compression to extension between localities f and g in specimen 14 (Figure 19). The same situation is indicated by the calcite data from specimen 3 (Figure 21) and by the fact that type B fractures in this specimen die out at a depth of about 1.3 cm from the upper surface of the fold. The crossover between localities f and g (specimen 14) also coincides with the marked change in trajectories of type A fractures. The observation that in specimen 12 (Figure 13a), deformed under lower confining pressure, a corresponding change occurs closer to the central fiber supports our view that changes in fracture orientations are significant. Since the ultimate strength increases with confining pressure (Figure 3), there is less axial stress difference and hence less shift in the position of the apparent neutral fiber in specimen 12 than in specimen 14. Finally, the twin-lamella indices decrease from a maximum, where the maximum bending compression is superposed on the axial compressive prestress, to a minimum, where maximum extension occurs.

Initiation of folding. Chronologically, the fabric studies of specimen 14 were completed before the sequential tests at 1- and 2-kb confining pressure were made (Table 1). These tests demonstrated that the folding begins at very small axial strains. This fact had already been

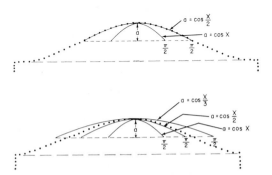

Fig. 27. Profiles of folded beams of Indiana limestone specimens **83** (top) and **100** (bottom) with aspect ratios of **32** and **13** in comparison with ideal cosine curves. Dotted line, contour of the specimen; solid line, theoretical curve.

anticipated from results of the study of calcite twin lamellas.

At the uppermost fiber in the extension zone of the central anticline (Figure 20, location d) very few of the compression axes derived from the twin lamellas reflect the initial axial compression. A similar observation was made in specimen 3 (Figure 21). This fact is contrary to what would be expected if the specimen had undergone appreciable inelastic shortening before folding. At least the uppermost fiber at the anticlinal crest was extended perpendicular to the fold axis throughout the history of folding. This conclusion is supported by the very low twin-lamella index at the same location (Figure 16). Evidently the early superposition of stresses in the uppermost fiber was such that the net fiber stress remained minimal throughout the deformation.

Experimental and natural macrofractures. The geometric and genetic implications of macrofractures on natural folds have been treated extensively in the literature. For example, *Stearns* [1964, 1968] has recognized five different arrays of conjugate systems of shear and extension fractures (Figure 28). The orientation of each conjugate array relates to an orientation of the principal stress axes in the rocks at the time of that fracturing.

Elements of three of these five fracture assemblages have so far been recognized in the experimental folds. The type B fractures correlate with those of assemblage 2 on natural folds, the extension fracture in the type B array is indistinguishable from that of assemblage 3 in the extended portion of the fold, type D fractures are parallel to those of assemblage 3 in the compressive part, and type C fractures correlate with those of assemblage 4. In Coconino sandstone beams, fractures parallel to elements of assemblage 3 fractures are developed in both the compressive and the extensive parts. Counterparts of natural assemblage 1 and 5 fractures are missing in the experiments. Fractures parallel to those of assemblage 5 are not expected, since they are conjugate to shear displacements along bedding planes. The fact that fractures corresponding to those of assemblage 1 have not been found in any of the experiments is not understood. Since fractures of assemblage 1 are ubiquitous and often the best developed fractures of all five assemblages on all natural folds, their absence from experimental folds may imply that they are formed by a process that is not operative in the laboratory, e.g., erosional unloading, as *Price* [1959] has suggested, or as a result of different loading conditions, e.g., plane stress versus plane strain.

The correlation between type B and assemblage 2 fractures is particularly detailed. In both natural and experimental folds the angle between these conjugate shear fractures is less by 10°–15° than that of shear fractures of other arrays. This fact suggests that fractures are formed under a stress state in which σ_3 is close to 0 or even tensile, as is readily visualized on a Mohr diagram with a curved failure envelope [*Muehlberger*, 1961]. The fact that both strike-slip and dip-slip displacement took place on the type B shear fractures suggests that σ_1 and σ_2 were close to being equal in magnitude at the time of fracturing. In the experiments, type B fractures form only when the strain parallel to the specimen surface and normal to the fold axis

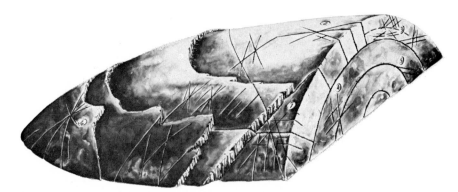

Fig. 28. Schematic illustration of fracture patterns on folds [after *Stearns*, 1968, Figure 10].

is an elongation, a fact that is consistent with the idea that the smaller dihedral angle is associated with lower mean pressure.

Dynamic analysis of calcite twin lamellas. Correlations between calcite compression and extension axes, twin-lamella spacing indices, surface strains, and the controlled loading conditions for the experimental folds substantiate previous interpretations of petrofabric data on natural folds, and they provide new insight for improving these interpretations.

As was summarized by *Friedman* [1964], *Carter and Raleigh* [1969], and *Friedman and Sowers* [1970], previous workers, with only one significant exception, have found that in folds calcite compression axes invariably form concentrations subparallel to the bedding plane and at high angles to the nearest fold axis and that extension axes are either normal to bedding or are distributed in a girdle about the compression-axis concentration. The single exception is *Conel*'s [1962] study of deformed calcite in the trough of a small syncline in a 10-cm-thick limestone bed. He has found that the compression axes are subparallel to the bedding in the compressive part of the layer and at high angles to it in the extension zone. His diagrams [*Friedman*, 1964, Figure 30; *Carter and Raleigh*, 1969, Figure 7] are very similar to ours for locations *f* and *g* in an experimental fold (Figure 19). This similarity extends even to the minor concentrations of compression axes parallel to the layering in the extended zone. *Friedman and Stearns* [1970] have shown that for superposed deformations the calcite compression axes derived from e_1 lamellas tend to cluster around the compression axis associated with the largest compressive strain, because by definition e_1 lamellas are the best-developed set in each grain and because the lamellas increase in abundance as the strain increases (Figure 18). Friedman and Stearns demonstrate that the original e_1 lamellas resulting from natural deformation are demoted to e_2 when a specimen is redeformed experimentally such that the axis of shortening is oriented 90° to the original axis of compression. Hence, in natural folds, calcite compression axes have not been found parallel to fold axes, because the compressive strains parallel to the fold axes are not as great as those normal to it. The results from the study of specimens 3 and 14 substantiate this inter-

pretation. First, without exception, the compression axes are locally subparallel to the direction of greatest shortening as determined from the surface-grid distortions. Second, except in zones of extension, the compression axes are oriented at high angles to the fold axes, and they either lie parallel to the upper and lower surfaces, i.e., the 'bedding planes,' or are inclined at low angles to them.

The distribution of twin-lamella indices throughout a folded limestone beam (Figure 16) is also consistent with a stress pattern that results from the superposition of bending stresses on the axial compressive prestress.

The following new facts may be useful to those who apply these techniques.

1. Minor concentrations of compression axes, if they are consistent and reproducible, may be used as evidence for superposed deformation, as in diagrams *f* and *g* of Figure 19 and diagram *d* of Figure 20. This practice is consistent with the numerical modeling of *Dieterich* [1970, p. 472].

2. Where compression axes form a point maximum, the extension axes are distributed in a complete or partial girdle around it. Concentrations of extension axes within the girdles are not necessarily parallel to the direction of greatest elongation.

3. Where the extension axes form a point maximum, the compression axes are distributed in a complete or partial girdle around it. Here the maximum of extension axes does coincide with the direction of greatest elongation, and concentrations of compression axes within the girdle do seem to coincide with the local direction of maximum shortening. Points 2 and 3 are consistent with the orientation of calcite compression and extension axes after the method of *Turner* [1953] for states of stress defined by $\sigma_1 > \sigma_2 \simeq \sigma_3$ and $\sigma_3 < \sigma_1 \simeq \sigma_2$, respectively.

CONCLUSIONS

Folding of single-layer beams of limestone and sandstone has been achieved under moderate confining pressures in the laboratory. From the experimental data and the associated petrofabric studies we draw the following conclusions, which we regard as geologically significant.

1. End-loaded thin beams of rock and of brass (for comparison) with aspect ratios of about 30 or more are geometrically unstable.

They buckle elastically, and their critical buckling stresses are well predicted by the Euler formula for an unsupported beam fixed at both ends.

2. Thick beams of the brittle sandstone do not fold; they first fail by shear fracturing near their ends, because the breaking strength is exceeded by both the yield stress and the critical buckling stress. Thick beams of the ductile limestone do fold, even for an aspect ratio as low as 4.4. Their axial stress-strains curves rise monotonically, and thus they reflect no instability. Nevertheless, these beams must also be geometrically unstable, since short columns do not fold; they flow homogeneously in conventional triaxial compression tests.

3. Once a thin beam has buckled, a physical instability arises from failure in the crestal region. In the brass, a nearly perfectly elastico-plastic material that yields by intracrystalline slip, plastic flow occurs in the hinge, the limbs are rotated rigidly, and a chevron fold develops. In the sandstone, an essentially brittle material that behaves triboplastically under our test conditions, cataclastic flow occurs in the hinge, and the folds that develop are again V shaped. For both materials the half wavelength of the fold is equal to the original length of the beam.

4. The limestone beams are ductile, and their inelastic deformations are certainly due to intracrystalline slip, yet no physical instability is obvious from either the axial stress-strain curves or the ultimate shape of the fold. Only at low confining pressures or large deflections does the stress-strain curve fail to signal the instability that is due to the onset of fracturing. The profiles of the central anticlines that develop are closely approximated by cosine curves; the half wavelength is equal to about three quarters of the original length independent of confining pressure and aspect ratio, and the central anticline and two adjacent synclines always form. Our data are insufficient to establish the generalized rheologic law for limestone, but this material is evidently not simply elasticoplastic.

5. Our petrofabric analyses of the development and orientation of fracture arrays, of surface strains, and of calcite twin-lamella spacing indices and derived compression-extension axes are all consistent with the concept that the net fiber stress in the folded layer is closely approximated as the sum of the axial compressive prestress and the superposed bending stresses (2), even though the deformations are finite.

6. We have identified the counterparts of three of the four fracture arrays that are commonly observed in the field and that can be expected in a single layer lacking interfacial shear stress. These are (1) an array of conjugate shears and associated extension fractures related to extension normal to the fold axis where σ_1 is parallel to that axis, σ_2 is normal to bedding, and σ_3 is parallel to dip, (2) an array of conjugate shear and extension fractures related to elongation in the extended portion of a single layer where σ_1 is normal to bedding, σ_2 is parallel to the fold axis, and σ_3 is parallel to dip, and (3) an array of conjugate shear fractures related to thrust faulting on the flanks in the inflection region between anticline and syncline where σ_1 is horizontal, σ_2 is parallel to the fold axis, and σ_3 is vertical. The ubiquitous natural flank assemblage that relates to a σ_1 normal to the fold axis and a σ_2 normal to bedding is not observed in the laboratory, and we suggest that the origin of this array is not yet understood.

7. The petrofabric techniques developed by *Turner* [1953] for the dynamic interpretation of data on calcite twin lamellas are clearly applicable to the mapping of complex stress fields in folded limestones. In particular, we find that compression-axis maximums are definitely skewed to the bedding, as the axes of quadric shortening are parallel to them, so that the similar skewing often observed in diagrams from natural folds should not be ignored. Furthermore, this skewing is consistent with the prediction of *Dieterich* [1969, 1970] and of *Spang and Chapple* [1970] from their numerical modeling of folding.

Acknowledgments. We wish to gratefully acknowledge the inspiration of Professor David Griggs. In particular, the entire field of experimental rock deformation, now more viable than ever, has sprung largely from his pioneering and always first-rate research.

Our research on folding has been generously supported by the National Science Foundation under grant GA-23332.

Plasticity of Single Crystals of Synthetic Quartz

B. E. HOBBS,[1,2] A. C. MCLAREN,[3] AND M. S. PATERSON[1]

Single crystals of synthetic quartz containing amounts of OH in the range 700–6000 ppm of H/Si have been deformed plastically at 3-kb confining pressure, and the influence on their behavior of temperature, crystal orientation, OH content, and strain rate has been examined. Three stages can generally be distinguished in the stress-strain curve after initial yield, but whether, or to what extent, each is developed is strongly dependent on the parameters above. The initial or yield stage often shows a yield point drop, the magnitude of which increases with OH content. The following stage, stage 1, increases in extent with temperature and is associated with a low rate of strain hardening. Then stage 2 enters, showing a high rate of strain hardening that decreases as the temperature increases. Dislocation structures for each stage have been established by using transmission electron microscopy, and they can be correlated with strain-hardening behavior. There is considerable correspondence with the behavior of diamond structure materials, and the microdynamical theory of the yield behavior of these materials can be applied to quartz, its application indicating a similarly low stress dependence of dislocation velocity. An influence of OH on the dislocation velocities is seen as the principal role of OH in the initial yield behavior, and the seemingly paradoxical observation that at high temperatures OH-rich crystals have a higher yield stress than OH-poor crystals can be rationalized. However, the hydrolytic weakening phenomenon of Griggs and Blacic (1965) also involves changes in the strain-hardening behavior, probably including recovery processes.

In 1964 D. T. Griggs and J. D. Blacic demonstrated that the introduction of trace amounts of OH into the quartz structure has a great effect on the mechanical properties, resulting in a decrease in strength by an order of magnitude at high temperatures. *Griggs and Blacic* [1965] and *Griggs* [1967] further showed that, below a critical temperature T_c, quartz remains strong, its strength decreasing only slowly as the temperature increases up to T_c. Above T_c, quartz is weak, and a relationship between T_c and the OH content of the quartz was established [*Griggs*, 1967, Figure 9]. In those experiments, T_c was defined in a stress relaxation experiment by plotting $\Delta\sigma/\sigma$ against temperature, where $\Delta\sigma$ is the change in the differential stress that the crystal can support when the temperature is changed by a prescribed amount. In general, there is a discontinuity of slope in such a plot, and T_c is defined as the temperature at which this discontinuity occurs [*Griggs*, 1967, Figure 8].

This phenomenon of weakening, dependent on OH content, is known as hydrolytic weakening, and T_c is the hydrolytic weakening temperature [Griggs, 1967]. The explanation advanced at that time (jointly by D. T. Griggs and F. C. Frank) was that Si-O-Si bridges become hydrolyzed and that this process enables slip to take place relatively easily once the temperature T_c is reached where diffusion of OH takes place rapidly enough to hydrolyze the Si-O-Si bridges adjacent to moving dislocations. The dislocations were considered to propagate by developing kinks in regions of high OH concentration; thus some parts of dislocation lines would be relatively mobile, whereas adjacent parts in regions of 'dry' quartz would be relatively immobile. Thus the important processes in hydrolytic weakening were considered to be the diffusion of OH through the quartz structure to dislocations and the breaking of the hydrolyzed bonds. This

---

[1] Department of Geophysics and Geochemistry, Australian National University, Canberra, A.C.T., Australia.
[2] Now at the Department of Earth Sciences, Monash University, Clayton, Victoria, Australia.
[3] Department of Physics, Monash University, Clayton, Victoria, Australia.

hypothesis was in accord with all observations made up to mid 1966.

Further experiments by D. T. Griggs and J. D. Blacic (personal communication, 1966), using synthetic crystals of high OH content, established that much of the strain in what had been thought to be the elastic range was nonrecoverable on unloading. Similar results were described by *Hobbs* [1968]. Furthermore, *Mc-Laren and Retchford* [1969] found that the crystals loaded a little below T_c with an approximately linear stress-strain curve had an amazingly high density of dislocations, in excess of 10^{11} cm^{-2} for strains of about 3%. It appeared then that what had been thought to be 'strong' quartz was actually strain-hardened quartz and that the slope of the strain-hardening curve was similar to the elastic slope; it had not been possible to demonstrate clearly the transition from elastic to plastic behavior.

McLaren and Retchford [1969] examined a number of specimens deformed above and below T_c and specimens that had been given various types of heat treatment above T_c after deformation at a lower temperature. They found that specimens deformed below T_c had a high density of tangled dislocations typical of a heavily cold worked metal, whereas specimens deformed above T_c had low dislocation densities with many obvious signs of recovery. They concluded that T_c is the temperature at which the rate of recovery balances the rate of strain hardening, and they derived an expression relating T_c to the concentration C of H atoms in the quartz structure for a constant strain rate:

$$C = (W/A) \exp (E/kT_c)$$

where W is the rate of strain hardening, A is a constant, E is the activation energy for recovery, and k is the Boltzmann constant. This expression is consistent with the relationship between C and T_c observed by *Griggs* [1967, Figure 9]. The essential point of the hypothesis put forward by McLaren and Retchford is that climb of dislocations rather than slip is the important process in hydrolytic weakening.

The hydrolytic weakening effect was also demonstrated in synthetic crystals by *Heard and Carter* [1968], using a gas apparatus. Further studies of hydrolytically weakened synthetic quartz were made by *Baëta and Ashbee*

[1970], whose experiments were done at atmospheric pressure. This work, published after most of our experiments had been made, demonstrates some of the orientation effects reported here, but it is mainly complementary to the present work because of the different conditions (the different OH content, the absence of confining pressure, the consequently more limited range of temperatures, and the other orientations).

This paper describes the results of experiments in which the effects of temperature, crystal orientation, OH content, and strain rate on the mechanical properties of single crystals of synthetic quartz were investigated in more detail than previously. Finer details of the stress-strain curves were revealed, and a correlation between the mechanical properties and the dislocation structure established. The experiments were conducted at 3-kb confining pressure, and the results are not complicated at the higher temperatures by the growth of metastable polymorphs of silica, as they were in the experiments at atmospheric pressure of *Baëta and Ashbee* [1970].

EXPERIMENTAL DETAILS

The specimens were deformed by axial shortening in the apparatus described by *Paterson* [1970], which employs Ar as the pressure medium and has an internal load cell. The specimen was an oriented cylinder 12 mm in length (ends flat and parallel to within 10 μm) and 5 mm in diameter, cored from a single crystal of synthetic quartz. The specimens were sleeved in annealed Cu tubing to produce an external diameter of 1 cm to suit the specimen arrangements in the apparatus. The Cu-sleeved specimen was then assembled between the deforming pistons and sealed from the Ar gas by an overlapping thin Cu jacket. Carbide rings were used to make push-fit seals on the carbide pistons.

TABLE 1.   Composition of Crystals

Crystals	Initial dislocation density N_0, 10^3 cm^{-2}	OH content, atomic ppm H/Si	Na content, weight ppm
S1	10	6100	280
W1	1.2	1850	120
W2	~1	1600	60
W4	1	730	70

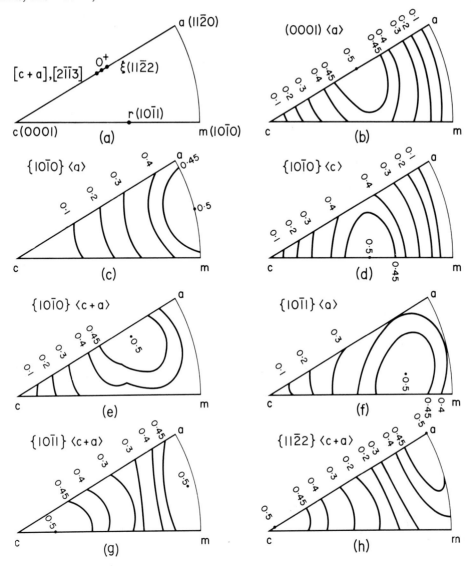

Fig. 1. Crystallography of slip systems and resolved shear stress coefficients cos χ cos λ, where χ and λ are the angles from compression axis to slip plane normal and slip direction, respectively. Equal-area projection is used. (*a*) Projection of planes and directions. (*b*)–(*h*) Resolved shear stress coefficients for the most highly stressed system with indices shown. (In some cases, different combinations of equivalent planes and directions may appear as the most highly stressed system in different parts of the projection). If the equivalence of r and z, ξ and ξ', and $\langle c + a \rangle$ and $\langle c - a \rangle$ is assumed, other sectors of the whole projection can be obtained by reflections across the $c - a$ and $c - m$ boundaries.

The specimen itself was vented to the atmosphere through the top hollow carbide piston, through which thermocouples could be inserted [*Paterson*, 1970, Figure 3].

All experiments were performed at 3-kb confining pressure, at temperatures in the range 350°–900°C, and, unless otherwise specified, at an average strain rate of 10^{-5} sec^{-1}. In each run a moveable thermocouple was placed just above the top of the specimen, and during the run the temperature gradient above the specimen was monitored by moving the thermocouple through

2 cm. The temperature gradient can be altered by changing the current in the windings of the upper furnace [*Paterson*, 1970, Figure 2], and previous calibration runs are used to establish the gradient necessary over these 2 cm so that the gradient along the specimen is small. In this way the gradient can be flattened to within 5°C or better.

Stress-strain curves have been derived from the load-displacement records after correcting for the elastic distortion of the apparatus and for the load supported by the thick Cu sleeve (which is often as much as one-third of the load supported by the quartz, although it is always substantially less at the higher stress levels). Both load and displacement were recorded continuously on a strip-chart recorder. The load was measured with an internal load cell that was calibrated by comparing load increment readings with those from an external

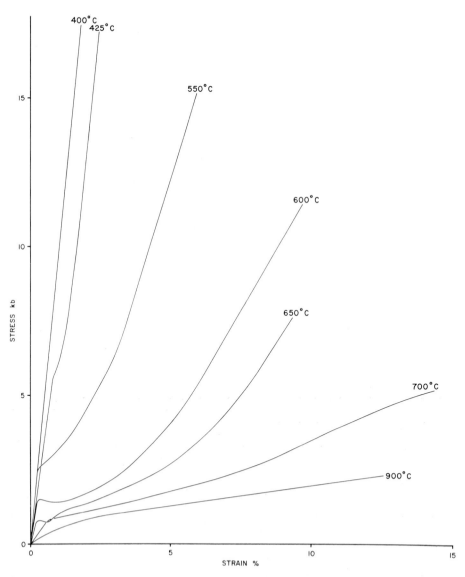

Fig. 2. Stress-strain curves for the crystal $W4$ loaded at 3-kb confining pressure at various temperatures. The strain rate is 10^{-5} sec^{-1}; the orientation is perpendicular to r.

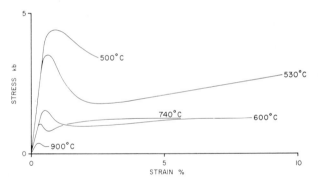

STRESS kb

STRAIN %

Fig. 3. Stress-strain curves for the crystal *W1* loaded at 3-kb confining pressure at various temperatures. The strain rate is 10^{-5} sec^{-1}; the orientation is perpendicular to r.

load cell during loading within the elastic range of a strong specimen. The displacement was measured with a linear variable differential transformer. The apparatus distortion was calibrated from the apparent load-displacement curves obtained from strong dummy specimens (carbide or natural quartz) of known lengths and Young's moduli. The correction was found to be linear, except at low loads, with a value of 0.1 mm per 1000 kg of load; this figure has been used throughout in the reduction of the stress-strain curves. At piston loads of less than 10^3 kg the applied correction should have been slightly greater, but some lack of reproducibility made a single correction unrealistic. It is estimated that, at stress levels in the ranges 0–2 kb and above 2 kb on the specimens used here, the indicated elastic slope could be in error by factors of the order of 10 and 2, respectively.

The reduction of the stress-strain curves has been programmed for an IBM/360 computer, and the final stress-strain curves were generally based on 20–30 calculated points. The computed stress is calculated on the current cross-sectional area of the specimen, when a homogeneous finite strain and no volume change are assumed, and is thought to be accurate to within about 5% except at very low loads. The strain is based on the length at room temperature and pressure and is thought to be accurate to within about 0.2% strain. In all about 100 experiments were performed, and of these about 20 runs were repeats; a particular stress-strain curve appears to be reproducible within 10%.

All specimens were heated at 600°C and 3-kb confining pressure for 1½ hours before deforma-

tion. There is some evidence that this treatment tends to homogenize the OH distribution in synthetic quartz crystals [*Hobbs*, 1968].

The material used was supplied by Drs. Baldwin Sawyer (crystal *S1*) and D. W. Rudd of Western Electric (crystals *W1, W2, W4*) in the form of optically clear synthetic quartz crystals. The OH and Na contents of these crystals are given in Table 1; the OH content was determined from the 3-μm infrared absorption spectrum measured at the Spectroscopic Laboratory of the Chemistry Department of Monash University, Clayton, Victoria, Australia, by using crystal X_0 [*Griggs*, 1967] as a standard, and the Na content was determined by B. W. Chappel of the Geology Department of the Australian National University, Canberra, A.C.T., Australia, by flame photometry. The intrinsic defects present in these crystals before deformation were examined by X-ray topography, and the initial dislocation densities are given in Table 1. These defects are similar to those reported by *Lang and Miuscov* [1967], *McLaren and Retchford* [1969], and *McLaren, et al.* [1971] for other synthetic quartz crystals.

The orientation conventions used throughout this paper are those described by *Carter et al.* [1964]. Figure 1 indicates the resolved shear stress coefficients for each of the slip systems thus far recorded for quartz at high confining pressure [*Christie and Green*, 1964] and for all orientations of the compressive stress in relation to the crystal. Orientations were checked by taking back-reflection Laue exposures from the ends of some cores before deformation.

Specimens for transmission electron micros-

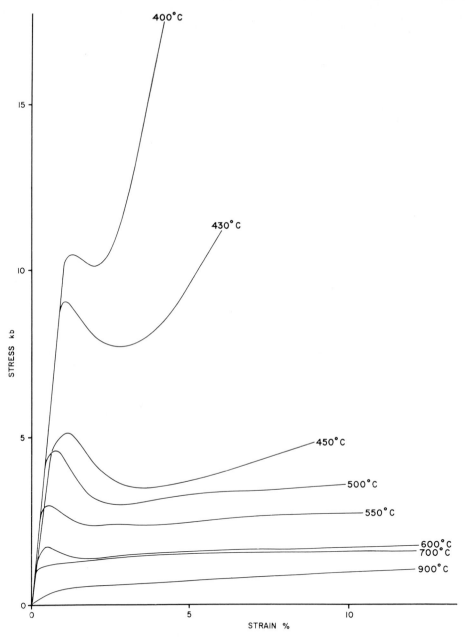

Fig. 4.   Stress-strain curves for the crystal *S1* loaded at 3-kb confining pressure at various temperatures. The strain rate is 10^{-5} sec^{-1}; the orientation is perpendicular to r.

copy were mainly prepared by crushing [*Mc-Laren and Phakey, 1965a*], but late in the study an ion bombardment technique was used [*Gillespie et al., 1971*]. The specimens were examined in a JEM-6A electron microscope operating at 100 kv.

EXPERIMENTAL RESULTS

Influence of temperature: orientation perpendicular to r.   Figures 2, 3, and 4 show examples of the stress-strain curves for the crystals *W4*, *W1*, and *S1*, respectively, loaded in the $\perp r$ orientation at temperatures in the range 400°–

900°C. Each curve shows an initial elastic part with a slope of approximately 10^{12} dynes cm^{-2}. Departures from 10^{12} dynes cm^{-2} occur, but, in view of the uncertainties regarding the initial slope of the stress-strain curve at stresses less than about 5 kb (see the section on experimental details), no significance can be attached here to this variation. Following the elastic part of the curve the crystal gradually yields; this yield region is often associated with a decrease in stress (yield point). (Yield is the first onset of substantial plastic flow. However, when yielding is accompanied by a drop in stress with further straining, the term 'yield point' is often used in the material sciences to describe this part of the stress-strain curve. The maximum stress reached before the drop is called the 'upper yield stress,' and the minimum to which the stress falls is called the 'lower yield stress.') The prominence of the yield point increases in going from the crystal $W4$, where it is only developed at 600° and 650°C (Figure 2), to the crystal $S1$, where well-developed yield points are present in the range 400°–600°C (Figure 4). The yield point is not sharp, as it is in α-Fe [*Cottrell and Bilby,* 1949], but it is rounded and gradual, as in LiF [*Johnston,* 1962] or in materials with the diamond structure [*Alexander and Haasen,* 1968]. Following the yield region the slope of the stress-strain curve tends to be relatively low. Later, after a range of strain that increases as temperature increases, the slope may increase again to relatively high values, even comparable at the lower temperatures to the elastic slope. Thus one can distinguish four stages in the stress and strain curve, as are depicted in Figure 5, although one or more of the stages may be absent for a given specimen. The over-all trend is for strain-hardening rates in all the stages to be lower at higher temperatures.

At high stress levels (generally greater than 25-kb differential stress) there is often a stress drop associated with fracture. At low temperatures this drop is sudden and accompanied by a loud report. Thin sections show a single fracture plane. At intermediate temperatures the stress drop is small and quite gradual; afterward the deformation may continue to large strains (in excess of 20% shortening) at fairly constant stress. Thin sections show a wide zone through which many fractures are distributed.

At high temperatures, deformation continues to 25% shortening without fracture.

Influence of crystal orientation. The stress-strain curves for specimens loaded in the orientations O^+, $\perp m$, and $\|c$ at various temperatures show a range of forms similar to that of those curves given for the $\perp r$ orientation, as is shown in Figures 6, 7, and 8. Only crystal $W4$ was investigated in detail. From the remaining fragments of the $S1$ crystal, two cores of other orientations were obtained, and thus the stress-strain curves of Figure 9a were measured. There is a general trend to higher stress levels in going from $\perp r$ through $\perp m$ and O^+ to $\|c$.

Influence of strain rate. The effects of changes in strain rate have not been systematically studied here, but Figure 10 is presented to give an indication of the rapid way in which the mechanical properties vary with strain rate. This figure refers to the crystal $W4$ loaded in the $\perp m$ orientation at 600°C and at strain rates of 10^{-4}, 10^{-5}, and 10^{-6} sec^{-1}. An extraordinarily rapid variation of both the initial flow stress and of the strain-hardening characteristics is indicated, although this sensitivity to strain rate

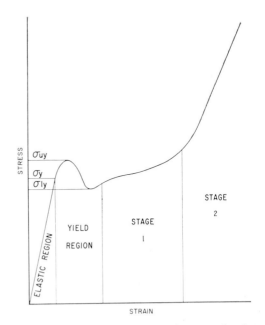

Fig. 5. Generalized stress-strain curve showing the elastic and yield regions and stages 1 and 2. The yield stress, at which departure from ideal elastic behavior is first observed, is σ_y, the upper yield stress is σ_{uy}, and the lower yield stress is σ_{ly}.

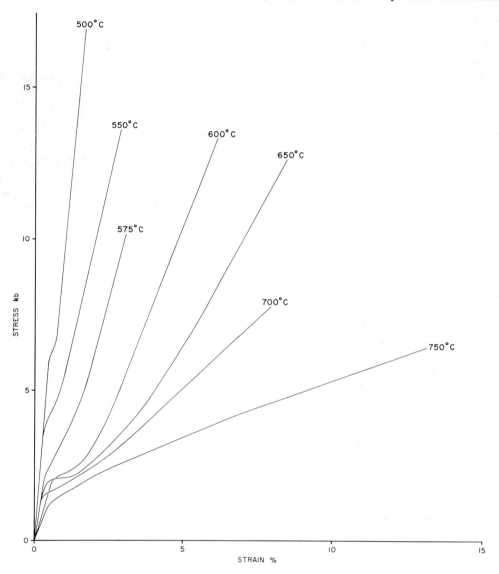

Fig. 6. Stress-strain curves for the crystal *W4* loaded at 3-kb confining pressure at various temperatures. The strain rate is 10^{-5} sec^{-1}; the orientation is O^+.

may be accentuated by the fact that the temperature is near the hydrolytic weakening temperature.

This strong dependence of mechanical properties on strain rate apparently explains, at least in part, why some of the stress-strain behavior reported here was not observed in crystals of high OH content by *Heard and Carter* [1968], who initially loaded their specimens rapidly before continued straining at 10^{-5} sec^{-1}.

OPTICAL AND ELECTRON MICROSCOPE OBSERVATIONS

Since the primary aim of this study was to determine the characteristics of the stress-strain curve, most specimens were loaded to high strains or high stresses. This situation means that little detail can be determined from these experiments concerning the development of deformation features with increase in strain. However, there are enough specimens deformed to various strains at different temperatures to permit a few general comments.

Before yielding. As is indicated in Table 1 and illustrated by Figure 11, the initial dislocation density of all crystals is low (10^3–10^4 cm^{-2}), increasing as the OH content increases. This low density is not grossly affected by the preliminary heat treatment. Most specimens were taken from the Z-growth regions, where dislocation lines are approximately parallel to c and their Burgers vectors are predominantly parallel to a or to $\langle c + a \rangle$ [cf. *McLaren et al.*,

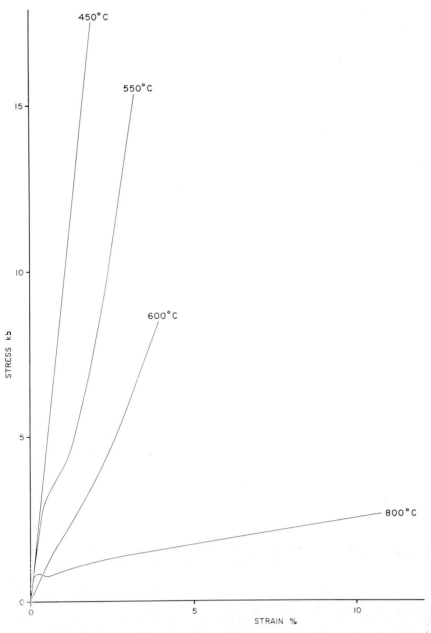

Fig. 7. Stress-strain curves for the crystal *W4* loaded at 3-kb confining pressure at various temperatures. The strain rate is 10^{-5} sec^{-1}; the orientation is perpendicular to *m*.

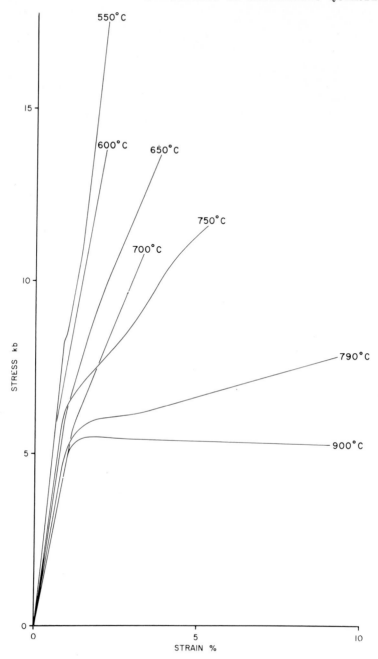

Fig. 8. Stress-strain curves for the crystal $W4$ loaded at 3-kb confining pressure at various temperatures. The strain rate is 10^{-5} sec^{-1}; the orientation is parallel to c.

1971]. However, some specimens from all crystals, in particular $S1$ (Figure 11b), were taken from the X-growth regions, where the initial dislocation lines trend approximately parallel to an a axis. There appears to be little difference in the mechanical behavior of specimens taken from these two regions, this finding indicating that the intrinsic defects have little influence on the mechanical properties. Specimens that have been heat treated at 3-kb confining pressure (see the section on experimental details) and loaded in the elastic range show no change in the optical microscope, and no dislocations or bubbles are observed in the electron microscope. Thus the dislocation density is less than 10^6 cm^{-2}. However, the specimens are too distorted because of elastic strains associated with cracking to be successfully examined by X-ray topography.

Yield region. The development of large-amplitude yield points in crystal *S1* loaded in the $\perp r$ orientation appears to be associated with the pervasive development of deformation lamellas throughout the entire specimen. No large kinks or regions of grossly inhomogeneous strain are developed. On the other hand, in *W4* loaded in the $\perp r$ orientation at 600° and 650°C, in which a small-amplitude yield point is developed (Figure 2), the optical microscope shows a development of lamellas in a wide kink band approximately parallel to (0001). This feature consistently appears at one endpiece and extends across the crystal, producing a large kink offset in the specimen margin. It presumably develops at the yield point. In all cases the lamellas are

8°–10° from the $\{10\bar{1}0\}$ plane with highest resolved shear stress, toward the axis of the specimen. The type of behavior in *W4* is to be expected from analogy with diamond structure materials [*Alexander and Haasen*, 1968]. However, the behavior of crystal *S1* indicates that the yield point is not necessarily associated with the development of large-scale inhomogeneities in strain.

In *S1* loaded in the $\perp r$ orientation at 450°C to just past the yield point (about 1% permanent strain) the dislocation density has risen to 5×10^9 cm^{-2} (Figure 12a). The dislocation lines are wavy but show an over-all trend parallel to the $\langle c + a \rangle$ or $\langle c - a \rangle$ directions with high resolved shear stress. The dislocation structure at this stage is remarkably similar to that in other specimens that have been strained to 10% in stage 1 of the stress-strain curve (Figure 12b).

From the present observations it is difficult to draw unequivocal conclusions about active slip planes, but on the following grounds it would appear that $m = \{10\bar{1}0\}$ is the predominant slip plane in the $\perp r$ experiments.

1. The lamellas are nearly parallel to $\{10\bar{1}0\}$,

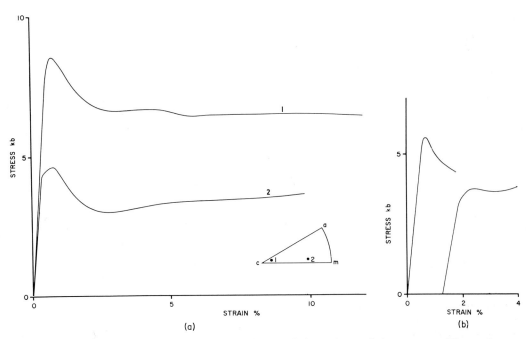

Fig. 9. Stress-strain curves for the crystal *S1* loaded at 3-kb confining pressure. The strain rate is 10^{-5} sec^{-1}. (*a*) Orientations shown in the pole figure. The temperature is 500°C. (*b*) The orientation is perpendicular to *r*, loaded to just past the yield point and then reloaded. The temperature is 480°C.

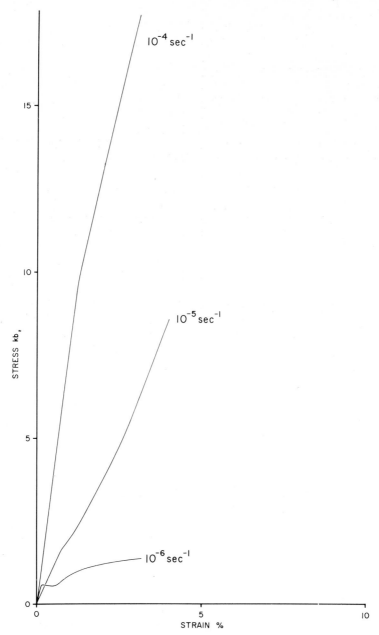

Fig. 10. Stress-strain curves for the crystal *W4* loaded at 3-kb confining pressure at various strain rates. The temperature is 600°C; the orientation is perpendicular to *m*.

and thus it is suggested that $\{10\bar{1}0\}$ is the slip plane by analogy with the slip trace observations of *Christie et al.* [1964]. Christie et al. established a correlation between lamellas and slip planes, although their observations were on basal slip in natural quartz, in which later observations

have also revealed Brazil twinning [*McLaren et al.*, 1967].

2. The large-scale kinking parallel to the basal plane around an axis close to *a* is consistent with slip on $\{10\bar{1}0\}$.

However, there are many loops and cusps

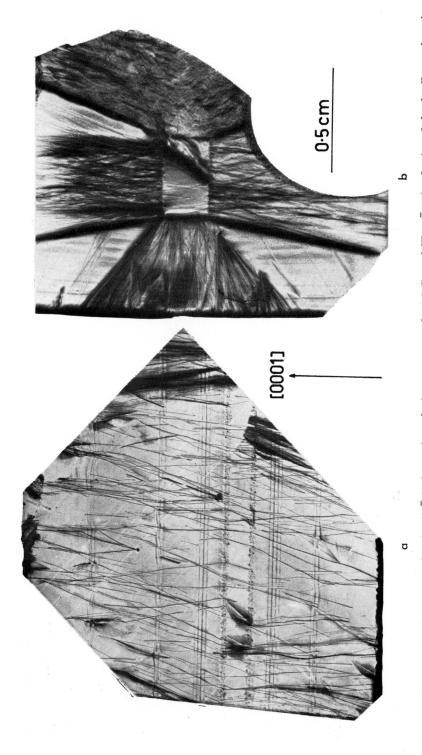

Fig. 11. X-ray diffraction topographs of $m(10\bar{1}0)$ plates of synthetic quartz crystals. (a) Crystal $W4$; ($\bar{1}101$) reflection. Only the Z-growth region is shown, and the seed crystal, passing horizontally across the topograph, is easily identified. (b) Crystal SI; ($1\bar{2}10$) reflection. Both X- and Z-growth regions are shown, and the seed crystal can be seen near the center of the topograph.

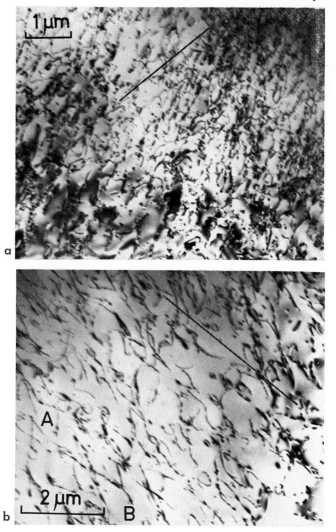

Fig. 12. Transmission electron micrographs of deformed synthetic quartz crystals. The trace of the predominant slip plane ($10\bar{1}0$), which is normal to the crystal slice, is marked. Note that the dislocations tend to lie parallel to this trace. (*a*) The crystal *S1* was loaded in the $\perp r$ orientation at 450°C to just past the yield point. (*b*) Similar to the top micrograph except that the crystal was deformed to 10% strain in stage 1. Note the cusps and loops, such as A and B, respectively.

(Figure 12) that do not lie in $\{10\bar{1}0\}$, and thus some climb or cross slip and the possibility of significant activity of other slip planes are indicated.

Stage 1, lamellas. All specimens loaded in the $\perp r$ orientation and deformed into the stage 1 region show the development of a broad deformation band that begins at or near one endpiece and extends diagonally across the specimen approximately parallel to (0001). In *S1* this band is rather diffuse (Figure 13*a*) and develops during stage 1. On the other hand, in *W4* the band has sharp boundaries (Figure 14) and develops at yield (cf. above). If the same argument as that given above for the yield region is followed, it appears (cf. Figure 12*b*) that the predominant slip plane continues to be $\{10\bar{1}0\}$ throughout stage 1.

Differences in the degree of development of lamellas in the typical *S1* specimen illustrated

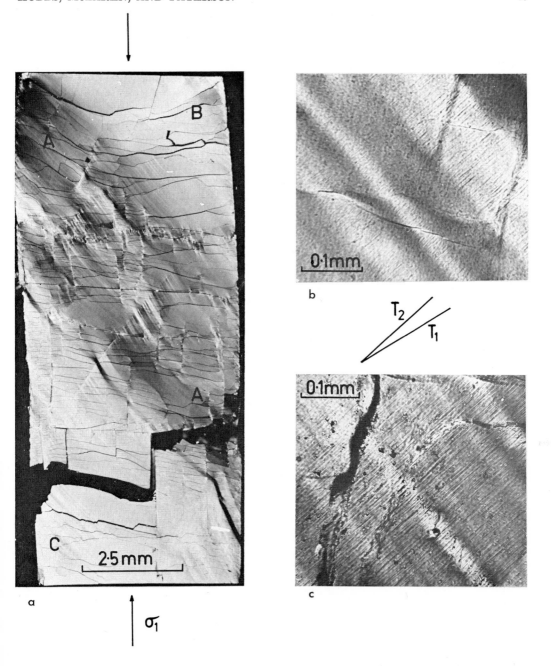

Fig. 13. Optical micrographs (crossed polarizers) of a thin section of crystal *S1* loaded in the orientation $\perp r$ and deformed to 7% strain at 450°C. (*a*) The whole of the thin section showing the broad deformation band AA that consists of narrow diffuse deformation bands. (*b*) Irrational deformation lamellas within the broad deformation band AA. Note that these lamellas develop preferentially in the (bright) regions that have suffered less lattice rotation than the adjacent (dark) regions suffered. T_1 and T_2 mark the traces of (1010) and the lamellas, respectively. (*c*) Strongly developed irrational deformation lamellas in the triangular end regions B and C. Note that the lamellas are farther from rational orientation in the bright band, which has suffered less lattice rotation than the surrounding dark region suffered.

in Figure 13 can be correlated with differences in lattice reorientation, the lattice rotation being less in the triangular end regions B and C than in the band AA. Closer examination of the broad deformation band AA shows that it consists of many narrow diffuse deformation bands (Figure 13b) and that deformation lamellas are confined to those zones (bright) in which the lattice rotation is much less than in the adjacent bands (dark). Also, in relation to the band AA, deformation lamellas are generally much more strongly developed in the end regions B and C. Here there are weakly developed narrow deformation bands, and the lamellas now appear both in and between the bands, but again the lamellas are more prominent where there is less lattice rotation; they also depart further from the

rational $\{10\overline{1}0\}$ orientation where there is less lattice rotation. Thus the smaller the amount of lattice rotation, the greater the tendency for lamellas to form and the greater the tendency to depart from rational orientation.

In $W4$ specimens there is some tendency for lamellas to develop throughout the specimen as they do in $S1$, not just to develop in the sharply defined deformation band, although their development may not be as marked as it is in $S1$. However, there are often also secondary, weakly developed, conjugate deformation bands, an example being the band BB cutting across the stronger band AA at C in Figure 14a. In the intersection region C, lamellas develop almost parallel to the axis of loading, whereas away from C they are 5°–10° from $\{10\overline{1}0\}$. The kink-

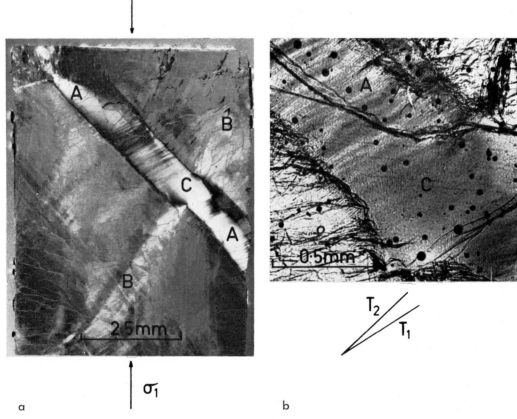

Fig. 14.   Optical micrographs (crossed polarizers) of a thin section of crystal $W4$ loaded in the $\perp r$ orientation and deformed to 13% strain at 600°C. (a) The upper half of the specimen showing the very strong deformation band AA due to prismatic slip, cut at C by the weaker band BB due to basal slip. (b) The region C at higher magnification showing deformation lamellas almost parallel to the axis of loading. T_1 and T_2 mark, respectively, the trace of (10$\overline{1}$0) and the direction of the lamellas in the matrix away from the deformation bands, respectively.

like nature of the band indicates its formation by basal slip, although basal lamellas have not been seen, so that lamellas in region C may have resulted from operation of the two intersecting slip systems {1010} and (0001), on both of which there is high resolved shear stress. It seems unlikely that these lamellas could result from operation of a single slip system; also, their orientation is inconsistent with simple rotation of {1010} lamellas by slip on (0001).

The strong development of lamellas in regions B and C of Figure 13 is also consistent with the formation of lamellas by multiple slip. Because these regions are clearly deformed but rotated a negligible amount, slip systems other than {1010} have presumably operated there.

The optical characteristics of all deformation lamellas are those discussed by *McLaren et al.* [1970]. Transmission electron microscopy shows that all of the deformation lamallas developed in these experiments consist of narrow slabs containing a dislocation density in excess of 10^{10} cm^{-2}; this finding is true independent of the orientation of the lamellas. The dislocations are tangled [see *McLaren et al.*, 1970, Figure 4], and the structure within each slab resembles that developed throughout the entire crystal during stage 2 at low temperatures (see below). The tangled nature of the dislocations within the lamellas is also suggestive of multiple slip. The regions between the lamellas have the structure illustrated in Figure 12b. Thus deformation has occurred between as well as in the lamellas. In fact, the tangled nature of the dislocations within the lamellas may indicate that these are regions in which not much movement of dislocations has been possible. That is, deformation lamellas may be regions of relatively low deformation in comparison with the intervening regions of lower dislocation density.

As the temperature is increased, the prominence of deformation lamellas decreases, and the subgrain structure becomes apparent optically. There is, however, no evidence of recrystallization in any of the specimens.

Stage 2. In specimens that have been strained well into stage 2, transmission electron microscopy shows very high dislocation densities, in excess of 10^{11} cm^{-2}, even in specimens that have been shortened only 3% at low temperatures [cf. *Hobbs*, 1968, Plate 11a; *McLaren and Retchford*, 1969, Figure 3]. The dislocation-generating

mechanism, then, is capable of rapidly producing high dislocation densities throughout the volume of the crystal. In $\perp r$ and O^+ specimens the dislocations are tangled, and there are no dislocation-free areas. In $\perp m$ specimens the dislocations tend to be straight and lie in the unit rhomb and {1010} planes that have high resolved shear stress, this finding thus suggesting that they are slip planes. There is an indication then of slip on at least three planes in this orientation. In ||c specimens, dislocations are again straight, and their orientations suggest slip on the unit rhomb planes. Specimens loaded in the ||c orientation also contain a high density of Dauphiné twins with composition planes parallel to {1011} and {1010}. In all orientations there is a considerable amount of strain associated with the dislocation image, and thus much interaction between dislocations is suggested.

The specimens strained into stage 2 also contain a high density of optically visible closely spaced cracks oriented normal to the axis of shortening when taken down from pressure [cf. *Hobbs*, 1968, Plate 1b]. These cracks are apparently not present at pressure after the load has been removed, since they do not appear when pressure is released in specimens with identical deformation history but annealed at high temperature at pressure. This treatment reduces the dislocation density to of the order of 10^8 cm^{-2}. The cracking must, therefore, be a response to high internal stress concentrations generated by the deformation.

DISCUSSION

General. The present experiments again show that quartz crystals containing small quantities of bound OH can be plastically deformed at elevated temperatures under moderate stresses, much below those needed for dry natural quartz. If differences in OH content and strain rate are allowed for, the stress-strain curves are generally in agreement with those measured on other samples of synthetic quartz by previous workers, at least within a factor of 2 and often closer [*Griggs*, 1967; *Heard and Carter*, 1968; *Hobbs*, 1968; *Baëta and Ashbee*, 1970]. The only marked discrepancy is with the earlier single measurement by *Hobbs* [1968] on a $\perp r$ specimen at 900°C, which gave a strength that was higher by roughly an order of magnitude than that found in the present experiments; the reason for this discrepancy is not known. Since the confining pressure has ranged from atmospheric pressure to 15 kb in the various studies, there is also evidently no marked influence of pressure on the stress-strain curves. However, the present results reveal much more detail in the stress-strain behavior than the earlier experiments in solid-medium apparatus revealed, and they cover a wider range of conditions than the experiments at room pressure covered.

The considerable behavioral differences between the individual crystals used here can be correlated, to a first approximation, with differences in OH content. The few analyses made indicate that the amounts of other impurities are relatively minor, although the Na contents do also increase as OH content increases (Table 1), and the possibility that Na plays an active role should be considered. The higher initial dislocation content in crystal S1 is not thought to contribute greatly to the differences in behavior (see the section on the yield region below). Therefore, in the ensuing discussion we treat OH content as the principal variable.

All the stress-strain curves conform in some degree to the generalized form shown in Figure 5, although whether, or to what extent, an individual region is developed depends on temperature, orientation, OH content, and strain rate. The distribution between regions of low hardening rate (stage 1) and high hardening rate (stage 2), and their correlation with the predominance of a single slip system and the marked activity of several intersecting slip systems, respectively, corresponds to what has been recognized in a number of materials of different crystal structure; these materials include face-centered cubic metals, body-centered cubic metals, hexagonal close-packed metals, diamond structure materials, and alkali halides [Nabarro et al., 1964]. (A stage 3, in which a lower rate of work hardening succeeds the high rate in stage 2, has not been observed in the quartz, possibly because stresses rise to a level at which fracture intervenes in experiments at 3-kb confining pressure before this stage is reached.) However, the two stages may not have the same physical significance in terms of hardening mechanisms in all cases [cf. Frank and Seeger, 1969]. Thus special characteristics of the mechanisms in quartz are suggested by its following properties.

1. The extent of stage 1 where it is well developed increases as temperature increases, a characteristic that is opposite the trend observed in most other materials.

2. The slope of stage 2 where it is well developed decreases as the temperature increases, whereas in other materials it is commonly independent of temperature.

3. During stage 2 the mean dislocation density rises quickly to a level (10^{11}–10^{12} cm^{-2}) greatly in excess of that characteristic of stage 1 (10^8–10^9 cm^{-2}), and the rate of strain hardening in stage 2 is abnormally high.

Despite the over-all conformity to the general characteristics just outlined, there are striking differences in detail between the stress-strain curves for the different crystals. Thus comparison of the $\perp r$ orientation results for crystals S1 and W4, the most comprehensive series, shows the following observations.

1. The stress-strain curves cannot be correlated in terms of homologous temperatures. For example, although at small strains the 600°C curve for W4 is lower than the 500°C curve for S1, they cross over at 5% strain to a reversed relationship at higher strains (Figures 1 and 3). Thus, although the stress level at 5% strain is about the same for both W4 and S1, the mechanisms of deformation must differ substantially in detail to account for the much higher rate of strain hardening in W4; this difference is presumably related to the different OH contents. Generally, higher rates of strain hardening, associated with lower stress levels, are found in the OH-poorer W4 crystal than are found in the OH-richer S1 crystal.

2. The yield point is more prominent in the OH-rich S1 specimens, but heterogeneity of deformation is, paradoxically, more marked in the OH-poor W4 specimens. Since the W4 specimens have the lower initial dislocation density N_0, the second trend may be compared with that in Ge and Si, where greater heterogeneity of deformation in yielding has been related to lower N_0 [Schröter et al., 1964; Siethoff, 1969], although the range of N_0 involved there was much greater than that involved here. However, the same analogy does not extend to the prominence of the yield point, since Ge and Si give a sharper yield point when N_0 is lower [Patel and Chaudhuri, 1963].

3. The yield stress is generally less temperature dependent in the *S1* specimens than it is in the *W4* specimens (Figure 15a).

There is no indication of any marked difference in behavior in the α and β stability fields. Assuming that the OH content does not significantly affect the α-β transition temperature, the transition will occur somewhat above 650°C, depending on the differential stress, since the confining pressure of 3 kb will raise it by about 78°, to about 652°C, and the compressive differential stress will raise it a further amount, which will vary with orientation from 5°C/kb for $\|c$ to 10°C/kb for $\perp m$ [*Coe and Paterson*, 1969]. Therefore all the stress-strain curves up to and including 650°C refer to α quartz, and all those 700°C and above refer to β quartz (except that, near the end of the 700°C curves for the *W4* specimens of $\perp r$, O^+, and $\|c$ orientations, the differential stress may have become just large enough to take the quartz back into the α field). However, there is a steady progression in behavior through the transition. The suggestion of *Baëta and Ashbee* [1970] that there is a significant change in deformational behavior at the α-β transition is therefore not confirmed. In fact, it does not seem to be supported by their results, since, according to the above figures for the influence of nonhydrostatic stress on the transition, all their illustrated stress-strain curves correspond to the β field, except the upper half of the 600°C $\perp B$ curve (their 650°C tests for this orientation would be well within the β field).

The present results are consistent with the role of OH in assisting to break Si-O bonds in the slip movement of dislocations or in diffusion-controlled processes such as climb or jog dragging (see the introductory section). However, the details of the processes in which it is involved may differ in the different stages of the stress-strain curve. We now discuss these stages in turn.

Yield region. Within the range of observation here the yield stress for specimens of a given OH content depends exponentially on temperature, and there are no strong discontinuities. This dependence is illustrated in Figure 15, in which, for convenience, the upper yield stress σ_{uy} is plotted in cases where a yield point occurs and the initial flow stress σ_y is plotted in the other cases. (The initial flow stress is the stress at which departure from a linearly rising curve is first obvious (Figure 5); σ_y is only slightly below σ_{uy} in the first cases.) Thus σ_{uy} or σ_y can be expressed as $\sigma_0 \exp(A_0/kt)$, except possibly for *W4* below 600°C. The preexponential term σ_0 increases with OH content, but the apparent activation energy A_0 decreases (Figure 15a). A_0 and σ_0 also change markedly with orientation (Figure 15b), the value of A_0, and hence the degree of temperature sensitivity, decreasing and σ_0 increasing, in the order $\perp r$, $\perp m$, O^+, $\|c$.

In other materials, yield points have been attributed to either dislocations breaking away from the pinning effects of an impurity atmosphere [*Cottrell and Bilby*, 1949] or multiplication of dislocations from an initially low density to a density capable of producing the required strain rate under a lower stress [*Johnston*, 1962]. The latter appears the more likely explanation of the yield points in the quartz in view of the low initial dislocation density and the similarity in the shape of the stress-strain curve near the yield point to that in LiF and diamond structure materials, where dislocation multiplication is thought to be the important factor [*Johnston*, 1962; *Alexander and Haasen*, 1968]; this explanation implies that the dislocation velocity has relatively low stress sensitivity. We therefore now consider the microdynamical theory of yielding developed for these materials, following particularly the theory expounded by *Haasen* [1964] and *Alexander and Haasen* [1968], and discuss its applicability to quartz, at least qualitatively.

This theory is based on a model having the following properties.

1. Velocity law. The dislocation velocity v depends on the shear stress and temperature according to

$$v = B(T) \cdot \tau_{eff}{}^m = B_0 \tau_{eff}{}^m e^{-U/kT} \tag{1}$$

where B_0, m, and U are constants and τ_{eff} is the effective or local shear stress acting on the dislocation in the slip plane. This law is supported by observations where dislocation densities are small and the applied shear stress τ is identified with τ_{eff}. Typical values of m are 1 for diamond structure materials, 16 for LiF, and a much higher value for the common metals.

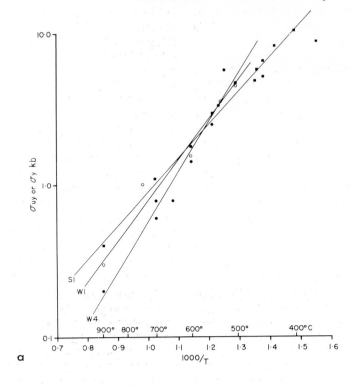

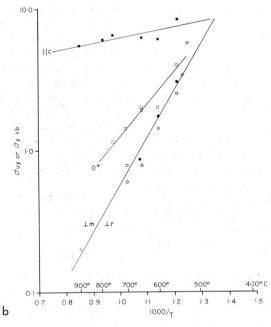

Fig. 15. Temperature dependence of yield stress (upper yield stress σ_{uy} when developed; initial yield stress σ_y when a yield point is absent). (*a*) Orientations perpendicular to r for crystals *S1* (squares), *W1* (open circles), and *W4* (solid circles). (*b*) Orientations perpendicular to r (open circles), O^+ (open squares), perpendicular to m (solid circles), and parallel to c (solid squares) for crystal *W4*.

No observations have been made on quartz, but it seems likely in view of the nature of the bonding that the value of m will be relatively low.

2. *Multiplication law.* The dislocation density N increases in time dt according to

$$dN = N_m v \, dt \, \delta \qquad (2)$$

where N_m is the density of mobile dislocations and δ is a multiplication parameter that is assumed to have the form

$$\delta = K \tau_{\text{eff}} \qquad (3)$$

where K is a constant. That is, the increase in dislocation density under a given stress is proportional to the moving length of dislocations present and to the distance traveled, and the number generated for a given dislocation movement is proportional to the stress. This situation leads to an exponential multiplication, which would help to account for the rapid increase in dislocation density observed in Ge and quartz.

3. *Strain-hardening law.* Strain hardening is introduced by using an effective stress, related to the applied stress by

$$\tau_{\text{eff}} = \tau - A N^{1/2} \qquad (4)$$

where N is the total dislocation density and A is a constant, $Gb/[2\pi(1 - \mu)]$, G being the shear modulus, b the Burgers vector, and μ the Poisson ratio. That is, the internal stress from the mean stress field of the other dislocations partly counteracts the applied stress, this partial counteraction thereby introducing a strain-hardening effect.

The theory of yielding is developed for a constant strain rate experiment in the following manner [*Haasen,* 1964, p. 186; *Alexander and Haasen,* 1968, p. 102]. The total strain rate is given by

$$\dot{\epsilon} = \dot{\epsilon}_{\text{el}} + \dot{\epsilon}_{\text{pl}} = \text{const}$$

The elastic strain rate $\dot{\epsilon}_{\text{el}}$ can be written as $\dot{\tau}/G$, where $\dot{\tau}$ is the rate of change of the applied stress. (Strictly, the appropriate Schmid factor (Figure 1) should be applied if axial strain is to be used instead of resolved shear strain on the slip system, but this refinement is unimportant in relation to the approximations involved in the theory.) The plastic strain rate $\dot{\epsilon}_{\text{pl}}$ is $N_m b v$, leading to

$$\dot{\epsilon} = (\dot{\tau}/G) + N_m b v \qquad (5)$$

Then, if we put τ_{eff} equal to τ and thereby neglect any strain hardening due to the $A N^{1/2}$ term in (4) for the small strains at first involved in yielding and if we assume that $N_m = N$, it follows that the flow stress will pass through a maximum (upper yield stress) as the second term becomes dominant. *Haasen* [1964] derives an approximate expression for this upper yield stress τ_{uy}, which can conveniently be written

$$\tau_{\text{uy}} = \left[(m + 2) G \cdot \frac{1}{K} \cdot \frac{\dot{\epsilon}}{B(T)} \cdot \ln \frac{N_{\text{uy}}}{N_0} \right]^{(1/m+2)} \qquad (6a)$$

$$= \left[(m + 2) G \cdot \frac{1}{K} \cdot \frac{\dot{\epsilon}}{B_0} \cdot \ln \frac{N_{\text{uy}}}{N_0} \right]^{(1/m+2)}$$

$$\cdot \exp \left[\frac{U}{(m + 2)kT} \right] \qquad (6b)$$

where N_{uy} is the density of dislocations at the upper yield stress.

At larger strains, as N increases further and the strain hardening due to the $A N^{1/2}$ term in (4) becomes important, the stress will rise again after passing through a minimum (lower yield stress τ_{ly}) given by

$$\tau_{\text{ly}} = \left[\frac{C(m) A^2}{b} \cdot \frac{\dot{\epsilon}}{B(T)} \right]^{1/(m+2)} \qquad (7a)$$

$$= \left\{ \frac{C(m) A^2}{b} \cdot \frac{\dot{\epsilon}}{B_0} \right\}^{1/(m+2)}$$

$$\cdot \exp \left[\frac{U}{(m + 2)kT} \right] \qquad (7b)$$

$C(m)$ is the number $(m + 2)^{m+2}/4m^m$, which is approximately 7 for $m = 1$, 66 for $m = 5$, and 218 for $m = 10$.

The occurrence of the yield point is, therefore, dependent on the relative strengths of the tendencies for the stress to fall as smaller velocities are required of the greater number of moving dislocations and to rise, owing to the $A N^{1/2}$ strain hardening. If the latter term is sufficiently prominent at small strains, rather than being negligible as was assumed in deriving (6), a yield point drop will not occur; and thus the absence of a yield point when N_0 is large would be explained. When the yield point is well developed, a measure of its prominence is given by

$$\frac{\tau_{\text{uy}}}{\tau_{\text{ly}}} = \left[\frac{(m + 2) G b}{C(m) A^2} \cdot \frac{1}{K} \cdot \ln \frac{N_{\text{uy}}}{N_0} \right]^{1/(m+2)} \qquad (8)$$

The quantity $\ln (N_{uy}/N_0)$ is a slowly varying function of N_0, K, $B(T)$, and $\dot{\epsilon}$. It follows from (1), (5), and (7b), with $\dot{\tau} = 0$, that

$$N_{uy} = \frac{\tau_{uy}^2}{C(m) A^2 \left(\dfrac{\tau_{uy}}{\tau_{1y}}\right)^{m+2}} \tag{9a}$$

and

$$\ln \frac{N_{uy}}{N_0} = \ln \frac{\tau_{uy}^2}{C(m) A^2} - \ln \left(\frac{\tau_{uy}}{\tau_{1y}}\right)^{m+2} - \ln N_0 \tag{9b}$$

We now apply this theory to quartz. This application can only be made tentatively here because of the paucity of data thus far available, and it should not be assumed that the following conclusions have more than qualitative value. If the 450°C curve in Figure 4 is taken to be typical of a well-developed yield point, (9a) leads to $N_{uy} = 9 \cdot 10^9$ cm^{-2} for $m = 1$, $4 \cdot 10^8$ cm^{-2} for $m = 5$, and $5 \cdot 10^6$ cm^{-2} for $m = 10$. Since a dislocation density of $5 \cdot 10^9$ cm^{-2} was observed a little past the yield point, the validity of *Haasen's* [1964] theory would therefore imply a value of m of probably around 5, intermediate between the values for Ge and LiF. If $m = 5$ is assumed, (8) leads to $K = 5 \cdot 10^{-5}$ cm dyne^{-1}, which corresponds to a multiplication rate substantially greater than that of Ge (for which $K \simeq 10^{-6}$ [*Berner and Alexander*, 1967]); $m = 10$ would be required to give a multiplication rate roughly equal to that of Ge.

When comparing the $\sigma_0 \exp(A_0/kT)$ fit to the data in Figure 15a with (6b), and assuming that all the high-temperature points for $W4$ correspond closely to the upper yield stress, even when a yield point drop is not clearly developed, it follows from the increase in σ_0 with OH content that the quantity KB_0 must decrease as the OH content increases. Also U, being equal to $(m + 2)A_0$, must decrease as the OH content increases. On the other hand, applying (8) to cases where a lower yield stress is developed indicates that change in OH content does not markedly affect K, at least within an order of magnitude. Therefore, the behavior of σ_0, reflected in KB_0, is mainly due to a decrease in B_0 as the OH content increases. A decrease in B_0 corresponds to a relative decrease in the dislocation velocities at high temperatures. This effect is revealed in the greater initial strength of the crystals of higher OH content at the higher temperatures, where the effect of the B_0

term outweighs the effect of lowering the activation energy. Since, according to (6), change in the strain rate would affect crystals of all OH contents in the same way, the crossover of the yield strengths shown in Figure 15a may also occur at geologically important strain rates.

The reason for the lesser development of yield points in crystal $W4$ is not clear. It cannot be correlated with changes in N_0, to which, in any case, the prominence of a yield point is relatively insensitive according to (8). (The suppression of a yield point by the presence of a large initial dislocation density is illustrated by the experiment in which a preyielded specimen was reloaded (Figure 9b).) Also, the relatively small changes in K appear inadequate to give an explanation. Thus the distinction may lie in other factors, such as the occurrence of a much more effective strain hardening than that which would correspond to the $AN^{1/2}$ law.

The decrease in apparent activation energy for the yield stress at higher OH content (Figure 15a) means on the above model (6b) that the activation energy U for dislocation motion is reduced by OH. This finding suggests that OH is effective in reducing the activation barrier involved in the Peierls stress. We may expect the Peierls stress to be the main factor resisting dislocation motion in the yield region in quartz on the grounds of (1) the nature of the bonding in quartz, (2) its initial low density of defects that might hinder dislocation motion, and (3) the observation that isolated dislocations are almost invariably straight and crystallographically aligned (or consisting of straight segments). See the review by *Guyot and Dorn* [1967] for a general discussion of the Peierls stress. The role of OH presumably involves the hydrolysis of Si-O bonds, if one follows the suggestion of *Brunner et al.* [1961] that OH can be incorporated in the quartz structure as

$$\text{Si} - \text{OH}$$

$$\text{HO} - \text{Si}$$

[cf. *Griggs and Blacic*, 1965]. The actual mechanism may be to assist the migration of kinks along the dislocation line, one bond at a time, in which case U would be the activation energy for kink migration. The measured values of A_0, which are in the range 10,000–15,000 cal mole^{-1} for the $\perp r$ orientations (Figure 15a), imply values of U in the range 70,000–100,000 cal moles^{-1} if $m = 5$ in (6b); such values are incompatible with the hypothesis that the diffusion of OH is the rate-limiting process [cf. *Kats et al.*, 1962].

The single straight-line fit to the yield stress plot (Figure 15) points to the fact that the rate-limiting step in dislocation motion is the same over the whole temperature range studied. At lower temperatures, however, a straight-line plot cannot be expected, even with a single activation energy, because the temperature dependence of $\ln(N_{uy}/N_o)$ in the pre-exponential term of (6b) is no longer negligible; this situation would tend to give somewhat lower yield stresses at low temperatures than would be predicted by extrapolation of the straight lines in Figure 15.

The change in yield stress trends with orientation may result, at least in part, from the change in dislocation dynamics in going to other slip systems, although there is insufficient data to analyze this change for other orientations in the same detail as for the $\perp r$ orientation. For the extreme case of the $\|c$ orientation the plot in Figure 15b corresponds to a higher σ_0 but lower A_0 than that for the orientation perpendicular to r. However, this situation does not necessarily mean lower values of KB_0 and U for unit rhomb slip, since a lower value of m could lead to the same effect.

In summary, the main aspects of the behavior of quartz in the yield region can be accounted for by a microdynamical theory. With *Haasen's* [1964] model the stress exponent m in the dislocation velocity law would have a low value, of between 1 and 10 and probably of about 5. The role of OH in yielding would then be seen as lying in two opposing areas of influence on the dislocation velocities under a given stress, namely, a relative decrease due to an increase in B_0, which predominates at high temperatures, and a relative increase due to a lowered activation energy, which predominates at lower temperatures. However, these conclusions are rather tentative, and more detailed studies, particularly direct measurements on dislocation dynamics, are needed to test the theory more stringently or to suggest modifications to the model. Alternative velocity, multiplication, and work-hardening laws, such as those proposed by *Gilman* [1965, 1966], *Chen et al.* [1964],

and *Gillis and Gilman* [1965], should be examined; in particular, a multiplication law that takes into account the loss or immobilization of dislocations, such as may be associated with recovery processes or the trapping of dislocations in tangles in the lamellas, merits special examination.

Stages 1 and 2. In stage 1 the specimen enters a regime in which, in contrast to the yield region, its properties change relatively little in correspondence with further strain, more closely approximating a steady state. The rate of strain hardening is small, the dislocation density remains approximately constant, and the strain apparently continues to result from the predominance of slip on a single set of planes. Explanation of this behavior on microdynamical theory using *Haasen's* [1964] approach would require modification or refinement of the model to account for the stabilization of the dislocation density. Since an active role of OH could be expected to facilitate climb as well as slip and thus to promote recovery processes, a likely modification would be the introduction into the multiplication law of a rate of annihilation of dislocations in addition to the rate of generation. A suitable form of such a modified law might then account for the behavior of stage 1, if it is seen essentially as a regime in which recovery processes are nearly balancing the tendency for dislocation multiplication and strain hardening; the higher the temperature, the more effective the recovery process would be.

Stage 2 of the strain-hardening curve is characterized by a very rapid increase in dislocation density. From the somewhat sketchy observations made so far the important factor in its onset is the appearance of significant amounts of slip on other slip systems. This secondary slip should lead to rapid hardening, since many of the expected dislocation interactions in quartz produce jogs that are relatively immobile, either because of low resolved shear stress or because the slip plane for the jog is not known to be an active one.

Delay in the onset of stage 2 hardening beyond that of stage 1 can probably also be understood by microdynamical theory. The activity and multiplication of dislocations in the other, less highly stressed, slip systems should be governed by the same laws as those governing the main slip system, with the appropriate values of the adjustable parameters. Therefore, during the earlier stages of deformation, when the density of dislocations in the alternative systems is still relatively low and interaction is not yet important, we may expect the dislocation density to build up in these slip systems in a manner analogous to that in the main system but more slowly because of the lower resolved shear stress. Thus, if the values of the parameters are suitable, significant dislocation densities for the production of appreciable slip

on the alternative systems and the occurrence of appreciable additional hardening due to interactions with dislocations in the main system may only be attained after substantial strain has occurred. However, a much more detailed knowledge of the dislocation dynamics is needed for the development of theory along these lines.

Hydrolytic weakening temperature. D. T. Griggs and J. D Blacic have defined a critical temperature for hydrolytic weakening from stress relaxation experiments [*Griggs,* 1967].

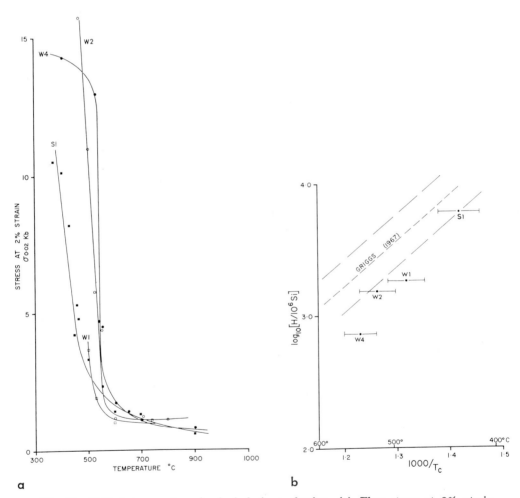

Fig. 16. Critical temperature for hydrolytic weakening. (*a*) Flow stress at 2% strain for crystals *S1* (solid squares), *W1* (open squares), *W2* (open circles), and *W4* (solid circles) as a function of temperature. (*b*) Relation between OH content and critical temperature for hydrolytic weakening derived from the diagram at the left; the dashed curve is the one obtained from stress relaxation experiments by *Griggs* [1967], for which the error limits are shown by the parallel broken lines. The error bars on the points from the present work correspond to an estimated ±20°C uncertainty in reading the temperature from the diagram at the left.

They measured the stresses that a given specimen would support at an increasing sequence of temperatures and defined the critical temperature as that at which the greatest rate of change of stress level occurred. Since significant plastic strain occurs during the relaxation following a temperature increment, the stresses measured are really flow stresses at points along a stress-strain curve of complicated history. If the machine were very stiff, the stresses would correspond approximately to intercepts on the normal stress-strain curves (Figures 1–9) at a strain abscissa reading of 4–5% (the initial elastic strain); the effect of machine compliance would tend to increase this total strain in the specimen as the stress relaxes. However, owing to the previous history of strain hardening at lower temperatures, the actual stresses will be somewhat higher.

Since we do not have data for plots of stress versus temperature exactly analogous to those from relaxation experiments, we have assumed that the stresses at 2% total strain indicate a comparable trend (2% was chosen rather than a higher value to give intercepts at the lower temperatures, where our curves have not been taken to very high stresses, but the exact choice of strain is not thought to be very critical). These stresses are plotted for $\perp r$ orientations in Figure 16a, and the inflexion point, or the point of most rapid decrease, is taken as a critical temperature of weakening. This plot leads to the dependence of the critical temperature of weakening on OH content that is shown in Figure 16b. The line from the relaxation observations of Griggs and Blacic is also shown for O^+ orientations.

A critical temperature for hydrolytic weakening, as determined by Griggs [1967] or as above, thus refers to the strength of the quartz after some strain hardening has occurred. That is, it reflects the influence of OH not only on the intrinsic or initial strength of the quartz but also on its subsequent strain hardening. Therefore, it presumably involves not only the influence on the Peierls stress but also that on the various factors that must affect the strain-hardening behavior, including any recovery processes (climb and so forth) that may be important, especially in stage 1, the stage to which our points in the plot of Figure 16b tend to refer at the higher temperatures. That the influence of OH on the Peierls stress alone does not lead to a sharply defined critical temperature of weakening is shown by the absence of an obvious bend in the yield stress plots of Figure 15 in the vicinity of the critical weakening temperature. Thus the Griggs-Frank explanation of hydrolytic weakening [Griggs, 1967] in terms of the influence of OH on the Peierls stress [Griggs, 1967] needs to be augmented by taking into account the additional roles of OH, probably including the role in recovery processes suggested by McLaren and Retchford [1969].

Acknowledgments. We thank G. T. Milburn for his assistance in preparing specimens for optical and electron microscopy, A. W. Geatly for keeping the deformation apparatus in running order, and B. Bryant for photographic work. Data from the crystal W2 were generously supplied by D. Morrison-Smith. The direction of this work has relied heavily on the accumulated experience in quartz deformation of D. T. Griggs and J. D. Blacic. We also thank them for their comments on an earlier version of this paper and are especially grateful to Griggs for pointing out some flaws in the discussion.

Transmission Electron Microscope Investigation of Some Naturally Deformed Quartzites

A. C. McLaren
Department of Physics, Monash University
Clayton, Victoria, Australia

B. E. Hobbs[1]
Department of Geophysics, Australian National University
Canberra, A.C.T., Australia

Transmission electron microscopy has been used to observe the dislocation substructures in a number of naturally deformed quartzites thinned by ion bombardment. The selected specimens exhibit the optical features that have been widely illustrated in the literature and that are characteristic of naturally deformed quartzites. By a comparison with experimentally deformed quartz and metals the observed substructures suggest that the deformation mechanism is more akin to cold work than to steady creep and that optical deformation lamellas are associated with sub-boundaries that do not necessarily define active slip planes.

Features such as deformation lamellas, Boehm lamellas, deformation bands, and subgrains have been described and measured in deformed quartzites since the pioneering work of *Sander* [1911, 1930], but there is still only limited knowledge of their detailed structure and origin. This paper examines by transmission electron microscopy a group of strongly deformed quartzites selected from regions with a detailed structural history that is well known. The aims are: (1) to describe the dislocation substructures developed in these rocks and, if possible, to correlate these substructures with the optically observed features; in particular, one aim is to describe the dislocation structure of naturally produced deformation lamellas, and (2) to comment on the conditions of deformation indicated by these substructures or on the thermal history experienced by these rocks subsequent to deformation.

In selecting the specimens an attempt was made to examine optical deformation features that are characteristic of deformed quartzites and that have been illustrated widely in the literature.

The existence of dislocations during the deformation and the recrystallization of quartz is now firmly established, as the result of studies originating in the experiments of Griggs and his co-workers [*Carter et al.*, 1964; *Christie et al.*, 1964]. These experiments established that natural quartz could be deformed plastically in the laboratory. At about the same time, dislocation arrays and networks were observed directly by transmission electron microscopy in natural vein quartz by *McLaren and Phakey* [1965b]. Shortly afterward the samples deformed experimentally by Griggs and his co-workers were examined with transmission electron microscopy by *McLaren et al.* [1967]. This examination confirmed the fact that dislocations were involved in the deformation mechanism and, in addition, showed that Brazil twinning by a shear mechanism was an important mode of deformation, particularly at low temperatures.

Subsequent transmission electron microscopy [*Griggs*, 1967; *Hobbs*, 1968; *McLaren and Retchford*, 1969; *Hobbs et al.*, this volume] has continued to emphasize the importance of dislocations in both the experimental deformation and the recrystallization behavior of quartz.

Until this stage all specimens for transmission

---

[1] Now at the Department of Earth Sciences, Monash University, Clayton, Victoria, Australia.

electron microscopy consisted of tiny randomly oriented fracture fragments prepared by crushing larger specimens [*McLaren and Phakey*, 1965b]. Application of this technique to strongly deformed natural quartzites, such as those from the Moine thrust [*Christie*, 1963], failed to reveal significant dislocation densities (J. M. Christie and A. C. McLaren, unpublished work, 1966). However, recently the technique of ion bombardment thinning, originally developed for metals by M. Paulus and F. Reverchon in 1961, has been applied with greater success to nonmetals [*Barber*, 1970; *Gillespie et al.*, 1971]. This paper is concerned with a study of the dislocation substructures in a number of naturally deformed quartzites thinned by the ion bombardment technique.

GEOLOGICAL SETTING FOR SPECIMENS

Specimens for this study were selected from three localities, the Mount Isa fault of Mount Isa, north Queensland, the Atnarpa thrust nappe complex in central Australia, and the Chewings Range quartzite of the Arunta complex in central Australia.

Mount Isa fault. The Mount Isa fault is one of the many zones of intense deformation that cut across the regionally metamorphosed rocks of the Mount Isa area [*Bennett*, 1965; *Wilson*, 1970, unpublished manuscript, 1971]. These zones consist of rocks with chlorite and white mica abundant in the pelitic varieties, and, although breccias may be present in local areas, most of the rocks are blastomylonites [*Christie*, 1960] with a steeply dipping foliation bearing a strong down-dip lineation. The Mount Isa fault itself is a feature about 70 km long and is associated with a zone of deformation varying in width from a few to 50 meters. Although truncation of stratigraphic markers is proved [*Wilson*, 1970], the magnitude of the displacement associated with this structure is unknown. In crossing the fault a complete progression from weakly deformed sediments to strongly deformed and recrystallized quartz blastomylonites is displayed. These zones are similar to blastomylonite zones from other parts of the world [*Christie*, 1963; *Johnson*, 1960; *Hobbs*, 1966b] and display microstructures and preferred orientation patterns that are typical of these low-grade blastomylonites.

The microstructure of quartz-rich rocks in the Mount Isa fault zone changes progressively toward the center of the zone in a manner similar to that described from the Moine thrust zone [*Christie*, 1963, p. 397]. The deformed detrital grains away from the zone of intense deformation show undulatory extinction and deformation lamellas and become progressively elongate and flat in the plane of the new foliation as they approach the center of the zone in correspondence with the appearance of numerous new grains (Figure 1a). Deformation lamellas are generally absent in the areas containing abundant new grains. In the more intensely deformed quartzite there is little evidence for the presence of the strongly oriented old grains, and the rock consists of numerous fine new grains. The preferred orientation has been discussed by *Wilson* [1970, and unpublished manuscript, 1971]. These rocks are Precambrian in age and were deformed in the Precambrian [*Farquharson and Wilson*, 1971].

Atnarpa nappe complex. This complex consists of a series of thrust nappes in a structurally low part of the Arltunga nappe complex [*Forman*, 1971]. The stratigraphic sequence consists predominantly of bedded and massive quartzites of Proterozoic age that have never been metamorphosed to grades higher than lowermost greenschist facies. This sequence is duplicated a number of times by a complicated group of thrusts (M. Yar Khan, unpublished manuscript, 1972). The quartzites between the thrusts are only weakly deformed, but, as each thrust is approached, the rocks become progressively more deformed and finally recrystallize to form a polygonal aggregate of new grains. The only rock (AR-1) selected for study here consists of a weakly deformed quartzite in which the grains have more or less their original detrital shapes. All grains show undulatory extinction, and deformation lamellas are very well developed (Figure 2a). There is no evidence of recrystallization in this rock. These rocks were deformed in the Carboniferous [*Stewart*, 1971].

Chewings Range quartzite. This unit is a major quartzite mass in the Archean Arunta complex [*Forman et al.*, 1967] in central Australia. The quartzite is strongly deformed throughout, the last period of deformation being that responsible for nappe structures in the overlying Proterozoic rocks. This deformation is dated as Carboniferous [*Stewart*, 1971]. The quartzites examined here were collected from a locality known as the Fish Hole. One quartzite (FH-1; Figure 2b) consists of grains with

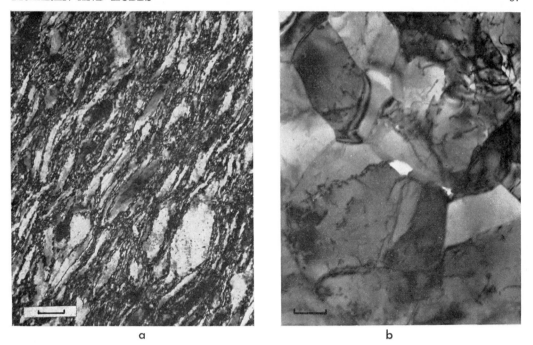

a   b

Fig. 1.   Mount Isa quartzite. (*a*) Optical micrograph (crossed polarizers) showing the structure of the more intensely deformed rocks of the Mount Isa fault. Specimen MI-1. The scale
mark indicates 0.1 mm. (*b*) Transmission electron micrograph showing the substructure most
commonly observed in the more intensely deformed rocks of the Mount Isa fault. Specimen
MI-1. The scale mark indicates 1 μm.

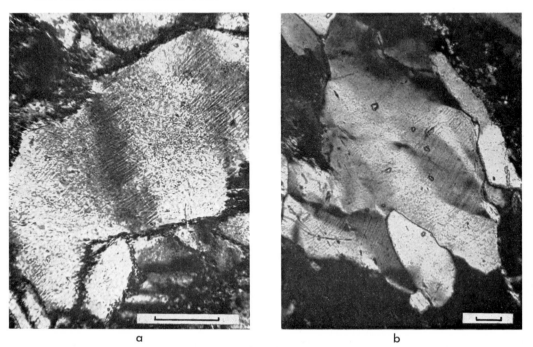

a   b

Fig. 2.   Quartzites from the Atnarpa nappe complex and the Chewings Range. (*a*) Optical
micrograph (crossed polarizers) of specimen AR-1 from the Atnarpa nappe complex. The
scale mark indicates 0.1 mm. (*b*) Optical micrograph (crossed polarizers) of specimen FH-1
from the Chewings Range. The scale mark indicates 0.1 mm.

TABLE 1. Comparison of Observed and Calculated Object Widths for Various Numerical Apertures

Objective Lens		Apparent Object Width, μm	
Magnification	Nominal NA	Observed	Calculated
×4	0.10	7	6.7
×10	0.25	4.2	2.7
×25	0.50	2.3	1.3

abundant deformation lamellas, whereas the other (FH-2; Figure 3) consists of markedly elongate quartz grains in which both deformation lamellas and polygonal subgrains are well developed. These grains are recrystallized along their margins. The two specimens were collected within about 1 meter of each other.

The important features of these areas, as far as this study is concerned, are (1) that detailed structural work has been carried out in each so that the structural history of each specimen is known in detail and (2) that the thermal history of each area has been such as to not disturb the Sr-Rb or the K-Ar isotopic constitution since the Precambrian (in the Mount Isa examples) or the Carboniferous (in the other examples).

OPTICAL MICROSCOPY

In detrital grains away from the zone of intense deformation in the Mount Isa fault the grain size is of the order of 0.2 mm, and many of the grain boundaries are decorated with mica. Plastic deformation is revealed by undulatory extinction and the associated slightly misoriented subgrains. Both undecorated grain boundaries and sub-boundaries that are approximately normal to the thin section are seen as fine black lines in cross polarizers (Figure 1a). Boundaries that are not normal to the thin section produce the same contrast if they are brought to a near-vertical orientation on a universal stage. These subgrain boundaries are generally irregular in shape, being curved and frequently changing their orientation along their length. Deformation lamellas are present but not common, and a detailed examination shows that their image contrast is consistent with the fact that the lamellas are submicroscopic phase objects in the form of a narrow wall approximately normal to the thin section, as was discussed previously by *McLaren et al.* [1970]. They are nearly basal in orientation (5°–15° from (0001)). The apparent object width of the lamellas decreases markedly with increasing numerical aperture (NA) of the objective lens, as in Table 1, this decrease thus indicating that the optical system is not resolving the actual structure of the lamellas.

Also included in that table are the apparent object widths calculated by using the theory given by *McLaren et al.* [1970] from the stated NA of the objectives for an object whose true width is less than 0.5 μm. For the ×4 objective there is very good agreement between the observed and calculated values. However, for the other two objective lenses the observed values are significantly larger than those calculated, owing to the fact that the full NA of these objectives was not being utilized, since the condenser diaphragm was closed to obtain maximum contrast from these weak phase objects. By using a Bertrand lens to observe the illuminated area in the back focal plane of the objective lens, it was seen that the diameter of the condenser diaphragm had practically no effect on the effective NA of the ×4 objective. However, when the condenser diaphragm was closed, the effective NA of the ×10 and ×25 objectives were reduced to about 0.16 and 0.3, respectively, this reduction giving calculated apparent object widths of 4.5 and 2.5 μm, which are in good agreement with the observed values. It was also found that near-vertical subgrain boundaries produced identical optical images in bright-field illumination. It is clear, therefore, that the objects that give rise to the images of these deformation lamellas and subgrain boundaries are phase objects less than 0.5 μm wide and that the observed image is characteristic of the microscope, giving no information from which the detailed structure of the object may be inferred. In cross polarizers there is no clear evidence for orientation changes across the deformation lamellas.

As the center of the zone of intense deformation is approached, the deformed detrital grains become progressively more elongate, and the average grain size decreases. The grain boundaries in these rocks are seen with the same contrast as described above, and there is no evidence of grain boundary decoration. In the more intensely deformed quartzites there is little evidence for the presence of the original grains, and the rock consists entirely of fine new grains, the maximum diameter of which is about 5 μm. These grains are too small to allow the definite observation of subgrain features by optical microscopy.

The detrital grains of the Atnarpa rocks contain abundant deformation lamellas (Figure 2a) that also display the optical characteristics of submicroscopic phase objects. Often there are multiple sets of lamellas present, and, for the most part, lamellas lie within 40° of (0001). Occasional lamellas near $\{10\bar{1}0\}$, as established from the electron diffraction of some thinned grains, are also present. Zones of undulatory extinction cross the lamellas (Figure 2a), and these zones are often subdivided into diffuse subgrains having boundaries that trend approximately parallel to $\{10\bar{1}0\}$ and (0001). Orientation differences across deformation lamellas and sub-grain boundaries are generally of the order of 5°.

Rocks from the Chewings Range quartzite show more obvious signs of deformation. Grains from FH-1 (Figure 2b) are frequently elongate and always contain well-developed deformation lamellas having orientations that lie within 30° of (0001). These lamellas also have the characteristics of submicroscopic phase objects. Diffuse elongate subgrains cross the deformation lamellas at right angles, the boundaries being approximately parallel to $\{10\bar{1}0\}$. Orientation changes across lamellas and subgrain boundaries are 1°–5°. In FH-2 (Figure 3) the grains are markedly elongate, and their margins are recrystallized. Within the grains a mosaic of equant subgrains is developed, the boundaries of subgrains being predominantly parallel to (0001) and $\{10\bar{1}0\}$. Crossing this mosaic are continuous deformation lamellas that lie within 30° of (0001). They also have the optical characteristics of submicroscopic phase objects. Orientation differences across both lamellas and subgrain boundaries are 1°–5°.

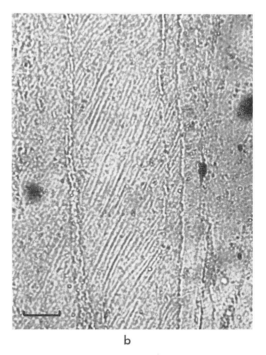

a   b

Fig. 3. Chewings Range quartzite. (a) Optical micrograph (crossed polarizers) of specimen FH-2, showing the mosaic of equant subgrains within the elongate grains. (b) Optical micrograph (bright-field, plane polarized light) of the same areas as shown in the left micrograph, showing near-basal deformation lamellas. The lamellas are made visible by a phase contrast mechanism due to a slight overfocus. Note that the optical images of the lamellas and the north-south grain boundaries are identical. The scale mark indicates 0.1 mm.

ELECTRON MICROSCOPY

Experimental details. Specimens thin enough for 100-kv transmission electron microscopy were prepared by ion bombardment thinning of standard 30-μm thin sections cut from the rocks discussed above. With this technique, regions (about 0.5 mm in diameter) containing individual optical features may be selected and thinned. Before or after investigation in the electron microscope the thinned specimen may again be examined optically, and the position of various optical features of interest noted, with respect to the thinned region.

With the ion beam at normal incidence (as described previously by *Gillespie et al.* [1971]), fine cracks in the specimen tended to open up more rapidly than the specimen thinned. This effect was largely overcome by bombardment at a glancing angle of 20° to the specimen. This change from normal to glancing incidence did not change the thinning rate from that observed with uncracked specimens (about 1 μm/hour), nor did it change the surface texture of the thinned specimen. Although no

residual ion bombardment damage was observed, the specimens, in common with all quartz, became damaged very rapidly (about 15 sec) in the 100-kv electron beam. We have evidence that for some quartz samples the damage is significantly reduced for accelerating voltages of 200 kv.

Considerable difficulty was also experienced in orienting the specimens to give good approximate two-beam diffracting conditions, especially in the finer-grained specimens. These conditions are essential for obtaining clear sharp diffraction contrast images of dislocations and other crystal defects. This difficulty appeared to result from distortion associated with long-range stresses locked in the crystal and the common difficulty of obtaining two-beam conditions in randomly oriented crystals with large d spacings.

Rocks of the Mount Isa fault. Figure 1*b* illustrates the substructure most commonly observed in the more intensely deformed rocks of the Mount Isa fault (specimen MI-1). The grain boundaries tend to be curved (Figure 4*a*), and the boundary intersections do not appear to

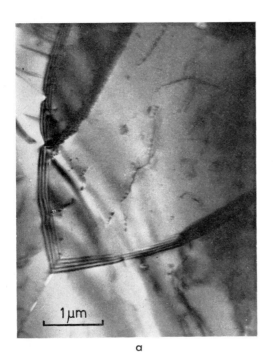

a   b

Fig. 4. Electron micrographs of the Mount Isa quartzite. (*a*) Transmission electron micrograph showing the intersection of curved grain boundaries in the more intensely deformed rocks of the Mount Isa fault. Note the voids along the boundaries and the curved subboundary. Specimen MI-1. (*b*) An area of extremely elongate subgrains whose boundaries are approximately parallel to the trace of (0001). Specimen MI-1.

have attained an equilibrium configuration typical of a well-annealed specimen. The grain boundaries often have voids along them. It is probable that these voids are characteristic of the grain boundaries in the bulk rock, but they may have been formed by preferential attack of impurities during the ion bombardment thinning process. The dislocation density within the grains varies considerably from grain to grain. The grains frequently contain curved sub-boundaries that almost invariably consist of an array of near-parallel dislocations. No clear examples were observed of dislocations interacting to form threefold nodes or dislocation networks, which are characteristic of recovery after deformation. The arrangement of grains and subgrains frequently gives an orthogonal pattern, the grain boundaries being approximately parallel to the traces of $\{10\bar{1}0\}$ and (0001). Occasionally the subgrains are quite elongate. An extreme example is shown in Figure 4b, in which the subgrain boundaries are approximately parallel to the trace of (0001). The substructure observed in the deformed detrital grains away from the zone of intense deformation consists mainly of an apparently random arrangement of dislocations similar to those in Figure 6a. However, arrays of approximately parallel dislocations forming low-angle boundaries that define a cell-like substructure were also observed. It was not possible to examine grains in which deformation lamellas were definitely present.

Atnarpa nappe complex. Transmission electron microscopy shows that the large grains of this rock, which have abundant near-basal deformation lamellas (Figure 2a), characteristically contain elongate subgrains with voids distributed along the boundaries (Figure 5a). These features may be traced up to 30 μm before they disappear in the thick parts of the specimen. Most boundaries are approximately parallel to the trace of (0001), but orientations near the trace of $\{10\bar{1}0\}$ are also observed.

Figure 5b shows an electron diffraction pattern taken from a region about 400 μm^2 in area containing elongate subgrains. This pattern indicates that within this region an *a* axis is constantly oriented in each subgrain and that [0001] varies over the range 20°–25°. The

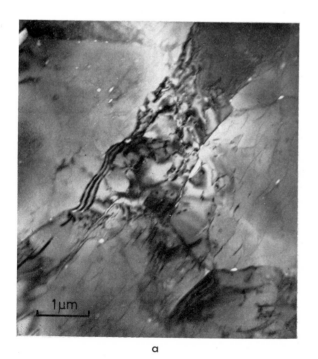
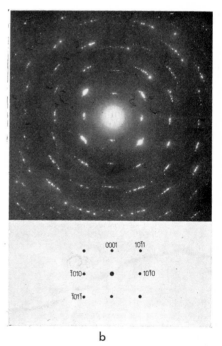

Fig. 5. Atnarpa nappe quartzite. (a) Transmission electron micrograph showing an elongate subgrain (parallel to the trace of the basal plane) characteristic of specimen AR-1. Note the voids along the sub-boundaries. (b) Electron diffraction pattern from an area of specimen AR-1 with elongate subgrains. See the text for a full description.

 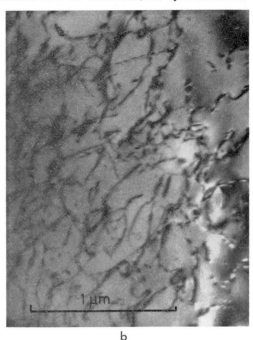

a   b

Fig. 6. Electron micrographs of the Chewings Range quartzite. (*a*) A micrograph illustrating the tangle of dislocations that are distributed fairly uniformly throughout specimen FH-1. Note also the subgrain boundary with voids along it. (*b*) An example of the localized regions of dislocation tangles that were commonly observed in specimen FH-2.

dislocation lines are randomly oriented within the subgrains.

Chewings Range quartzite. The specimen FH-1 of this rock (Figure 2*b*) generally shows a high density (about 5×10^9 cm^{-2}) of tangled dislocations distributed fairly uniformly. Within this distribution of dislocations, localized regions forming subgrain boundaries within 30° of the trace of (0001) are also observed (Figure 6*a*). Selected area diffraction shows that these boundaries are generally tilt boundaries associated with a misorientation of up to 5°.

Only regions of the large elongate grains of FH-2, which show the well-developed subgrains (Figure 3), were examined in the electron microscope. These specimens show to a varying degree all the features observed in MI-1, AR-1, and FH-1 as well as a limited number of dislocation arrangements normally associated with recovery.

Some subgrains contain a fairly high density (about 5×10^9 cm^{-2}) of dislocations (Figures 6*b* and 7*b*). On the other hand, the dislocation density in some grains can be lower by at least a factor of 10 (Figures 7*a*, 8, and 9*a*). Some subgrain boundaries are fairly straight and are

arranged in an orthogonal pattern (Figures 7*b* and 8*b*), whereas others are very curved (Figure 7*a*). The sub-boundaries usually consist of an array of near-parallel dislocations that are like those in Figure 9*a* but closer together. However, some sub-boundaries appear to consist of a much more complicated arrangement of dislocations (Figures 7*b*, 8*b*, and 9*b*).

Dislocation interactions are often observed in FH-2, and a well-developed network can be seen in Figure 8*b*. Figure 9*a* shows a regular array of dislocations abutting a well-defined subgrain boundary that is approximately parallel to the trace of (0001). The individual dislocations in the array are approximately parallel to an *a* axis, and the array produces a tilt of the *c* axis. If it is assumed that the dislocations are predominantly edge, their spacing indicates a tilt of the order of 0.5°. This indication is consistent with the observed change of contrast across the array. The individual parallel dislocations in the sub-boundary shown in Figure 8*a* are spaced much more closely, and the orientation change is very marked. This sub-boundary also has well-developed ledges along its length. Figure 9*b* shows a near-basal sub-boundary consisting of a

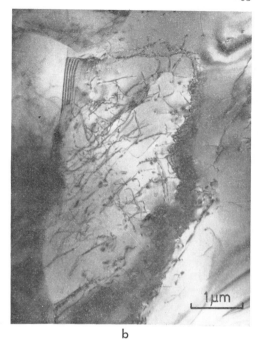

a   b

Fig. 7. Electron micrographs of the Chewings Range quartzite. (a) A micrograph illustrating the curved subgrain boundaries frequently observed in specimen FH-2. It can be
seen that in places the sub-boundaries consist of a simple array of parallel dislocations,
but in other places the dislocation density is too high for the dislocation arrangement to
be resolved. Note the comparatively low dislocation density within the subgrains. (b)
A micrograph showing a subgrain with a comparatively high dislocation density (about
$5 \times 10^9 \mathrm{cm}^{-2}$) as well as a sub-boundary consisting of a very high density of tangled dislocations. Specimen FH-2.

complicated arrangement of dislocations. It
terminates near a low-angle boundary, parallel
to the trace of $\{10\bar{1}0\}$, composed of dislocations
parallel to an a axis.

Again, many of these planar features, such as
the one illustrated in Figure 8a, may be traced up
to 30 μm before they vanish into thick parts of
the specimen. These long continuous boundaries
have variable orientations, but all lie within 30°
of (0001).

DISCUSSION

Conditions of deformation. The substructures observed by transmission electron microscopy in all the specimens examined indicate
that the rocks have been plastically deformed
by mechanisms involving generation, motion,
and interaction of dislocations. All samples contain regions of dislocation tangles in which
the localized dislocation density can be as high
as 5×10^9 cm^{-2} (Figure 6b). There is no evidence from these observations that brittle frac-

ture (crushing, milling, granulation) is active
as a mode of deformation. In FH-1 these
tangles are more or less uniformly distributed
throughout the specimens; in all other rocks
the tangles are localized although still widespread. This type of substructure is similar
to metals and quartz [*Hobbs et al.,* this volume]
that has been experimentally strain hardened.

The rocks examined in this study were selected from regions in which detailed structural
analyses and radiometric age determination
studies had been carried out. This work indicates that all of these rocks were deformed
more than 3×10^8 years ago, and there is no
evidence of subsequent small deformation
events. Also, the radiometric work indicates
that the temperature has not been high enough
to disturb the isotopic constitution for at least
this period of time.

Although some degree of recovery has occurred in FH-2 (see below), the existence of
dislocation tangles itself implies that the rate

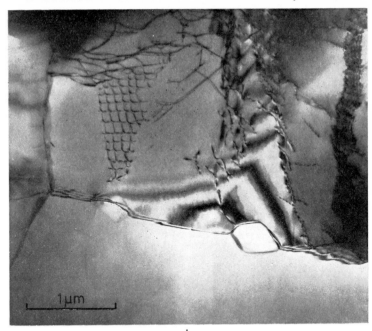

a   b

Fig. 8. Electron micrographs of the Chewings Range quartzite. (*a*) A sub-boundary consisting of closely spaced parallel dislocations and exhibiting well-developed ledges. The sub-boundary also gives rise to diffraction contrast fringes parallel to its length. Specimen FH-2. (*b*) A micrograph illustrating the near-orthogonal arrangement of subgrain boundaries in specimen FH-2. Note also the well-developed dislocation network, which is indicative of recovery.

of recovery is very slow at the temperatures these rocks experienced after deformation. In addition, the subgrain boundaries are generally curved and meet other boundaries in non-equilibrium configurations. This finding and the observations discussed above suggest that the microstructures observed are representative of the deformation process and that they have been 'frozen in' since deformation ceased.

The slow strain rates postulated for geological deformations (10^{-12}–10^{-14} sec^{-1}) combined with the low differential stresses thought to exist in the earth (≤ 1 kb) suggest that deformation by some steady creep mechanism should be common in natural deformation. If this suggestion is accurate, the deformation mechanisms must be such as to not generate obstacles such as tangles or other pinning points on dislocations. If such obstacles are created, the stress must rise to overcome them, and the process is no longer one of steady creep. Hence, in any material that has been deformed by steady creep, high densities of tangled dislocations should be absent.

Clauer et al. [1970] found that, in single

crystals of molybdenum, glide dislocations form sub-boundaries consisting of arrays of parallel dislocations in the very early stages of creep and that these arrays continue to move as units. *Boland et al.* [1971] found the same substructures in naturally deformed olivine, this finding suggesting creep deformation. Some examples of similar sub-boundaries have been observed in the quartzites studied here (Figure 8*a*). The ledges of these sub-boundaries imply that they may have moved as units by the sideways motion of the ledges in the manner described by *Gleiter* [1969].

Clear evidence of recovery (an over-all lowering of the dislocation density and interaction of dislocations to form threefold nodes and networks) is observed only in FH-2. The arrays of parallel dislocations (Figure 9*a*) could also be recovery features. However, networks, such as those in Figure 8*b*, are characteristic of recovery and not of creep.

The widespread occurrence of dislocation tangles in all the rocks examined is indicative of cold working. It is suggested, therefore, that these rocks have not been deformed by steady

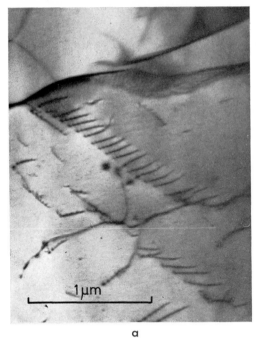 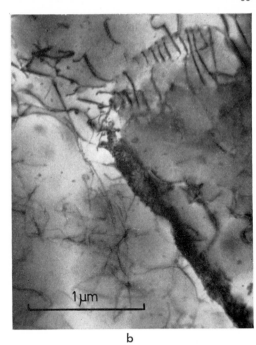

a   b

Fig. 9. Electron micrographs of the Chewings Range quartzite. (a) Low angle sub-boundaries consisting of arrays of dislocations approximately parallel to an a axis, producing a tilt of the c axis. (b) A sub-boundary (approximately parallel to the trace of the basal plane) consisting of a dense tangle of dislocations. Note also the simple array of dislocations parallel to an a axis, in the upper right-hand corner.

creep even though there is a strong resemblance optically in thin sections between the Mount Isa blastomylonites and the peridotite blasto-mylonite described by *Boland et al.* [1971]. The deformation processes apparently involved some strain hardening.

The evidence for strain hardening accompanying deformation and low recovery rate subsequent to deformation is based on the geological evidence that these rocks were last deformed in the Precambrian (Mount Isa) or the Carboniferous (Atnarpa and Chewings Range). The remote possibility that there has not been a recent deformation event in all three areas cannot be overruled, so that features observed in both optical and electron microscopy may have been imposed more recently. However, it is emphasized that there is no geological evidence to support this possibility.

Structure of deformation lamellas. Because of the great differences between the magnifications that are employed in the electron (about ×20,000 in these examples) and in the optical (about ×500) microscopes, until now it has not proved possible to obtain a direct one-to-one correlation between an optical deformation lamella and some feature observed in the electron microscope. This difficulty is enhanced by the extreme thinness of the regions in the immediate vicinity of the region examined in the electron microscope. Such regions may be only 5–10 μm thick, so that it is not possible to carry optical features precisely into the region that is examined in the electron microscope.

However, in the examples reported here it was always possible to make a careful optical study of the vicinity of the thinned region both before and after examination in the electron microscope and to note the trend and spacing of the deformation lamellas and the other optical features present. The only features that can be correlated with the deformation lamellas observed in the optical microscope are the boundaries of elongate subgrains, which have the same orientations and spacings as the lamellas observed optically in the same vicinity. In many examples these boundaries have such a high density of dislocations that their structure cannot be resolved, and they give rise to diffraction contrast fringes characteristic only of planar defects (Figure 7b). In other examples the boundary is seen to be composed of dense tangles (also see Figure 7b). Often voids are distributed along boundaries (Figure 6a). Such features are capable of acting as phase objects and thus of producing the optical characteristics observed. Therefore in these specimens it would appear that the deformation lamellas are associated with sub-boundaries and hence generally do not define active slip planes.

The deformation lamellas described here have dislocation structures different from those observed by transmission electron microscopy [McLaren et al., 1967] in the specimens deformed by Carter et al. [1964], where the lamellas were parallel to the active slip plane [Christie et al., 1964]. Many lamellas have the characteristics of those observed in experimentally deformed synthetic quartz [McLaren et al., 1970; Hobbs et al., this volume]. Compare, for instance, Figures 4 and 5 of McLaren et al. [1970] with Figure 6a. Other lamellas are similar to those described by Boland et al. [1971], where the lamellas are normal to the active slip planes (for example, Figure 8a).

All the deformation lamellas reported here from naturally deformed quartzites have the optical characteristics of submicroscopic plane phase objects [McLaren et al., 1970]. That is, they represent the optical image of a planar object that possesses an optical path difference different from that of its surroundings but having a width that is so small that it is not resolvable by the optical system. This lack of optical resolution arises from the diffraction pattern of the object, which has a width that is inversely proportional to the object width, being wider than the aperture at the back focal plane of the objective lens [McLaren et al., 1970, Figure 6d]. In the particular examples reported here this lack of optical resolution means that no information about the structure of the object can be obtained from detailed optical examination of the image. In particular, the optical phase changes to be expected on either side of a submicroscopic planar phase object [McLaren et al., 1970, Figure 7 a, b) that is not precisely vertical will be artifacts, and measurement of these phase changes by using either a compensator or an interference microscope will provide no information on the structure of the object. In addition, since the phase contrast microscope introduces additional phase and intensity changes at the back focal

plane of the optical objective, any truncation of the diffraction pattern arising from the object again results in a lack of optical resolution and means that the same limitations are inherent in this technique. These same statements are true of the deformation lamellas described from experimentally deformed synthetic quartz [McLaren et al., 1970; Hobbs et al., this volume] and from a naturally deformed peridotite mylonite [Boland et al., 1971].

However, it is not to be implied that no information can be obtained from detailed optical examination of other deformation lamellas. For the lamellas produced experimentally in natural quartz by Christie et al. [1964], image widths of 5 μm are recorded [Christie et al., 1964, Figure 2]. Without a statement of the NA of the objective lens used, one cannot say if these particular features have been resolved. For instance, in the image details reported in this paper, widths of 7 μm are recorded, yet Table 1 indicates that these features are not being resolved by the optical system. However, J. M. Christie (personal communication, 1971) has reported image widths of 10 μm for these lamellas, and these widths would appear to be within the resolution of most optical microscope objectives used. In such cases, where the optical feature is resolved by the microscope, detailed measurements of phase changes across lamellas by compensator, interference, or phase contrast techniques clearly give information on the detailed structure of the object.

In the particular examples of naturally produced deformation lamellas examined here, optical examination provided little detailed information about the deformation process, since the optical features characteristic of deformation lamellas appear to have arisen from a wide range of different dislocation arrangements that may have developed in a variety of ways.

Acknowledgments. We thank Mr. Gordon Milburn for the preparation of optical thin sections and Mr. R. L. Bryant for his very careful photographic work. We also thank C. J. L. Wilson and M. Yar Khan for the supply of specimens. P. F. Williams read the manuscript and offered many helpful comments.

We thank the Australian Research Grants Committee for financial assistance.

Preferred Orientation of Quartz Produced by Mechanical Dauphiné Twinning: Thermodynamics and Axial Experiments

JAN TULLIS AND TERRY TULLIS

Brown University, Providence, Rhode Island   02912

Mechanical Dauphiné twinning of quartz exchanges positive and negative trigonal forms and thus is capable of producing a preferred orientation of these forms in aggregates. Because Dauphiné twinning is a coherent transformation involving elastic but no permanent strain, it is easily amenable to thermodynamic analysis. Such analysis predicts that the favored twin member is the one that has the higher elastic strain energy calculated for a given stress or the lower elastic strain energy calculated for a given strain. These predictions are confirmed by measurements of the complete preferred orientation in quartz aggregates loaded with little or no permanent deformation in axial compression experiments. The amount of twinning is insensitive to the temperature above 200°C and the duration of the loading. The variation in the amount of twinning with crystal orientation and stress may be explained by inhomogeneities in stress and strain in the aggregate and/or untwinning during unloading. The methods for identifying the part of the preferred orientation due to mechanical Dauphiné twinning in natural rocks are described. Such identification allows recognition of whether some other mechanism besides mechanical Dauphiné twinning has acted to orient the positive and negative trigonal forms, although at present it is unclear how useful twinning itself will be in practice for determining paleostresses and paleostrains.

Recent analyses have shown that a part of the preferred orientation found in all experimentally deformed quartz aggregates is due to mechanical Dauphiné twinning [*Tullis*, 1970]. The twinning operates independently of recrystallization processes and of other orienting mechanisms such as intragranular slip. There is no permanent strain involved in Dauphiné twinning, and thus the patterns of preferred orientation resulting from this easy and well-defined process depend only on the orientation and relative magnitudes of the principal stresses. Hence it is potentially of some use in the interpretation of natural quartz preferred orientations. This paper discusses the thermodynamics of mechanical Dauphiné twinning, presents expressions for the energy change involved in twinning under general and axial states of stress and strain, shows how such twinning operates in experimental polycrystalline samples where the deformation has axial symmetry, and shows what methods should be used in identifying the preferred orientations due to twinning in rocks in which the deformation has been more general.

A Dauphiné twin is related to the host by a 180° rotation about the c axis; thus the crystal axes remain parallel, but the electrical polarities of the a axes are reversed. A positive trigonal form, such as r, $\{10\bar{1}1\}$, is exchanged for a negative one, such as z, $\{01\bar{1}1\}$, and vice versa. (See a crystal projection such as that shown by *Baker et al.* [1969, Figure 12].) The operation of Dauphiné twinning may be represented on an inverse pole figure by a change in the orientation of the reference axis from one side of the c-a line to the symmetrical position on the other side.

Almost 40 years ago it was first discovered that Dauphiné twinning could be mechanically induced in quartz single crystals [*Zinserling and Shubnikov*, 1933]. This mechanical twinning is quite different from that usually encountered in metals and other minerals, as can be seen from the fact that the sense of stress makes no difference to the sense or the amount of Dauphiné twinning [*Wooster et al.*, 1947, p. 932]. Only a slight displacive rearrangement of the atoms occurs during mechanical Dauphiné twinning (see the illustrations of *Frondel* [1945, p. 449] and *McLaren and Phakey* [1969, p.

725]), and there is no permanent macroscopic strain involved. However, the two members of a twin pair have different elastic properties (Figure 1) and thus will strain different amounts elastically under the same stress. It is the difference in elastic strain energy between original and twinned orientations that determines the stable member.

Extensive work was done 20–30 years ago on the methods of eliminating the network of Dauphiné twin boundaries commonly found in natural quartz crystals [*Frondel*, 1962, p. 81]. It was found that movement of the twin boundaries could be achieved in various ways but that torsion of cut plates or bars at moderate temperatures was the most effective and reliable method [*Wooster et al.*, 1947; *Thomas and Wooster*, 1951]. The direction of movement of the twin boundary was found to depend sensitively on the orientation of the crystal with respect to the applied stress [*Wooster et al.*, 1947, p. 930]. However, these authors did not attempt experiments on polycrystalline aggregates, nor did they use their single-crystal results to predict how twinning might produce a preferred orientation in stressed aggregates. It is such experiments and predictions that form the basis for this paper.

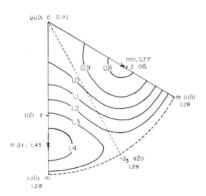

Fig. 1. Inverse pole figure showing the variation of S_{11}', the reciprocal of Young's modulus, with crystal direction for α quartz at room temperature and atmospheric pressure, in units of 10^{-12} cm²/ dyne. The figure was calculated from the isothermal elastic constants determined by *McSkimin et al.* [1965, p. 1632]. Negative crystal forms (to the right of the dotted line) consistently have compliances lower than those of the corresponding positive forms in twinned orientations.

THERMODYNAMICS

Thomas and Wooster [1951, p. 51] observed that, in their single-crystal torsion experiments, twinning acted in a way that maximized the elastic strain energy in the crystals. However, this statement by itself is misleading, since it is only strictly true for the case of constant external load, as in their experiments. A more careful analysis that predicts the twin member that will be in stable thermodynamic equilibrium under any given imposed set of stresses or strains is presented below. These thermodynamic criteria can then be used to predict the course of twinning in aggregates and thus the development of a preferred orientation.

An increment of twinning occurs by small movements of twin boundaries; a crystal containing such boundaries is not homogeneous, and no forces or displacements applied to its margins can produce homogeneous stresses or strains. If a crystal contains a statistically homogeneous distribution of twin boundaries, its total elastic strain energy can be determined from the average stress or strain and the average elastic constants [*Hill*, 1963, pp. 358–359, 362–363] in the same way that the total strain energy of a polycrystalline aggregate can be determined. Thus, as far as the total elastic strain energy is concerned, a representative volume of a twinned crystal can be imagined to be perfectly homogeneous. In the discussion presented below we assume that an increment of twinning can be represented by an incremental change in the elastic constants of a homogeneous crystal, the constants changing toward those of the growing twin member. This assumption is central to the validity of our derivations. Although an investigation of the general validity of the assumption is beyond the scope of this paper, it appears to be reasonable even for certain cases that lack statistically homogeneous distributions of twin boundaries. One example is a crystal under uniaxial stress in which a single planar twin boundary that is perpendicular to the axial stress migrates while maintaining its orientation (B. Kamb, personal communication, 1971).

Of the two criteria that *Gibbs* [1906, p. 56] established for stable equilibrium, the second one is more useful for this problem. He stated it in a variational manner: 'For the equilibrium of any isolated system it is necessary and sufficient that in all possible variations in the state of the system which do not alter its entropy, the variation of its energy shall either vanish or be positive.' In other words, at a state of equilibrium the internal energy assumes a stationary value, and for stable equilibrium the internal energy is a minimum [*Gibbs*, 1906, p. 57].

We should specifically state at this point that in this paper the phrase 'stable thermodynamic equilibrium' refers only to relative stability with respect to the specific process of an alteration in elastic constants of the type involved in Dauphiné twinning. Other mechanisms by which a crystal may change its state, such as plastic deformation or dissolution and redeposition, are specifically excluded from consideration. Thus such mechanisms are not included in the 'possible variations of the state of the system,' to use Gibbs's phrase.

To apply Gibbs's criterion to Dauphiné twinning, it is convenient to imagine a system composed of two parts enclosed in a rigid adiabatic envelope. One of the parts is the quartz crystal, which we imagine to be homogeneous. The other part of the system is a loading apparatus that surrounds the quartz and subjects it to homogeneous stresses and strains by application of suitable boundary displacements and forces. Whether the existing member of the twin pair is the thermodynamically stable member is only a function of the orientation of the crystal and the stresses or the strains; the nature of the loading apparatus is not important. However, to test for the conditions of strain energy that accompany the stable thermodynamic equilibrium of the crystal, it is instructive to consider two variational approaches involving different types of loading apparatus. Both approaches predict the same twin member to be the stable one, and they are entirely equivalent.

In the first approach, termed the constant stress approach, the loading apparatus is capable of maintaining the stresses within the crystal constant and thus can do work on the crystal. We prove below that for constant stress any variation of the elastic constants from the values at stable thermodynamic equilibrium results in a decrease in the elastic strain energy of the crystal. In symbols, at stable thermodynamic equilibrium,

$$\delta W < 0$$

when $\delta\sigma_{ij} = 0$ and $\delta S = 0$, where

$$W = \int_V \tfrac{1}{2} S_{ijkl}\sigma_{ij}\sigma_{kl} \; \delta V$$

is the elastic strain energy of the crystal, σ_{ij} is the stress tensor, S is the entropy of the system, V is the volume of the crystal, and S_{ijkl} are the elastic compliances.

The proof follows: the variation in the strain energy density that accompanies a variation in the elastic compliances is given by

$$
\begin{aligned}
\delta W_d &= W_d\big|_{S_{ijkl}+\delta S_{ijkl}} - W_d\big|_{S_{ijkl}} \\
&= \tfrac{1}{2}(S_{ijkl} + \delta S_{ijkl})\sigma_{ij}\sigma_{kl} \\
&\qquad - \tfrac{1}{2}S_{ijkl}\sigma_{ij}\sigma_{kl} \qquad (1) \\
&= \tfrac{1}{2}\delta S_{ijkl}\sigma_{ij}\sigma_{kl} \\
&= \tfrac{1}{2}\delta\epsilon_{kl}\sigma_{kl}
\end{aligned}
$$

where ϵ_{ij} is the strain tensor. The work done by the loading apparatus δL is equal to the integral over the surface Σ of the quartz crystal of the forces T_i times the displacements δu_i, or

$$\delta L = \int_\Sigma T_i \; \delta u_i \; d\Sigma \qquad (2)$$

Using $T_i = \sigma_{ij}\nu_j$, where ν_j is the outward unit vector, transforming with Gauss's theorem, neglecting body forces, and employing the symmetry of the stress tensor, we have

$$
\begin{aligned}
\delta L &= \int_\Sigma \sigma_{ij}\nu_j \; \delta u_i \; d\Sigma \\
&= \int_V (\sigma_{ij} \; \delta u_i)_{,j} \; dV \\
&\qquad\qquad\qquad\qquad (3) \\
&= \int_V \sigma_{ij,j} \; \delta u_i \; dV + \int_V \sigma_{ij} \; \delta u_{i,j} \; dV \\
&= \int_V \sigma_{ij} \; \delta\epsilon_{ij} \; dV
\end{aligned}
$$

The variation of the internal energy of the system is the variation of the strain energy of the crystal minus the work done by the loading apparatus during the variation in displacement or, when (1), (2), and (3) are used,

$$\delta U = \int_V \delta W_d \, dV - \delta L$$

$$= \frac{1}{2} \int_V \sigma_{ij} \, \delta\epsilon_{ij} \, dV - \int_V \sigma_{ij} \, \delta\epsilon_{ij} \, dV$$

$$= - \int_V \delta W_d \, dV$$

$$= - \delta W$$

For the system to have been at stable equilibrium, Gibbs's condition is that $\delta U > 0$ and thus $\delta W < 0$ or, in other words, the strain energy was at a maximum. Hence, with the assumption that a gradual change in the elastic constants can be used to represent Dauphiné twinning, the stable twin member is the one for which the elastic strain energy is greater when the energies are compared for the same stress.

This proof is essentially the same as that of *MacDonald* [1960, p. 2–3], but our starting point is distinctly different because unlike MacDonald, we carefully specify the process under consideration. *Kamb* [1961a, p. 262; 1961b, pp. 3985–3986] has convincingly argued that thermodynamic treatments must be limited to a specific orienting process and that a preferred orientation produced by one orienting mechanism may be quite different from that produced by a different mechanism, even under the same conditions of stress or strain. *Kamb* [1961a, p. 264] states that MacDonald erred in considering his $\delta\epsilon_{ij}$ to be an actual deformation and that they were instead only a change in the strain tensor components (expressed in a fixed coordinate system) that resulted from a rotation of the body (and a consequent deformation under constant stresses). In our case the specific mechanism of Dauphiné twinning provides a basis for us to assume that the elastic constants change in the fixed coordinate system without a rotation of the body. Thus the strains $\delta\epsilon_{ij}$ in (1) are an actual deformation of the body.

Kamb [1961a, p. 263] further states that 'MacDonald's derivation starts out along lines identical to the variational derivation given by *Sokolnikoff* [1956, pp. 382–386] of the equations of equilibrium of elastic bodies,' but that his error led him to a false conclusion [*Kamb*, 1961a, p. 265]. It is important to clearly under-

stand that our derivation is conceptually distinct from proofs of the theorem of minimum potential energy given, e.g., by *Sokolnikoff* [1956, pp. 382–386] and *Fung* [1965, pp. 284–288], because different processes are being considered. Proofs of this theorem show the equivalence between the conditions of stress equilibrium and a minimum in the potential energy. However, we are using the conditions of stress equilibrium and Gibbs's condition for stable thermodynamic equilibrium to determine the favored combination of elastic constants, our choice of combinations being limited to the two possibilities offered by the members of a Dauphiné twin pair. If there were no mechanism whereby the elastic constants could change, the body would be in mechanical equilibrium under homogeneous stress with any set of elastic constants [*Kamb*, 1961a, p. 263]. If variations of the elastic constants of the body are permitted, however, only some combinations of constants correspond to thermodynamic equilibrium.

To clarify the difference between a process in which the elastic constants actually change (such as Dauphiné twinning) and the process considered by *Sokolnikoff* [1956, pp. 382–385] and *Fung* [1965, pp. 284–287], it is convenient to consider a simple situation of uniaxial stress. In uniaxial stress the energies involved may be represented by areas on stress-strain diagrams such as Figure 2.

For cases in which the elastic constants may change, consider Young's modulus E and a variation in it, δE. The crystal with Young's modulus E has strain energy equal to areas a plus c of Figure 2. If the variation occurs under constant stress, the new strain energy is $a + b$, and thus the variation in strain energy of the crystal (a decrease) is $(a + b) - (a + c) = b - c$. The work done on the loading apparatus is $c + d$. Hence the energy of the system changes by $(b - c) + (c + d) = b + d$. The fact that the energy of the system increases demonstrates that, with respect to processes that can increase Young's modulus, E is the stable equilibrium value. (If it were physically possible to have variations that decreased Young's modulus, E would not be the equilibrium value, because the variation in the energy of the system would be negative.)

For elastic solids like those considered by *Fung* [1965] and *Sokolnikoff* [1956], the stress and the strain are always related by Young's modulus E. (Actually *Fung* [1965, p. 286] does not restrict himself to linearly elastic bodies; as long as there is a one-to-one functional relationship between stress and strain, the arguments hold.) Here the initial strain energy is $a + c$, the new strain energy is a, and the variation in strain energy is $a - (a + c) = -c$. The work done on the loading apparatus is again $c + d$, and thus the variation in energy of the system is $-c + (c + d) = d$. In this second case, if variations in the strain of the opposite sign are considered, the energy of the system will again be found to increase. Hence, as long as the elastic constants cannot change, the body is in equilibrium under stress σ and strain ϵ.

Let us now more briefly consider the second approach (termed the constant strain approach). In this case the loading apparatus is regarded as being rigid and is therefore not capable of doing any work on the crystal. Thus virtual changes in the elastic constants from the equilibrium values will not result in an additional strain of the crystal, but they will result in a change in the stresses experienced by it and in the surface tractions applied to it. We prove below that, for constant strain, any variation of the elastic constants from the values at stable equilibrium results in an increase in the elastic strain energy of the crystal. In symbols, at stable thermodynamic equilibrium,

$$\delta W > 0$$

when $\delta\epsilon_{ij} = 0$ and $\delta S = 0$, where

$$W = \int_V \tfrac{1}{2} C_{ijkl}\epsilon_{ij}\epsilon_{kl}\ dV$$

is the elastic strain energy of the crystal, ϵ_{ij} is the strain tensor, S is the entropy of the system, and C_{ijkl} are the elastic stiffnesses.

The proof is very simple. Since the apparatus does no work, it contributes nothing to the change in the internal energy of the system. The change in the strain energy of the crystal on twinning is thus equal to the change in the internal energy of the system, which, according to Gibbs's criterion, must be >0 for variations from stable equilibrium at constant entropy. Thus the strain energy was a minimum, and, under the assumption that a gradual change

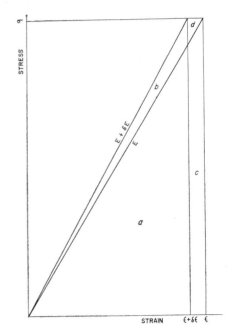

Fig. 2. Stress-strain curves showing the difference between variations in Young's modulus and variations in the strain with fixed Young's modulus. For an explanation, see the text.

in the elastic constants can be used to represent Dauphiné twinning, the stable twin member is the one for which the elastic strain energy is less when the energies are compared for the same strain.

Let us again consider the situation of uniaxial stress, this time to gain some intuitive understanding of the equivalence of the two variational approaches just considered. In this instance we consider the spontaneous twinning of a crystal rather than incremental variations in stress or strain from a state of stable equilibrium.

If a crystal is loaded with a constant force, as by placing a heavy weight on it (or by hanging one from it), intuition tells us that it will deform elastically as much as possible and that, if some mechanism exists for reducing the stiffness of the crystal so that it may deform more, this mechanism will tend to operate. Alternatively, if the crystal is strained and then has its ends pressed against (or bonded to) a rigid frame, intuition says that, if a mechanism exists for the force to relax, it will tend to do so. These statements represent an application of

LeChatelier's principle: on application of uni-axial stress to a stress-free crystal in equilib-rium, the process that will tend to reduce the stress will occur. Whether the process is able to actually reduce the stress depends, of course, on the manner of the application of the stress.

The essential point is that the direction in which the process will operate is the one that reduces the stiffness (or equivalently increases the compliance) of the crystal. Hence in Figure 3 the process of twinning should reduce Young's modulus from E to $E + \Delta E$ (where ΔE is <0). If this process occurred under constant stress, the strain energy would change from areas $a + b$ on the diagram to areas $a + c + d$. Hence the energy increases during the constant stress twinning process by the amount $a + c + d - (a + b)$ or $c + d - b$, which is equal to $b + e$. If the process occurred under constant strain, the strain energy would go from $a + b$ to a. Hence the energy decreases during the constant strain twinning process by the amount b. (This discussion is therefore analogous to a discussion of

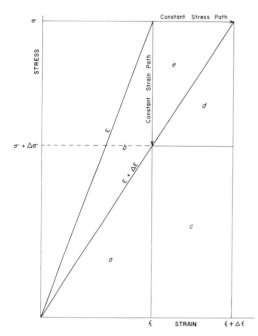

Fig. 3. Stress-strain curves showing that a decrease in Young's modulus results in an increase in elastic strain energy if the stress is constant or a decrease in elastic strain energy if the strain is constant. The indicated fractional change in Young's modulus is equal to the greatest change that can occur by Dauphiné twinning.

the growth of a Griffith crack given by *Cottrell* [1964, p. 344].)

The examination of twinning in the case of uniaxial stress presented above makes the physics of the process seem intuitively reasonable, because only one component of stress and one component of strain need to be considered. For this reason also, LeChatelier's principle can be easily applied. To gain an intuitive understanding of twinning under general states of stress or to apply LeChatelier's principle, it is necessary to think of the energy of the system rather than of the individual stress and strain components. Under constant strain, for example, twinning may result in an increase in the magnitude of some stress components accompanied by a decrease in the magnitude of others. The energy of the system would, of course, decrease during the transition to stable equilibrium, but this decrease would result from the summation involved in the expression $\Delta\sigma_{ij}\epsilon_{ij}$ and not necessarily from a negative value for each $\Delta\sigma_{ij}$.

The whole system must be included in one's consideration of both the uniaxial and the general cases. If one's intuition predicts the correct answer for the uniaxial case of constant stress discussed above, it is because one considers not only the crystal but also the reduction in potential energy of the heavy weight in the gravity field.

EXPRESSIONS FOR ENERGY DIFFERENCE

Having demonstrated the criteria for stability with respect to the process of mechanical Dauphiné twinning in terms of elastic strain energy, we now present expressions for the change in strain energy accompanying twinning in terms of the elastic constants of quartz and the stresses or the strains. This presentation will first be made for general states of stress or strain. The general result will then be used to obtain a convenient expression for axial stresses.

Our crystal coordinate axes are chosen in accordance with the standards of the *Institute of Radio Engineers* [1949]: they form a right-handed orthogonal set with positive X_1 in the same direction as a positive crystallographic a axis, X_2 in the basal plane, and positive X_3 in the same direction as the positive crystallographic c axis. We designate quantities applying to untwinned and twinned orientations (which differ by a 180° rotation about the c axis) by single and double primes, respectively. It is convenient to refer the stresses experienced by the two orientations of crystal to crystal coordinates in each case. In crystal coordinates the strain energy density is given in matrix notation [Nye, 1957, p. 136] as

$$W' = \tfrac{1}{2}S_{ij}\sigma_i'\sigma_j'$$

for the untwinned crystal and by a similar expression using double primes for the twinned crystal. We are interested in the energy change on twinning or

$$\Delta W = W'' - W' = \tfrac{1}{2}S_{ij}(\sigma_i''\sigma_j'' - \sigma_i'\sigma_j') \tag{4}$$

It is easy to show that $\sigma_i'' = \sigma_i'$, except for $i = 4$ and 5, and that, in these cases, $\sigma_i'' = -\sigma_i'$. With this result, the symmetry of the S_{ij}, and the form of the S_{ij} for class 32 [Nye, 1957, pp. 141–142], (4) becomes

$$\begin{aligned} \Delta W &= W'' - W' \\ &= -2S_{14}(\sigma_1'\sigma_4' - \sigma_2'\sigma_4' + 2\sigma_5'\sigma_6') \end{aligned} \tag{5}$$

(This expression is the same as that given by Klassen-Neklyudova [1964, p. 154, equation 15]. McLellan's [1970, p. 454, equation 51] analogous expression is missing two factors of 2.) The same procedure can be followed to show that the relationship expressed in terms of the strains referred to crystal coordinates for the untwinned crystal is

$$\begin{aligned} \Delta W &= W'' - W' \\ &= -2C_{14}(\epsilon_1'\epsilon_4' - \epsilon_2'\epsilon_4' + \epsilon_5'\epsilon_6') \end{aligned} \tag{6}$$

Equations 5 and 6 readily show an important characteristic of the strain energy change involved in mechanical Dauphiné twinning, namely, that the energy change is independent of the mean stress or the dilatation. This fact, together with the independence of the energy on the sign of the stresses or strains, can be used to show the equivalence of various different states of stress or strain as far as Dauphiné twinning is concerned.

Since, under the assumption of constant stress, twinning will tend to maximize the strain energy in the crystal, crystal orientations for which ΔW of (5) is negative should not twin, whereas those for which ΔW is positive will tend to twin. Alternatively, since, under the assumption of constant strain, twinning will tend to minimize

the strain energy in the crystal, crystal orientations for which ΔW of (6) is positive should not twin, whereas those for which ΔW is negative will tend to twin. (Since the signs of S_{14} and C_{14} are opposite, these two statements are compatible.) These statements constitute predictions of the form of the preferred orientation to be expected from twinning.

For uniaxial stress σ, the change in strain energy in going from one Young's modulus E^a to another E^b is clearly

$$\Delta W = \tfrac{1}{2}\sigma^2[(1/E^b) - (1/E^a)]$$

or, if S_{11}' is used to denote the reciprocal of Young's modulus in the direction of the uniaxial stress [Nye, 1957, pp. 143–144], we have

$$\Delta W = \tfrac{1}{2}\Delta S_{11}'\sigma^2 \tag{7}$$

where S_{11}' is the change in the reciprocal of Young's modulus on twinning. However, since we have just demonstrated the fact that ΔW is independent of the mean stress, (7) can be written as

$$\Delta W = \tfrac{1}{2}\Delta S_{11}'(\sigma_1 - \sigma_3)^2 \tag{8}$$

in which the quantity $(\sigma_1 - \sigma_3)$ is the differential stress.

Figure 4 is an inverse pole figure showing the distribution of $\Delta S_{11}'$. As (8) shows, those orientations for which $\Delta S_{11}'$ is positive should twin and those for which it is negative should remain untwinned. (Note that the signs of $\Delta S_{11}'$

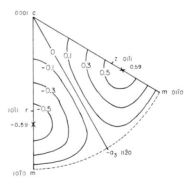

Fig. 4. Inverse pole figure showing the variation with crystal direction of the difference in compliance S_{11}' between original and Dauphiné twinned orientations, in units of 10^{-12} cm²/dyne. Under axial stress the orientations for which $\Delta S_{11}'$ is positive should twin and those for which it is negative should remain untwinned.

in Figure 4 are opposite from those given by *Tullis* [1970, p. 1343, Figure 2*b*], who took the difference to be in the opposite sense.) Thus this figure constitutes a prediction of the form of the preferred orientation to be expected from mechanical Dauphiné twinning in an axially stressed aggregate.

EXPERIMENTAL RESULTS

To test the prediction embodied in Figure 4 for the preferred orientation produced by mechanical Dauphiné twinning in aggregates, several samples of fine-grained Dover flint and Arkansas novaculite were subjected to axial compression at low temperature and with little or no permanent deformation, so that no other known orienting mechanisms would operate. In these experiments (using a scaled-up version of the solid confining medium piston cylinder apparatus described by *Griggs* [1967]) the samples were loaded at a strain rate of 10^{-4} sec^{-1} and a confining pressure of 4 kb. When the desired differential stress level was reached, the motor was turned off, and the sample was allowed to remain stressed for a certain time, during which the stress usually decayed somewhat. The conditions of these experiments are given in Table 1. Only one sample (GB-356) was loaded beyond the yield point, and it was loaded just barely beyond that point.

Five samples of flint were tested. Specimen GB-352 (400°C, 3.1 kb) showed no preferred orientation and thus no evidence of twinning. Specimen GB-354 (400°C, 15.8 kb) showed a very slight preferred orientation of the positive and negative forms and no accompanying preferred orientation of the *c* axes (Figure 5*a*), consistent with the operation of mechanical Dauphiné twinning. Specimen GB-365 (450°C, 12.7 kb) developed a much stronger preferred orientation of the same type (Figure 5*b*). The strength of the preferred orientation in this sample (and in all the others discussed) decreases in intensity away from the hotter center; there is none at all in the cold (roughly 200°C) ends of this sample. The two specimens loaded at 500°C, GB-353 and GB-356, are somewhat anomalous, since the one loaded at a lower differential stress and for a shorter length of time (GB-353, Figure 5*d*) developed a somewhat stronger preferred orientation of the positive and negative forms than the other (GB-356, Figure 5*c*).

The results from the flint experiments were puzzling, because for a given temperature a much higher differential stress was required to produce significant twinning than had been required for single crystals [*Wooster et al.*, 1947, p. 931]. This result might be due to complex effects in an aggregate, to some effect of mean

TABLE 1.   Experimental Conditions for Samples Loaded To Produce Preferred Orientations Only from Mechanical Dauphiné Twinning

Material	Sample	Temperature, °C	Maximum Differential Stress, kb	Time, hours	Stress Decay, kb
Flint	GB-352	400	3.1	0.3	0.25
	GB-354	400	15.8	4.5	7.2
	GB-365	450	12.7	3.0	6.4
	GB-353	500	9.4	2.5	6.4
	GB-356	500	12.9	3.25	7.4
Novaculite	GB-413	25	15.9	4.5	2.0
	GB-403	100	17.1	3.0	0.24
	GB-396	200	15.9	3.0	0.43
	GB-390	300	16.3	4.25	0.80
	GB-376	400	15.9	6.0	0.87
	GB-377	500	16.0	6.0	0.92
	GB-380	600	16.4	6.5	1.25
	GB-397	300	16.3	0.0	
	GB-407	300	8.7	0.08	0.24
	GB-412	300	2.2	0.08	0.0
Quadrant quartzite	GB-399	400	15.3	4.5	0.59
Black Hills quartzite	GB-404	400	15.4	3.0	0.90
Simpson sandstone	GB-405	400	15.2	3.5	0.92

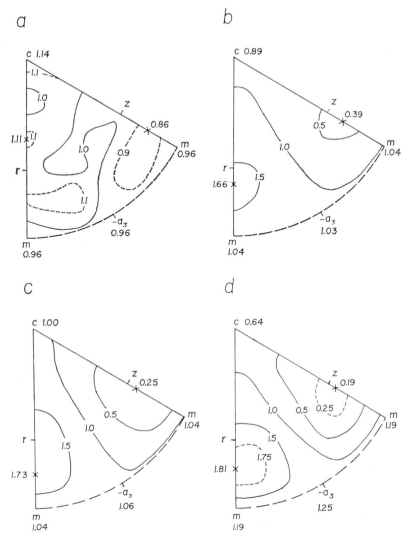

Fig. 5. Inverse pole figures showing preferred orientations produced in flint samples loaded with little or no permanent deformation. The concentration of the compression direction is contoured in multiples of a uniform distribution in these and all the following inverse pole figures. (For a discussion of inverse pole figures, see *Baker et al.* [1969, pp. 158–162]). (*a*) GB-354, 400°C, 15.8 kb; very little preferred orientation of the positive and negative forms. (*b*) GB-365, 450°C, 12.7 kb; a very definite preferred orientation, with the poles to positive forms preferentially aligned parallel to σ_1 as expected. (*c*) GB-356, 500°C, 12.9 kb; a somewhat stronger preferred orientation of the same type. (*d*) GB-353, 500°C, 9.4 kb; an even stronger preferred orientation of the same type.

stress, or to the presence of impurities. To test the latter possibility, a series of similar experiments was done on novaculite, which has nearly the same grain size as the flints but contains much less water and other impurities [*Green et al.*, 1970].

The inverse pole figures for the first series of experiments, loaded at different temperatures but at the same differential stress and for approximately the same time, are shown in Figure 6. The novaculite sample loaded at room temperature shows the same amount of twinning as the flint loaded at 450°C shows at almost the same differential stress. Thus impurities ap-

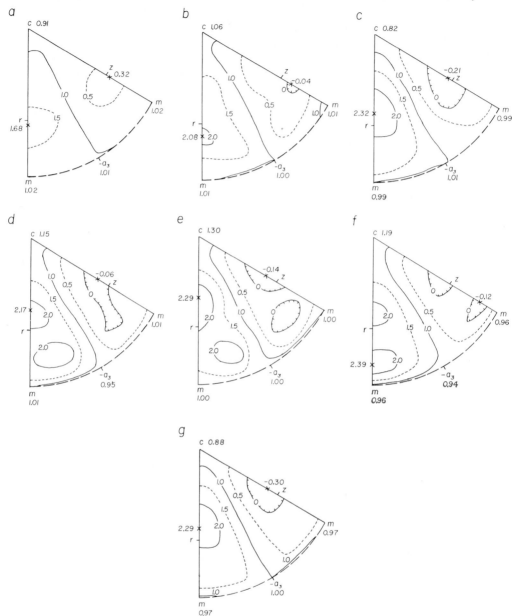

Fig. 6. Inverse pole figures showing the amount of mechanical Dauphiné twinning as a function of the temperature for seven novaculite samples loaded to approximately the same differential stress (16 kb). The negative areas (and the corresponding areas of concentration of >2.0) on some of the figures are physically impossible; see *Baker et al.* [1969, p. 165–166] for a discussion of their origin. (*a*) GB-413, 25°C. (*b*) GB-403, 100°C. (*c*) GB-396, 200°C. (*d*) GB-390, 300°C. (*e*) GB-376, 400°C. (*f*) GB-377, 500°C. (*g*) GB-380, 600°C.

parently play a large role in impeding the movement of the twin boundaries, as was suspected by *Wooster et al.* [1947, p. 937] from their single-crystal experiments. In this series of novaculite experiments the amount of twinning increases with temperature up to a certain point; although there are distinct increases from 25° to 100°C and from 100° to 200°C, there is little or no further increase up to 600°C.

The second series of experiments was done to test the effect of varying the differential stress at constant temperature (300°C). The inverse pole figures for these samples are shown in Figure 7. Of the two samples loaded to 16.3 kb the one that remained loaded for 4.25 hours shows only slightly more twinning than the one that was immediately unloaded. The sample loaded to 8.6 kb has almost as much twinning as the one loaded to 16.3 kb for a similar period

of time. The sample loaded to 2.2 kb has much less twinning but still as much as the novaculite sample loaded to 15.9 kb at room temperature.

Three samples of quartzite were also loaded without permanent deformation to test for the occurrence of twinning. Although the X-ray counting statistics for these undeformed coarse-grained aggregates are very poor, the inverse pole figures (Figure 8) definitely show the same type of preferred orientation of positive and negative forms as that seen in the flints and the novaculites.

The patterns of preferred orientation of the

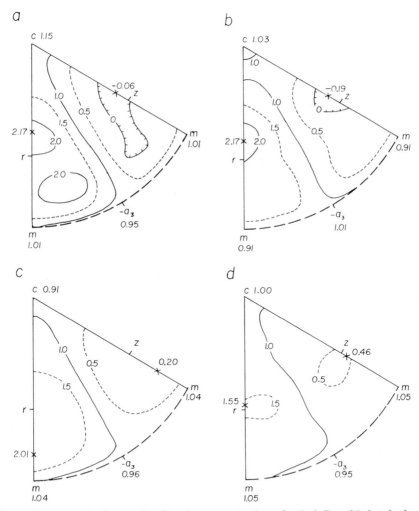

Fig 7. Inverse pole figures showing the amount of mechanical Dauphiné twinning as a function of differential stress and loading time for novaculite samples loaded at the same temperature (300°C). (a) GB-390, 16.3 kb for 4.25 hours. (b) GB-397, 16.3 kb but unloaded immediately. (c) GB-407, 8.7 kb for 5 min. (d) GB-412, 2.2 kb for 5 min.

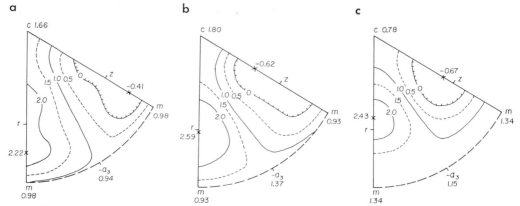

Fig. 8. Inverse pole figures showing mechanical Dauphiné twinning induced in three different quartzites by loading with little or no permanent deformation at 400°C and 15 kb for 3–4.5 hours. (*a*) GB-399, Quadrant quartzite. (*b*) GB-404, Black Hills quartzite. (*c*) GB-405, Simpson sandstone.

positive and negative forms observed in experimentally deformed quartzites (including those which are unrecrystallized through those which are completely recrystallized) and in syntectonically recrystallized flints (Figure 9) are essentially identical to the inverse pole figures shown above, except that the simultaneous development of a *c*-axis preferred orientation has shifted the area of maximum concentration along the *c-m* line of the inverse pole figure. Thus a large component of the preferred orientation in all experimentally deformed quartz aggregates is evidently due to mechanical Dauphiné twinning.

DISCUSSION OF EXPERIMENTAL RESULTS

We now discuss the relationship of the twinning in the simple loading experiments to the

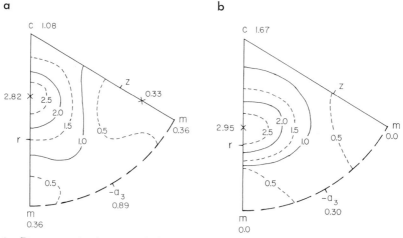

Fig. 9. Representative inverse pole figures for plastically deformed quartz aggregates showing that mechanical Dauphiné twinning contributes substantially to their total preferred orientation. (*a*) GB-171, Quadrant quartzite shortened roughly 50% at 850°C, 10^{-5} sec^{-1}, and 15-kb confining pressure. There is very little recrystallization in this sample; the preferred orientation has been developed in the original grains [*Tullis*, 1971, Figure 42]. (*b*) GB-84, Dover flint shortened roughly 31% at 850°C, 10^{-6} sec^{-1}, and 24-kb confining pressure. This sample was recrystallized during deformation, and the average grain size increased significantly [*Green et al.*, 1970, Figure 13].

temperature and the energy change. The amount of twinning varies with the temperature up to a certain point and appears to vary quite markedly with the energy difference involved in twinning. However, the absolute values of these energy differences are quite small. Evaluation of the contribution of twinning to the preferred orientation of plastically deformed samples is also discussed.

The variation of the amount of twinning with the temperature is similar to that found by *Wooster et al.* [1947, p. 931]; i.e., moderate temperatures (200°–300°C) aid twinning, but beyond them increases in temperature are not important. This result seems consistent with *Thomas and Wooster's* [1951, p. 44] statement that the atomic displacements necessary for twinning are little more than the amplitude of the thermal vibration at room temperature. A moderate increase in the temperature might increase the amplitude of the thermal vibrations to the point that twin boundaries would spontaneously move around if they were not pinned by impurities in the crystal. However, the process of twinning is diffusionless and does not seem to be thermally activated in the usual sense.

The apparent dependence of the amount of twinning on the energy difference involved is more problematical. This dependence shows itself in two ways. First, the amount of twinning increases with an increase in the differential stress (compare Figure 7d, c, and b). Second, the inverse pole figures for all of the specimens are similar in form (not just in sign) to the plot of the difference in compliance between the original and the twinned orientations (Figure 6). It would perhaps seem more reasonable if all the negative forms had twinned in the experiments, since they are all unstable under the macroscopically applied stress. Possible explanations for the apparent dependence of the amount of twinning on the energy difference will now be discussed. Several possibilities exist, and at present none can be proven to be the sole explanation.

One possibility might be that the rate of twinning is proportional to the energy difference and that the experiments have durations that are too short for twinning to be complete. However, this hypothesis is not confirmed by the available evidence of the effect of time on twin-

ning. In particular, consider the inverse pole figures of Figure 9a and b. The second sample experienced differential stress for a period that lasted at least 10 times as long as that of the first sample. Although there is a slightly greater amount of twinning in the second sample, it is by no means true that orientations that have twinned the same amount in the two samples had energy differences that differed by a factor of 10; instead, the factor appears to be ~1.5. In addition, some quartzite experiments that have lasted up to 3 months still do not show complete twinning, and there appears to be no systematic increase in the amount of twinning in these quartzites over such long time periods. In further support of the contention that the rate of Dauphiné twinning is very rapid, *Wooster et al.* [1947, p. 930] note that complete twinning occurred in a single crystal loaded for only a few seconds. Thus the twin boundaries may migrate at a speed close to the speed of sound, as in martensitic transformations.

Another possible explanation for the dependence of the amount of twinning on the magnitude of ΔW is that there could be a critical energy difference (the theoretical basis for which we do not specify) that must be exceeded for twinning to occur. An argument against this concept is that the energy difference for which 50% twinning occurred in the 2.2-kb novaculite experiment is about 20–40 times lower than that for which the same amount of twinning occurred in the 16.3-kb experiment at the same time and temperature (compare Figures 9b, and d and 6). If the energy change was near the critical energy in one experiment, it should have been above or below it in the other.

One reasonable explanation for the apparent dependence of the amount of twinning on the energy difference involved lies in the consideration of inhomogeneities in stress and strain that must exist within an aggregate of anisotropic crystals. If the actual stresses acting on a crystal differ from the macroscopically imposed stresses on the aggregate, the actual difference in strain energy on twinning for that crystal could differ in magnitude or even in sign from the nominal difference. Nominally unstable crystals for which the energy difference on twinning is greater can experience greater departures from the nominal stress while still remaining unstable than crystals with lower energy differences can experience. Since there should be a distribution in stress states centered in some sense about the macroscopically imposed stress, the lower the nominal energy difference, the more likely it is that a nominally unstable crystal will remain untwinned.

Although inhomogeneities in stress may result in some crystal orientations never becoming twinned, it is also possible that some crystals that twin during the loading may untwin during the unloading. This untwinning could occur if some crystals become plastically deformed during loading, and thus it would be impossible for any crystals to return to their initial stress-free state. The unloading would thus induce residual stresses, whose character could be such as to preferentially untwin crystals of low nominal energy change.

A fundamental aspect of the observed preferred orientation is that some crystals of a given orientation twin, whereas others of that orientation do not. Thus, in addition to systematic differences in the stresses and the strains experienced by crystals as a consequence of their differing orientations, individual crystals of a single orientation must also behave differently from one another as a consequence of their differing surroundings. Although this observation is perhaps self-evident, it would nevertheless be convenient for the analysis of inhomogeneities of stress and strain in an aggregate if all the grains of a given orientation could be treated alike. It appears that, at least for the elastic deformation of a quartz aggregate, this simplification cannot be made.

The effect of mechanical Dauphiné twinning on the preferred orientation of axially deformed samples can be removed from the inverse pole figures by averaging the concentrations in symmetrical positions on either side of the c-a line and plotting the averaged points again with the c-a line as a mirror plane. This averaging forces the preferred orientation due to all mechanisms except twinning to have mirror plane symmetry about the c-a line, when, in fact, this may not be the case, and it assumes that there is no other mechanism that operates in just the same way as twinning (i.e., that transforms a negative form into the corresponding positive form). However, this averaging is a simple way to find out if there is any preferred orientation of

the positive and negative forms that cannot possibly be accounted for by mechanical Dauphiné twinning. For the flints, novaculites, and quartzites loaded with no permanent deformation, such averaging removes all preferred orientation, as was expected. For the plastically deformed quartzites and the syntectonically recrystallized flints, there are several other components of the preferred orientation left after the averaging. These components have been produced by other orienting mechanisms and are discussed by *Tullis et al.* [1973] and *Green et al.* [1970].

METHODS OF APPLICATION TO NATURALLY DEFORMED ROCKS

In the experimentally deformed quartz aggregates, low differential stresses at moderate temperatures acting for very short times are sufficient to produce significant amounts of mechanical Dauphiné twinning. Thus twinning should be very common in naturally deformed rocks, including those which show little or no evidence of having been plastically deformed and those which are completely recrystallized. It is hoped that the study of mechanical Dauphiné twinning in rocks can be used to determine something about the stresses and the strains that caused the twinning, although there will be inherent ambiguities in the sign of the stresses or the strains resulting from the quadratic form of the energy expression.

The general method of approach presently envisaged is to compare the pattern of twinning derived from the measured preferred orientation with the patterns of twinning predicted from a number of different stress states and eventually to find one that is 'best.' The inverse problem, of determining directly from the observed twinning pattern the states of stress consistent with that pattern, is perhaps more desirable, but it is less straightforward and has not yet been attempted.

Baker and Wenk [1972] have described a method for representing the complete preferred orientation of a rock, i.e., the variation of the concentration as a function of crystal orientation within specimen coordinates. The same method can be used for representing the variation of ΔW as a function of crystal orientation with respect to a coordinate system in which the stresses or strains are given. Euler angles are used to represent different crystal orientations relative to the fixed coordinates.

To present the concentration of ΔW for a large number of crystals having orientations specified by the Euler angles, one can plot a value for each orientation in a Cartesian coordinate system in which the Euler angles are the coordinate axes. This representation, termed the 'orientation distribution function' (ODF) by *Bunge* [1968b], can be presented as a three-dimensional block diagram or as a series of sections normal to one of its axes.

In the choice of Euler angles made by *Baker and Wenk* [1972], two of the angles (ψ and θ) give the orientation of the crystal c axis with respect to the specimen coordinates, and the third angle ϕ gives the orientation of the crystal a axes about the c axis. Thus the operation of Dauphiné twinning is equivalent to going from a point in the ODF to a similar point displaced by 60° in ϕ. Thus, if Dauphiné twinning were the only orienting mechanism that had operated, a sum of concentrations at ϕ and $\phi + 60°$ would result in an ODF with uniform concentrations everywhere. If an ODF modified in this way still shows variations in intensity with ϕ, some mechanism other than mechanical Dauphiné twinning has oriented the positive and negative forms.

In many rocks a preferred orientation of the c axes will exist so that within a section through the ODF normal to the ϕ axis there will be ψ and θ variations in concentration in addition to those which could result from mechanical Dauphiné twinning. To determine the portion of the ψ, θ, and ϕ variation that results only from twinning, a 'fraction twinned ODF' may be prepared by replacing the concentration value at each point by a fraction having a numerator that is the concentration at the point displaced 60° in ϕ minus the concentration at the point of interest and a denominator that is the sum of these two concentrations. The values of this fraction will range from -1 to $+1$, negative values representing orientations depleted by twinning and positive values representing those enriched by twinning.

A fraction twinned ODF as defined above can be compared with a similar representation of ΔW for twinning (a 'ΔW function' or ΔWF). If the stresses or strains assumed in the calculation of ΔW produced the twinning, the fraction twinned ODF should match the ΔWF, just as the inverse pole figures for axial experiments match the inverse pole figure showing $\Delta S_{11}'$. Many different states of stress or strain could obviously be tried to find a ΔWF that matched the fraction twinned ODF. This search would be simplified if the specimen coordinates used in initially obtaining the ODF were known to be the principal directions of the stress or the strain that caused the twinning.

An attempt was made to interpret the twinning in the only quartz tectonite yet to have its complete preferred orientation determined [*Baker and Wenk*, 1972]. However, the analysis was unsuccessful, since the specimen, a recrystallized quartz mylonite [*Christie*, 1963, specimen F-6], is not really suitable for this type of analysis. It has a strong c-axis preferred orientation, has been subjected to several episodes of deformation, and has triclinic symmetry of its preferred orientation, although in the X-ray analysis, monoclinic symmetry was assumed and imposed. The rock shows a strong difference in the preferred orientation of the positive and negative forms, consistent with the operation of twinning, but the fraction twinned ODF is not similar in form to any of a series of ΔWF calculated for systematically varying deviatoric stress states. Overprinting of twinning from several episodes of deformation is the most likely explanation for this discrepancy. It is worthy of note that the ϕ variation for this specimen is very considerably reduced by removing the effect of twinning, this reduction indicating perhaps that no other mechanism besides twinning has operated to orient the positive and negative forms. However, we do not yet have enough experience to know how much significance to place on the remaining ϕ variation.

Although there are certain inherent ambiguities and complexities in using mechanical Dauphiné twinning for determination of the deformational history of a natural quartzose specimen, it does have some possibilities. For certain carefully chosen specimens, perhaps twinning can be used to infer directions and relative magnitudes of stresses for rocks that otherwise appear undeformed. Also, analysis of the complete preferred orientation of a rock for evidence of twinning may provide concrete evidence for the operation of some other orienting mechanism that has also acted to orient the positive and negative forms. In addition, analysis of patterns of preferred orientation produced by twinning in natural rocks may possibly yield information concerning the homogeneity of the stress or the strain within the rock.

Acknowledgments. This work is an outgrowth of the intensive program in Dave Griggs's laboratory dedicated to understanding the deformation of quartz. Dave's pioneering equipment design coupled with his penetrating insight and remarkable intuitive understanding of many problems has helped to create one of the most stimulating and productive laboratories. We are grateful to have had the opportunity to be his students, a role that Dave elevates to that of colleague. Many discussions with Dave about Dauphiné twinning have contributed substantially to the authors' understanding of the subject, and it is with pleasure that we dedicate this paper to him.

This work is strongly dependent on the techniques for measuring the complete preferred orientation of quartz aggregates developed by Dr. D. W. Baker; we are grateful to him for these contributions and for his help and cooperation at many early stages of this work. We are grateful to Dr. W. M. Chapple for many helpful discussions, for reading and suggesting improvements in the manuscript, and for pointing out the similarity between our treatment and that of Cottrell for Griffith cracks. We are also grateful to Dr. R. Shreve for offering valuable suggestions at several stages. In addition, we acknowledge helpful discussions with Drs. R. Coe, J. Christie, R. Fletcher, and M. Conway. Drs. B. Kamb and G. DeVore offered helpful criticism of the manuscript.

The experiments were done at UCLA and were supported by the National Science Foundation under grants GP-5575 and GA-1389.

Orientation Distribution Diagrams for Three Yule Marble Fabrics

H.-R. WENK AND W. R. WILDE

*Department of Geology and Geophysics, University of California
Berkeley, California 94720*

The orientation distribution has been derived for a specimen of undeformed Yule marble and two specimens of experimentally deformed Yule marble (Turner et al., 1956). In contrast to a fabric diagram, which describes the preferred orientation of only one crystallographic direction, hkl, the orientation distribution, which is in common use among metallurgists (Bunge, 1968a, b, 1969; Morris and Heckler, 1968, 1969), is a complete description of preferred orientation, and we advocate that it become the standard representation of orientation information in petrofabric analysis. It is a three-dimensional frequency distribution of the Euler angles Ψ, Θ, and Φ that relate the crystal coordinate system XYZ and the specimen coordinate system ABC. The orientation distributions of the three Yule marble fabrics are presented in three-dimensional stereoscopic Calcomp plots and two-dimensional contoured sections. An attempt is made to interpret the orientation distribution diagrams: $\Psi\Theta$ sections and projections onto the $\Psi\Theta$ plane give the orientation of the crystallographic c axes with respect to the specimen coordinates (fabric diagram), and $\Phi\Theta$ sections and projections onto $\Phi\Theta$ planes show the orientation of the specimen C axis with respect to the crystal (axis distribution chart). There is a strong difference in orientation for positive and negative forms in the experimentally deformed fabrics.

D. T. Griggs suggested that the orientation distribution function, a new way of representing the preferred orientation in crystal aggregates, using Euler angles, be applied to experimentally deformed marbles. The ODF, which was introduced by *Bunge* [1965] and *Roe* [1965], is a mathematical expression that gives a complete description of preferred orientation in crystal aggregates (in contrast to a fabric diagram, which just describes the orientation of one crystallographic direction) and which has become popular especially among metallurgists. The fundamentals are treated by *Bunge* [1969]; for applications see, for instance, *Morris and Heckler* [1968, 1969].

A first application of the orientation distribution in geology has been made by *Baker and Wenk* [1972]. They derive the orientation distribution function for a quartzite mylonite. Optical methods are insufficient for quartz, since only one crystallographic direction, the c axis, can be measured. Instead they used X-ray pole figure data coupled with sophisticated spherical harmonic analysis. The present study introduces the orientation distribution

to the geologist who does not have an X-ray texture goniometer and therefore has no opportunity to do the necessary analysis but is familiar with the standard universal-stage technique.

In calcite both the c axis and an e lamella $(01\bar{1}2)$ can be measured on the universal stage, and hence the orientation distribution can be derived directly from universal-stage data. To advertize this method of describing preferred orientation, we have derived the orientation distribution for three calcite fabrics. We have chosen from the literature three specimens of experimentally deformed Yule marble from a study that has contributed greatly to the understanding of deformation mechanisms in calcite [*Turner et al.*, 1956]. Table 1 gives the specifications of the three samples used in this study. The original universal-stage measurements of Roberta Dixon were used as input data. In developing one example we do not claim to make the orientation distribution easily understood. Indeed, from our own experience, we realize that it is not easy to visualize. However, by demonstrating its application and some of its possi-

TABLE 1. Specifications of Samples Used in This Paper [*Turner et al.*, 1956]

Specimen	Conditions of Deformation	Grains	Orientation (cf. Figure 1)
172	undeformed Yule marble	343	c-axes maximum at B
459	elongated 40%, 500°C, 5 kb	162	extension parallel to A
477	elongated 48%, 500°C, 5 kb	91	extension parallel to B

bilities, we hope to convince the reader that the orientation distribution is a better way of presenting orientation information than the ordinary fabric diagram and possibly to entice him to use this method in the future. Computer programs written in Fortran 4 for the CDC 6400 computer at Berkeley are available on request. We also offer to process data sets on the Berkeley computer.

PREFERRED ORIENTATION IN TERMS OF EULER ANGLES

The classical method of representing preferred orientation in crystal aggregates is the fabric diagram or pole figure [*Schmidt*, 1925]. The fabric diagram gives the orientation distribution of a crystallographic direction [*hkl*] with respect to the specimen coordinates *ABC* (Figure 1). The spherical coordinates ϕ, ρ (Figure 2b) of the direction [*hkl*] of each crystal are measured on a universal stage or by a suitable X-ray method, and the poles are plotted in equal-area projection. Areas of equal density, expressed as multiples of a uniform density distribution, are then contoured [see *Turner and Weiss*, 1963].

A fabric diagram is an incomplete description of preferred orientation, because it specifies the orientation of only one crystallographic direction, for instance, the *c* axis. Separate fabric diagrams are necessary to give information about the preferred orientation of other crystallographic directions, and, even with a whole set of fabric diagrams, it is not possible to determine directly, for example, the orientation of *e* lamellas of crystals having *c* axes that contribute to a certain maximum in the *c*-axis pole figure. If the orientation of at least two crystallographic directions can be specified simultaneously for a crystal, its orientation is established completely, and the relative orientation of the two coordinate systems, specimen and crystal, can then be described uniquely by the three Euler angles Ψ, Θ, and Φ. A right-handed rectangular coordinate system *XYZ* must be placed in the calcite crystal (symmetry $\overline{32}/m$); we have used the convention of *Baker et al.* [1969] (Figure 2a). Table 2 lists the spherical coordinates in the crystal system (Goldschmidt angles ϕ' and ρ') of these rectangular axes *XYZ*. Owing to the trigonal symmetry, there are three possible sets of axes for each crystal. Another rectangular

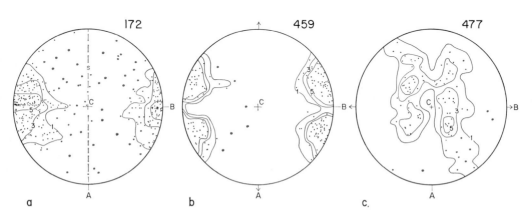

Fig. 1. The *c*-axis fabric diagrams of (*a*) a specimen of undeformed Yule marble and (*b*, *c*) two experimentally deformed Yule marbles. Data are from *Turner et al.* [1956]. Arrows indicate axis of extension. Equal-area projection is used on upper hemisphere. The contours are 1%, 3%, and 5%, per 1% area.

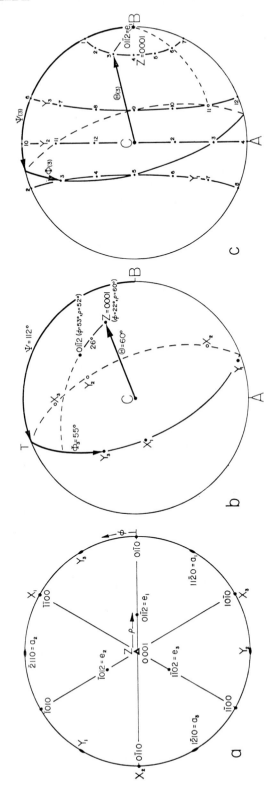

Fig. 2. (a) Orientation of an orthogonal crystal coordinate system XYZ in the trigonal calcite crystal. All diagrams are equal-area projections of the upper hemisphere. The dashed lines are on the lower hemisphere. (b) Derivation of the Euler angles Ψ, Θ, and Φ that relate the orientation of the crystal (XYZ) and specimen (ABC) coordinate systems from the position of the c axis [0001] and an e lamella $(01\bar{1}2)$. (c) A model for the orientation distribution if e $(01\bar{1}2)$ is aligned parallel to the specimen B axis. The c axes are on a small circle around B at 26°.

TABLE 2. Goldschmidt Angles (Spherical Coordinates) for Calcite

Crystallographic Direction	ϕ', deg	ρ', deg
$c = 0001 = Z$	0	0
$e_1 = 01\bar{1}2$	0	26.25
$a_1 = 11\bar{2}0$	−30	90
X_1	60	90
Y_1	150	90
X_2	180	90
Y_2	270	90
X_3	300	90
Y_3	30	90

coordinate system ABC is placed in the specimen (Figure 1). Possible specimen axes are: (1) geographic coordinates: south is A, east is B, and normal to the horizon is C, (2) mesoscopic fabric coordinates [*Sander*, 1948]: c (normal to foliation) is A, b (lineation) is C, and $\perp bc$ is B (these coordinates have been used by *Baker and Wenk* [1972]), and (3) symmetry axes of the fabric diagram.

In this study we chose two orientations to illustrate various aspects of preferred orientation. In the first setting all specimens are kept in their initial orientation. The mutually perpendicular directions of extension in specimens 459 and 477 (see *Turner et al.* [1956], Figure 1 and Table 1) are normal to the specimen C axis, and thus the effect of two different deformation experiments on the fabric of the undeformed marble (specimen 172) is illustrated. In the second setting the specimens are taken out of their original orientation to place the axis of highest symmetry in 172 and the extension axes in 459 and 477 parallel to C. In this position the symmetry of the fabric is easier to see, but the specimens are no longer in the same orientation (Figure 1. Specimen 172: $A = T$, $B = 1$, and $C = 2$; specimen 459: $A = 3$, $B = 2$, and $C = T$; specimen 477: $A = T$, $B = 2$, and $C = 3$). The assumption of $\bar{1}$ specimen symmetry had to be made because of the impossibility of distinguishing polar directions (i.e., between the upper and lower hemispheres) with a petrographic microscope.

The three angles $-\Phi$, $-\Theta$, and $-\Psi$ express a set of rotations about crystal axes that bring the two coordinate systems to coincidence. Angles Ψ, Θ, and Φ indicate the inverse operation, i.e., the rotations that bring XYZ from coincidence with ABC to a given orientation.

By convention a positive rotation is defined as a rotation that advances a right-handed screw in the positive direction along an axis (counterclockwise rotation if we look toward the origin). The angles are illustrated in Figure 2b: a first rotation $-\Phi$ about Z brings Y_3 to coincidence with T, a second rotation $-\Theta$ about T ($= Y_3'$) brings Z to coincidence with C, and a final rotation $-\Psi$ about Z' ($= C$) brings T ($= Y_3'$) to coincidence with B. A triplet of Euler angles can be derived for each crystal. Together they specify the orientation of a crystal in relation to the specimen unequivocally. Thus for each crystal in an aggregate we can plot a point in three-dimensional Euler angle space (Figure 3). The points in the Euler space that corresponds to the crystal orientation indicated in Figure 2b are shown in Figure 3a. The resulting distribution of Euler angles, called the orientation distribution [*Bunge*, 1965; *Roe*, 1965], gives a complete description of preferred orientation in a crystal aggregate. The three-dimensional point densities can be normalized and contoured in much the same manner as the two-dimensional densities in fabric diagrams after compensating for distortions in the surface area of the reference sphere (Figure 4).

It is customary to plot the orientation distribution in a Cartesian coordinate system with Ψ, Θ, and Φ as axes. *Baker* [1970] pointed out that such an orientation distribution has a translation periodicity of 2π along each of the axes, because a full rotation about any of the rotational axes brings a point to coincidence with itself. Therefore the ODF has a lattice character and a space group symmetry that is a function of the point group symmetry of the specimen and the crystal. *Baker* [1970] derived the space group symmetries for some important combinations. The mirror plane in calcite $(\bar{3}2/m)$ and the inversion center $(\bar{1})$ in the specimen, both enantiomorphous repetitions, would produce a left-handed coordinate system, which is not compatible with the definition of the Euler angles. The $\bar{3}2/m$ crystal symmetry thus becomes 32. The threefold axis parallel to Z causes a 120° repetition on Φ instead of a 360° one (Figure 2b). The unit cell of the ODF is therefore $\Theta = 360°$, $\Phi = 120°$, and $\Psi = 360°$. The twofold axis parallel to Y introduces an equivalent orientation $\Theta + \pi$, $-\Phi$, Ψ representing a Θ-glide plane perpendicular to Φ. No specimen symmetry has been assumed, but, because of the convention of choosing Y coincident with the twofold symmetry axis at a, no new crystal orientations are produced by the $\bar{1}$ specimen symmetry. An additional pair of equivalent positions is caused by the fact that each orientation can be described by two pairs of Euler angles: the orientation Θ, Φ, Ψ is identical to $-\Theta$, $\Phi + \frac{1}{3}\pi$, $\Psi + \pi$. An n-glide plane perpendicular to Θ at $\Theta = 0$ is thus introduced. The four equivalent positions, Θ; Φ, Ψ; $-\Theta$, $\frac{1}{3}\pi + \Phi$, $\pi + \Psi$; $\pi + \Theta$, $-\Phi$, Ψ; $\pi - \Theta$, $\frac{1}{3}\pi - \Phi$, $\pi + \Psi$, have been used to generate all equivalent orientations in the orientation distribution from one set of Euler angles. This configuration corresponds to space group $P\,n\Theta\,2_1$ (corresponding to space group 33 in the *International Tables for X-Ray Crystallography* [1952]). The origin is shifted by $\Theta = \frac{1}{2}\pi$, $\Phi = \frac{1}{6}\pi$, and $\Psi = 0$ in relation to the standard crystallographic setting. The reduction in the Φ axis results from the trigonal crystal symmetry, which causes a repetition of Φ after a $2\pi/3 = 120°$ rotation (Figure 2b). The symmetry elements of the orientation distribution and the size and the position of the asymmetric unit of the unit cell are shown in Figure 3a.

DERIVATION OF THE ODF

The following paragraphs describe briefly how the orientation distribution is derived for calcite from universal-stage measurements of the optic axis [0001] and an e lamella $(01\bar{1}2)$. Manipulation of the universal-stage data and the computation of Euler angles are done by a computer program that was modified from the programs Coordinate Transformation and Indirect Orientation of *Wenk and Trommsdorff* [1965]. (Note the unfortunate printing errors in their equations on p. 566.) Measurements for which the angle between [0001] and e deviates by more than 4° from the theoretical value of 26° are rejected.

After the conversion of universal-stage coordinates to spherical coordinates, all the specimens are rotated into the same orientation. A right-handed rectangular coordinate system is placed in each specimen, as is indicated in Figure 1 and Table 2, which gives other important information about the specimens.

Derivation of the Euler angles is illustrated by the example in Figure 2b. The spherical coordi-

nates (measured counterclockwise from B) are

$$[0001] \quad \phi\; 22.0° \quad \rho\; 60.0°$$

$$\perp(01\bar{1}2) \quad \phi\; 53.0° \quad \rho\; 52.0°$$

The spherical coordinates of the three symmetrically related pairs of orthogonal axes X_1Y_1, X_2Y_2, and X_3Y_3 are calculated.

	X		Y	
	ϕ	ρ	ϕ	ρ
1.	193.5°	30.3°	289.9°	86.3°
2.	305.5°	112.1°	65.9°	141.3°
3.	93.2°	119.2°	148.3°	44.3°

From these coordinates one set of axes with Y in the upper hemisphere (in this case X_3Y_3) is selected for the computation of the Euler angles:

$$\Psi = \phi_{0001} + 90° = 112.0°$$

$$\Theta = \rho_{0001} = 60.0°$$

$$\Phi = \arccos\,[\sin\rho_Y \cos(\phi_Y - \Psi)] = 55.7°$$

$$0 < \Phi < 120°$$

These calculations are carried out for every crystal, and the Euler angle triplet for each is punched out and used as input for a second program, designed to generate the three-dimensional orientation distribution diagram, which, because it is easy to visualize, is presented in a pair of stereoscopic figures. The first step is the calculation of the four symmetrically equivalent positions in the unit cell for each set of Euler angles. The complete unit cell is then rotated around the horizontal and vertical axes to an appropriate viewing position and finally plotted (in perspective) with an angle separation of 6° between the two stereoscopic images. In our example the following equivalent positions were calculated:

Θ, Φ, Ψ			
Θ, Φ, Ψ	60.0°	55.7°	112.0°
$-\Theta$, $\frac{1}{3}\pi + \Phi$, $\pi + \Psi$	300.0°	115.7°	292.0°
$\pi + \Theta$, $-\Phi$, Ψ	240.0°	304.3°	112.0°
$\pi - \Theta$, $\frac{1}{3}\pi - \Phi$, $\pi + \Psi$	120.0°	4.3°	292.0°

These positions are shown in Figure 3a in stereoscopic representation. Figure 3b–d shows the orientation distribution diagrams for the three marble specimens. As can be seen in these fig-

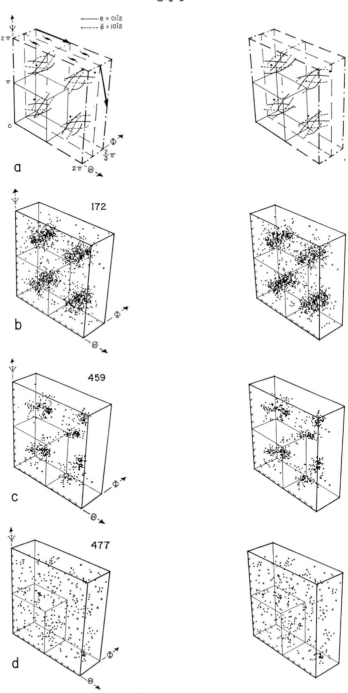

Fig. 3.

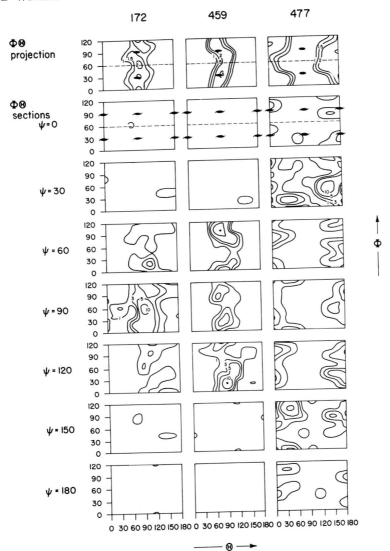

Fig. 4. Contoured ΦΘ projections and sections of the asymmetric unit of the orientation distribution for three Yule marble specimens. (Top panel) Contours of 0.5%, 1%, 2%, and 3% per 1% volume for Ψ projections. (Lower panels) Contours of 1%, 3%, 5%, and 10% per 1% volume for sections. (Note that the contours are not continuous across the unit cell boundaries. This discontinuity is an artifact of the contouring routine.)

Fig. 3. (Opposite) Stereoscopic pairs of orientation distribution diagrams. The asymmetric unit is indicated. Calcomp plots are used. The orientation of the specimen is indicated in Figure 1. (a) Unit cell of the orientation distribution for $\bar{1}$ specimen and 32 crystal symmetry. Symmetry elements are shown. Dots indicate equivalent positions for the orientation $\Theta = 60°$, $\Phi = 55.7$, and $\Psi = 112.0$. Solid lines represent orientations with e $(0\bar{1}12)$ parallel to the specimen axis B. Note that they follow high-density areas in the orientation distribution diagrams of 459 in Figure 3c. The dashed lines represent orientations with $\bar{e}$ $(10\bar{1}2)$ parallel to the specimen axis B. These lines follow the less dense areas in Figure 3c. (b) Undeformed Yule marble 172. Notice the uniform distribution parallel to Φ. There are 343 points in the asymmetric unit. (c) Experimentally deformed Yule marble 459. The extension axis is parallel to A. There are 162 points in the asymmetric unit. (d) Experimentally deformed Yule marble 477. The extension axis is parallel to B. There are 91 points in the asymmetric unit.

ures, 150–300 points seem to give the best stereoscopic effect; 100 points is a lower limit of adequate counting statistics in these fabrics.

The coordinates of all points within the asymmetric unit of the orientation distribution unit cell are punched out to be used by a third program designed to normalize and contour the distribution. Distortion in the surface area of the reference sphere caused by plotting Euler angles in Cartesian coordinates is corrected by converting Θ to cos Θ. The modified asymmetric unit is divided into equal rectangular volume elements. The points falling within each element are counted and normalized to a multiple of the uniform distribution. Cos Θ is then reconverted to Θ and the densities are contoured in any desired rectangular plane. Contouring is done with the plotter subroutine CONBVS [*Paradis and Hussey*, 1969]. Figure 4 gives contours in various planes perpendicular to Ψ for each of the three Yule marble fabrics shown in Figure 3b–d. Cells with sides cos $\Theta = \frac{1}{3}$, $\Phi = 20°$, $\Psi = 30°$ were used. Another method of numerically analyzing point densities has been mentioned by *Bunge* [1969, p. 23]. He expands the density distribution with a harmonic surface and obtains the smoothing effect by truncating the higher orders of the harmonic expansion.

INTERPRETATION OF THE ORIENTATION
DISTRIBUTION DIAGRAM

The orientation distribution is a representation of preferred orientation and exhibits many characteristics of a fabric. It contains substantially more information than a fabric diagram. But, since most geologists are not yet familiar with the orientation distribution, they have difficulties reading it. The following discussion is an attempt to explain some of the characteristics of the orientation distribution. We try to show many features that can also be seen in a fabric diagram and demonstrate how they are expressed in the orientation distribution.

The orientation distribution is difficult to read and, at least at first, requires a great deal of concentration to relate the three-dimensional Euler angle space to the actual orientations of crystals within a specimen. It is best to keep in mind the significance of the Euler angles implicit in their definition. Angle Ψ is the azimuth in the specimen system of the crystal c axis,

whereas angle Φ is the azimuth in the crystal system of the specimen C axis and corresponds to Goldschmidt's spherical angle ϕ (Figure 2). The angle Θ is a pole distance in both systems, being the angle between the crystal c axis and the specimen C axis.

If we consider Ψ and Θ to be independent of Φ (project the orientation distribution on to the $\Psi\Theta$ plane), we obtain the equivalent of the conventional fabric diagram in orthogonal representation, i.e., the distribution of crystal c axes in the specimen. Although we do not present such projections, they can be easily visualized in the stereoscopic figures (Figures 3 and 6) and correlated with the c-axis fabric diagrams (Figures 1 and 5).

On the other hand, we may consider Φ and Θ to be independent of Ψ. In this case we obtain the distribution of the specimen C axis in a fixed crystal system; this distribution is called the axis distribution chart for C. The top diagrams in Figures 4 and 7 show $\Phi\Theta$ projections for our specimens. They are left in orthogonal representation for easy correlation with the orientation distributions. Specimen 172 shows no significant variation with Φ, and thus it is indicated that the a axes are randomly distributed around the c axes within the specimen. In specimens 459 and 477, where the concentration varies markedly with Φ, the orientation of the a axis within the specimens was obviously affected by the deformation.

The deformed specimens show definitely trigonal symmetry with a strong difference in orientation for positive and negative forms. The twofold symmetry axes of calcite can be seen at $\Phi = 30°$, $90°$, and $\Theta = 90°$. In addition, the $\Phi\Theta$ projection nearly possesses a mirror plane at $\Phi = 60°$. This mirror plane corresponds to the mirror plane in the $\overline{3}2/m$ calcite symmetry. On geometrical grounds, because of the restriction that the coordinate system placed in the crystal must be right handed, the mirror plane was ignored in the derivation of the orientation distribution symmetry. Physically, however, calcite possesses $\overline{3}2/m$ symmetry, and the crystals respond to deforming forces in accordance with the true symmetry. The axis distribution chart contains a mirror plane at $\Phi = 60°$ by virtue of the statistical averaging of many grains.

Thus $\Psi\Theta$ and $\Phi\Theta$ projections of the orientation distribution contain the information of conventional fabric diagrams and axis distribution charts, respectively. Neither of these representations gives a complete description of the orientation of grains within the specimen. However, when they are combined and expanded into three dimensions, the resulting diagram does contain a complete description of grain orientation. Unfortunately, the three-dimensional nature of the distribution is not easily visualized or represented graphically. We have produced stereoscopic images of orientation distributions to aid their visualization. Another conventional method is to represent them in contoured sections at various levels in the unit cell of the orientation distribution. Figures 4 and 7 contain contoured sections perpendicular to Ψ in the orientation distributions depicted stereoscopically in Figures 3 and 6. Since all the essential information of the orientation distribution is contained in the asymmetric unit of the unit cell, only this unit has been contoured.

We have chosen two different specimen settings for our derivation of the orientation distribution. By placing all the specimens in the same orientation in the first setting (in relation to the fabric of the material before deformation), it becomes easy to visualize the change in the fabric during deformation. In this setting the c-axis maximum and the lack of a-axis control in the original fabric (specimen 172) gives rise to a roughly cylindrical concentration parallel to the Φ axis at $\Psi = 90°$ and $\Theta = 90°$.

In the extension experiment 459 this cylindrical concentration is split into two maximums, one at $\Psi = 110°$, $\Theta = 90°$, and $\Phi = 20°$ and the other at $\Psi = 70°$, $\Theta = 90°$, and $\Phi = 100°$. In specimen 477 the distribution is much more diffuse, and the main concentrations are shifted from the $\Psi = 90° \pm 20°$ plane to the $\Psi = 0° \pm 20°$ plane. This shift is the agreement with the change in the direction of extension from parallel to $A(\Phi = 90°)$ in specimen 459 to parallel to $B(\Phi = 0°)$ in specimen 477.

Since the orientation of each measured grain in a specimen is uniquely described in terms of Euler angles, the orientation distribution lends itself well to the testing of specific hypotheses regarding the nature of preferred orientation. It has been shown that, under certain experimental conditions of deformation, crystallographic directions other than the c axis (e.g., $e = \{01\bar{1}2\}$ and

$r = \{10\bar{1}1\}$) play important roles [*Wenk et al.*, 1972]. *Turner et al.* [1956] show for specimen 459 a high concentration of poles of e parallel to specimen direction B. By way of example, leaving genetic implications aside for the moment, let us test the hypothesis that this is the sole restriction on the orientation of the grains. We will begin by finding all possible orientations of grains having e poles that are exactly aligned with B (Figure 2c) and plotting them in the orientation distribution diagram (Figure 3a). (We also show the paths for poles of $\bar{e}$ parallel to B.) If our hypothesis is correct, paths for e parallel to B should follow high-density areas within the orientation distribution for specimen 459. The hypothesis can be tested quantitatively by integrating the point densities along these paths. Qualitatively, we can merely superimpose the paths on the orientation distribution diagram and notice whether they follow high-density areas.

For specimen 459 we find that the solid lines (poles of e parallel to B) do intersect high-density areas within the orientation distribution diagram and that the dashed lines (poles of $\bar{e}$ parallel to B) pass between high-density areas. We can deduce from this finding that poles of e do tend to fall parallel to B and furthermore that the direction $\bar{e}$ does not play an important role in the formation of preferred orientation. However, the lack of uniform density along these paths indicates that our hypothesis does not completely describe the preferred orientation. There is, in addition, a strong control on the orientation of the c axes. The same procedure can be applied to test other hypotheses about orientations of crystals in the specimen.

If the specimens are oriented with some unique axis parallel to C (second setting, Figures 5, 6, and 7), the symmetry of the fabric about this axis is much easier to visualize. For the undeformed specimen (172) we placed the single c-axis maximum at C. Points are nearly uniformly distributed in the $\Psi\Phi$ plane at $\Theta = 0°$ and $180°$, and thus true axial symmetry of the specimen is indicated. Specimens 459 and 477 were oriented with the axis of extension parallel to C. The strong variation with Ψ in both specimens shows a distinct departure from axial symmetry, which is not too surprising in 459, since the axis of extension is normal to the unique symmetry axis of the undeformed ma-

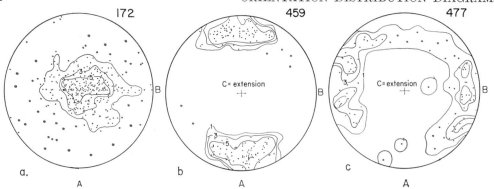

Fig. 5. The c-axis fabric diagrams of the three Yule marble specimens oriented such that the axis of highest symmetry (extension axis in 459 and 477) is parallel to C. Notice that in these settings all the specimens are not in the same orientation.

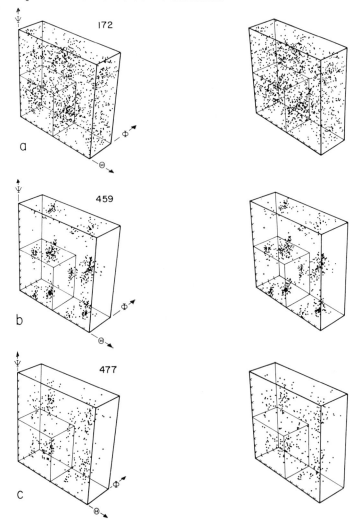

Fig. 6. Stereoscopic pairs of orientation distribution diagrams of the three Yule marble specimens in the second setting (Figure 5).

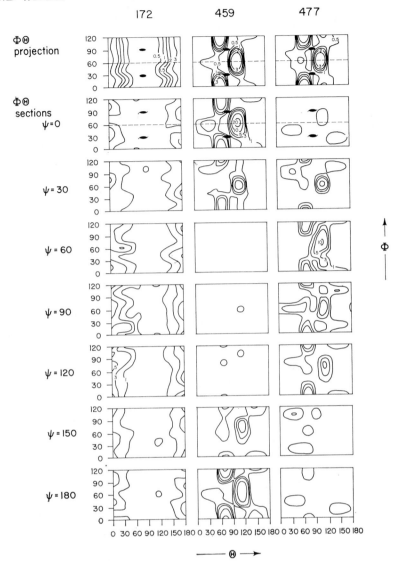

Fig. 7. Contoured ΦΘ projections and sections of the asymmetric unit of the orientation distributions in Figure 6 in a new setting. Contours of 0.5%, 1%, 2%, and 3% per 1% volume for projections and of 1%, 3%, 5%, and 10% per 1% volume for sections are given.

terial. However, in 477 both the unique symmetry axis of the starting material and the unique symmetry axis of the load coincide. This combination should produce an axially symmetric fabric. The departure from axial symmetry in this case is due to the difficulty of locating with the universal stage some c axes and e lamellas that have poles nearly normal to the plane of the thin section [*Turner et al.*, 1956]. Care should therefore always be taken

to remove any statistical bias from the measurements. Projections of the fabrics on to the ΘΦ plane (Figure 7) again show the same general characteristics as in the first setting, i.e., no Φ dependence in specimen 172 but definite trigonal symmetry for specimens 459 and 477.

In the derivation of the orientation distribution a specimen symmetry of $\bar{1}$ has been assumed. Coupled with the known symmetry of calcite

$(\overline{3}2/m)$, this symmetry produced an orientation distribution with the space group symmetry of $Pn\theta2_1$. As the specimen symmetry becomes higher, the orientation distribution also acquires higher symmetry. In specimens 459 and 477 the symmetry of the specimen approaches the symmetry mmm. The symmetry of the orientation distribution in turn approaches the space group symmetry $P2/\phi2_1/\theta2/\theta$.

It should be noted that the orientation distributions in both settings contain identical orientation information. It would be possible to draw the same conclusions from either distribution. The advantage of choosing one setting over another is merely that certain aspects of the preferred orientation can be made more immediately obvious.

CONCLUSION

There is no further need to mention the advantages of the orientation distribution over a classical fabric diagram. The orientation distribution is not the only complete description of preferred orientation, but it is a compact representation, well suited for mathematical analysis, which is essential if the orientation distribution must be derived from pole figure data [e.g., *Bunge*, 1969; *Baker and Wenk*, 1972]. The derivation from X-ray pole figure measurements is still a big project, since no standard computer programs are available. But it is quite easy to calculate the orientation distribution from universal-stage measurements of two crystallographic directions, which can be done in many important rock-forming minerals, such as calcite, dolomite, olivine, and feldspars. The calculations for an average specimen take 1-2 min on a CDC 6400 computer, including plotting of stereoscopic diagrams and contoured sections. The orientation distribution is necessary to calculate elastic properties of rocks with preferred orientation. The Euler angle representation can also be applied to systems other than deformed crystal aggregates. It is a suitable way to describe epitactic intergrowth, the host crystal representing one coordinate reference system. The orientation distribution might also prove to be useful in the representation of mesoscopic fabric coordinates (Y is the lineation or fold axis; Z is the pole to the foliation) in geographical space. Many of the characteristics of the orientation distribution will become familiar and easily visualized only if it has been applied many times. We therefore advocate that the fabric diagram, as the final step in representing preferred orientation and not as a preliminary description, be replaced by the orientation distribution.

Acknowledgments. Stimulating discussions with Dave Griggs and John Christie instigated this work. F. J. Turner kindly supplied the universal-stage data for three Yule marble specimens. We are thankful to D. Baker for his careful and critical reading of the manuscript.

Support from the Computer Center at Berkeley, the National Science Foundation under grants GA 1671 and GA 29294, and the Miller Institute for Basic Research at Berkeley is gratefully acknowledged.

Some Observations on Translation Gliding and Kinking in Experimentally Deformed Calcite and Dolomite

L. E. WEISS AND F. J. TURNER

Department of Geology, University of California
Berkeley, California   94720

Single crystals of calcite and dolomite were deformed at high lateral constraint. At 5-kb confining pressure and 300°C, calcite crystals jacketed in stainless steel were made to slip in a positive sense on $\{10\bar{1}1\} = r$ with $t = \langle 21.\bar{2} \rangle$. Thus, for example, if positive slip is taken to have occurred along $(10\bar{1}1) = r_1$, the glide line is $[r_1:f_2] = [21.\bar{2}]$. The possibility of such r glide suggests that the yield stress for the positive sense (hitherto obscured by twinning in the positive sense on $\{01\bar{1}2\} = e$) may possibly be lower than that for the negative sense under the physical conditions involved. Dolomite crystals were deformed in compression in a solid-medium apparatus at 10-kb confining pressure and room temperature. The σ_1 axis was inclined at small angles (generally 10°–20°) to the known translation plane $\{0001\}$ to induce kink band formation. Kink bands were formed with both $t = \langle 11\bar{2}0 \rangle$ and $t = \langle 10\bar{1}0 \rangle$, a previously unknown glide line.

D. T. Griggs's experiments on the deformation of calcite crystals led to the development of new microscopic techniques for identifying active systems of translation gliding. Any such system, identified by accompanying geometric rotational effects, is defined in terms of crystal morphology as summarized by *Buerger* [1930a]. For simple translation gliding a strained end state is related to an unstrained initial state by a finite simple shear. A strain of this kind conforms closely to the uniform pervasive mutual displacement of submicroscopic layers of the crystal structure parallel to a glide plane T in a constant direction t, the glide line. This description in no way implies that such simple displacements actually occur on the atomic scale. Atomic paths, which involve the motion of dislocations, are clearly more complex and are not specified.

In this paper we consider the following aspects of translation gliding in calcite and kinking in dolomite.

1. Experimentally demonstrated translation gliding on $r = \{10\bar{1}1\}$ in calcite and $c = \{0001\}$ in dolomite need not necessarily operate with equal ease in opposite senses. *Buerger* [1930a, p. 11] listed four symmetry criteria for T and t, any of which makes glide 'two sided' or symmetrical: (1) an even-fold symmetry axis parallel

to t, (2) a symmetry plane normal to t, (3) an even-fold symmetry axis normal to T, and (4) a symmetry plane parallel to T. Neither the translation planes nor the known glide lines ($\langle 21.\bar{2} \rangle$ in calcite and $\langle 11\bar{2}0 \rangle$ in dolomite) meet any of these criteria. Indeed, in calcite, r gliding hitherto has been produced in the negative sense only. (The sense is defined as negative when an upper layer parallel to, say, $(10\bar{1}1)$ is displaced downward with respect to an underlying layer as viewed along the $\langle 11\bar{2}0 \rangle$ direction in $(10\bar{1}1)$, in contrast with the unique positive sense of twin gliding on $e = \{01\bar{1}2\}$ [cf. *Turner et al.*, 1954, Figure 10]. Is translation in opposite senses possible for these two systems? And, if it is, does any geometric evidence of different behavior in opposite senses exist?

2. Translation gliding in laterally constrained crystal cylinders oriented with t inclined at small angles to the principal axis of compressive stress σ_1 is commonly accompanied by kinking. The ideal kink is a parallel-sided strain domain, in which the glide plane is externally rotated with respect to the unstrained host crystal about an axis in T normal to t, the rotation sense being opposite to that of translation. Evidence of translation in opposite senses in a single glide system is afforded by conjugate kinking. This fact is well known, e.g., in gypsum compressed

in the direction [001], where symmetry indeed demands such behavior: $T = \{010\}$, a symmetry plane, and $t = [001]$. Conjugate kinking is readily produced by basal translation ($t = [100]$) in muscovite even though not required by symmetry conditions [*Borg and Handin*, 1966, p. 280]. Dolomite crystals (cylinders) compressed parallel or at low angles to the potential translation plane $\{0001\}$ in a solid confining medium (talc) at room temperature tend to deform by a mechanism closely analogous to ideal kinking. This behavior, described in this paper, throws light on the sense of basal translation and reveals the capacity to glide normal to $\{10\overline{1}0\} = m[t = \langle 10\overline{1}0 \rangle]$ as well as in the familiar manner parallel to $\langle 11\overline{2}0 \rangle$, as was demonstrated by *Higgs and Handin* [1959].

SENSE OF r TRANSLATION IN CALCITE

Inhibiting influence of e twinning. The most common mechanism of translation gliding in calcite is translation on $\{10\overline{1}1\} = r$, with $t = \langle 2\overline{1}.\overline{2} \rangle$ [*Turner et al.*, 1954]. For example, for glide on $(10\overline{1}1) = r_1$ the glide line would be $[\overline{2}1.2] = [r_1{:}f_2]$, and the recorded sense invariably is negative, certainly mostly and perhaps entirely owing to the inhibiting effect of twin gliding on $\{0\overline{1}12\} = e$, sense positive. Calcite deformed by r translation (negative) is many times stronger (10 times at 300°C) than when failure is by e twinning. In triaxial axially symmetrical stress systems (compression with $\sigma_1 > \sigma_2 = \sigma_3$, extension with $\sigma_1 = \sigma_2 > \sigma_3$), a cylinder loaded to give a high positive shear stress on r automatically has a high stress in the same sense on one

or more e planes. It therefore fails by twin gliding on e.

Effect of lateral constraint on deformation by compression normal to $(10\overline{1}0)$. For compression normal to $(10\overline{1}0) = m_1$ the most highly stressed potential glide planes are r_1 ($S_0 = 0.5$) and $e_1(S_0 = 0.4)$. The sense of gliding in each case is positive. In all previous experiments, in which the test cylinder was sealed only by a thin weak jacket, the sole glide mechanism expectably was twinning on e_1. This twinning is carried to completion in a diagonal strain domain A, transverse to the trend of e_1, and there are lateral domains B of partial e_1 twinning (Figure 1, left). Small end domains C remain undeformed [cf. *Heard et. al.*, 1965, p. 88].

This pattern is modified when cylinders of the same orientation are laterally constrained by stainless-steel jackets (at 300°C and 5 kb; over-all shortening 10%), as was found in a series of experiments by M. S. Paterson in 1965. (Results of a series of extension experiments under similar conditions will be found in *Paterson and Turner* [1970], where details of experimental procedure are also recorded.) Strain again is concentrated in a broad diagonal domain transverse to e_1, and again twin gliding on e_1 has played a conspicuous role (Figures 1, right, and 2a). But twinning is concentrated in two lateral sectors X, in which it is about half complete ($L_{e_2}{\cdot}^{e_1}\Lambda e_2 = 10°$ (Figure 3, left), $S = 0.36$. (Mean values determined in specimen P. 303: $S = (\cot 45° - \cot 55°)/\sin 57° = 0.36$. For complete e twinning, $S = 0.688$.) In Y, the central sector, e_1 twinning is subordinate. The main glide mechanism here

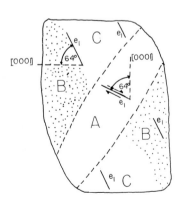
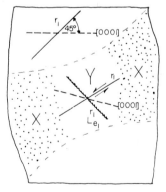

Fig. 1. Diagrammatic (simplified) sections of calcite crystals compressed (σ_1 is vertical) normal to $m_1 = (10\overline{1}0)$, showing distribution of strain domains: (Left) Specimen 212–190 deformed by D. T. Griggs [*Turner et al.*, 1954, p. 907; *Heard et al.* 1965, p. 88]. (Right) Specimens 212-P-274 and 212-P-303 constrained by stainless-steel jackets, deformed by M. S. Paterson (cf. Figure 2).

is r_1 translation in the positive sense. Geometric evidence is varied and conclusive (Figure 3, right):

1. Sparse early e_2 lamellas have been internally rotated counterclockwise through 12°–17° along the arc e_2-r_1. The maximum value of shear given for $e_2 \Lambda L_{e_2}{}^{r_1} = 17°$ is

$$S = (\cot 38° - \cot 55°)/\sin 51° = 0.74$$

(The relationship is [*Turner et al.*, 1954, p. 900]

Fig. 2a.Specimen 212-P-274.

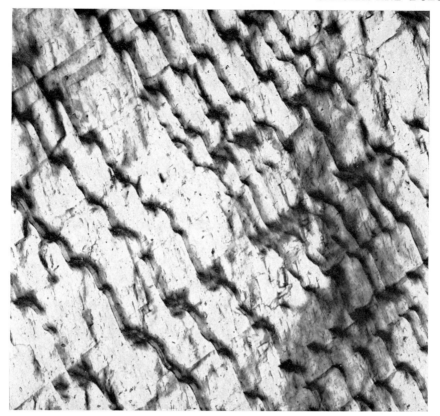

Fig. 2b.   Specimen 212-P-303.

Fig. 2.   Calcite crystals compressed normal to $m_1 = (10\bar{1}0)$ at 300°C, 5 kb. (The experiments were conducted by M. S. Paterson, Australian National University, Canberra.) The sections have been cut parallel to the deformation plane, the light is polarized, the polarizers are crossed, and σ_1 is vertical. (a) $\epsilon = 0.079$. (b) $\epsilon = 0.105$. This specimen is part of central domain Y (Figure 1, right), and r_1 is oriented as shown. The corrugated lamellas dip to the right arc $L_{e_1}r_1$.

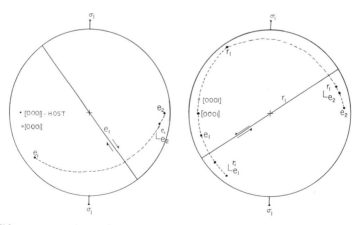

Fig. 3.   Glide systems on internal rotation of passive L_e lamellas in (left) domains X and (right) domain Y of specimen 212-P-303 (Figure 1, right).

$S = (\cot \alpha - \cot \beta)/\sin \gamma$, where α and β ($\alpha < \beta$) are angles between the internally rotated plane and the glide plane before and after rotation and γ is the angle between t and the axis of rotation.)

2. Numerous thin e_1 lamellas, conspicuous as narrow corrugated NW-SE bands in Figure 2b, have been internally rotated, also counterclockwise, through 30°–37° along the arc e_1-r_1. The corresponding maximum value of shear is

$$S = \cot 71° + \cot 72°$$

$$= 0.67$$

3. Parallel corrugations in successive L_{e_1}·r^1 lamellas (Figure 2b) show that r_1 gliding was concentrated in narrow packets a fraction of a millimeter thick. Axial planes of corrugations are precisely aligned parallel to r.

Identification of r translation (sense positive) in no way conflicts with obvious but very minor r twinning (sense positive) in the end sections of this specimen and in the many jacketed specimens deformed in extension [*Paterson and Turner*, 1970].

Implications of positive r translation. In specimens deformed under strong lateral constraint, the pattern of internal stress is likely to become markedly inhomogeneous. Adjoining strain domains might differ significantly with respect to the orientation of the principal stress axes as well as actual values of σ_1, σ_2, and σ_3. Nevertheless, high positive resolved shear stress on r_1 is difficult to visualize in the absence of significant positive stress on e_1 or one of the other two e planes. How then has positive r_1 translation occurred in preference to e_1 twinning? The answer may be partly in the fact that the known great strength of calcite in r translation refers to gliding in the negative sense. The anomaly would be more readily explained if r translation were notably easier in the positive than in the hitherto observed negative sense of gliding.

KINKING IN DOLOMITE

Experimental procedure. All experiments were made on cylindrical cores (5-mm diameter, 20-mm length) cut from a large (12-cm diameter) single crystal of dolomite from Ontario, Canada. In its unstrained state this crystal contains abundant thin $f = \{02\bar{2}1\}$ twin lamellas that were used as planar markers to determine internal and external rotations in the deformed specimens. All cylinders were shortened in compression from 5 to 10% at 10-kb confining pressure and room temperature in a triaxial tester using talc as a solid-pressure medium. Elevated temperatures were avoided to prohibit twin gliding on $\{02\bar{2}1\}$, which is favored more than basal glide above 400°C [*Higgs and Handin*, 1959]. The strain rate in all the experiments was 10^{-4} sec^{-1}.

The triaxial apparatus, designed by D. T. Griggs primarily for high-temperature creep and controlled-strain experiments, combines a solid-pressure medium with an external force gage and is not suitable for accurate determination of yield stresses. The present experiments suggest, however, that the yield stress for basal kinking where σ_1 is inclined at 20° or less to $\{0001\}$ is in excess of 5 kb at 10-kb confining pressure. No significant differences in yield stresses for specimens of slightly different orientations could be detected, although they undoubtedly exist.

The behavior of most crystals was largely ductile, although some cataclastic effects, notably fracturing on a variety of crystallographic planes, were always present in the sectioned specimens. But very little can be said about the nature of the 'yield point' at the onset of plastic behavior (kinking). Kinking of some crystals by compression at small angles to a glide plane is accompanied by a sharp yield followed by a large stress drop [e.g., *Klassen-Neklyudova*, 1964, p. 139]. No such behavior, if present, could be observed in the present experiments because of the design of the apparatus. Also, if some uniform translation gliding precedes kinking, as seemed to be the case in at least one specimen compressed at >25° to t, the yield is more gradual, and the sudden stress drops associated with rapid kink band nucleation need not occur.

The test cores were cut in a variety of orientations, but, because of some initial curvature in the crystal, these orientations could not be very closely prescribed. Most cores had axes lying between 10° and 20° of $\{0001\}$. To facilitate description of the core directions and particular crystallographic planes and lines, crystallographic axes were arbitrarily selected for the test crystal, and a system of notation as shown in Figure 4 (projected equal area, upper hemisphere on {0001}) was adopted. The shaded areas indicate three ranges in orientations of cores used: areas 1, cores in which initial resolved shear stress is higher in $[2\bar{1}\bar{1}0]$ than it is in $[10\bar{1}0]$; areas 2, cores in which initial resolved shear stress is higher in $[10\bar{1}0]$ than in $[2\bar{1}\bar{1}0]$; and areas 3, cores in which initial resolved shear stress is approximately equal in $[2\bar{1}\bar{1}0]$ and $[11\bar{2}0]$. Each pair of areas represents cores cut to favor positive and negative slip on {0001}, according to the convention given in Figure 5.

Identification of glide systems in kinking. The glide lines determined in the present experiments for kinking of dolomite by slip on {0001} are $t = \langle 11\bar{2}0 \rangle$ and $t = \langle 10\bar{1}0 \rangle$. The first glide line has been described by previous workers beginning with *Johnsen* [1902]; the second one is new. These two glide lines can be distinguished by the different patterns of internal rotation that they produce in inherited {02$\bar{2}$1} twin lamellas, as is shown in Figure 6. Translation gliding with $t = [2\bar{1}\bar{1}0]$ is shown in Figure 6 (left) for a shear strain of 0.4. Lamellas in orientations f_1 and f_2 are internally rotated through different angles, whereas lamellas in orientation f_3 are not rotated (they intersect the glide plane {0001} in the glide line). For the same shear strain with $t = [10\bar{1}0]$, f lamellas of all three orientations are internally rotated (Figure 6, right). Lamellas

with orientations f_2 and f_3 are rotated through the same angle, which is smaller than the angle of rotation of f_1.

In the present experiments we are concerned not with simple translation gliding but with kinking. Because the glide plane is reoriented during kink band formation, the external rotation of the slipped domain in relation to the unslipped domain can be measured directly. For dolomite with $T = \{0001\}$ the angle of external rotation is given by the angle between the optic axes in the host and the kink band, and the axis of external rotation, which in ideal kinking (see discussion below) should be at right angles to the glide line, is given by the pole of the plane in which the two optic axes lie. Both internal and external rotation effects were used to identify glide systems in the present study.

'Ideal kinking.' Many types of structures referred to generally as kink bands have been observed in naturally and experimentally deformed mineral crystals (see *Carter and Raleigh* [1969] for a general review), and very similar structures, commonly termed kink folds, have been observed in deformed pervasively foliated or laminated materials such as phyllite [e.g., *Paterson and Weiss*, 1966]. Study of all kinds of kink bands suggests that most kink bands can be idealized, as is shown in Figure 7. The

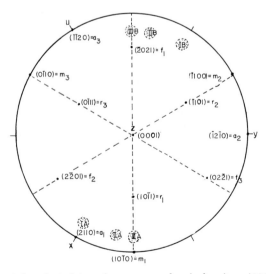

Fig. 4.   Dolomite crystal projected (equal area, upper hemisphere) on {0001}. Areas 1, 2, and 3 (A and B) show range of orientations of cores used in experiments.

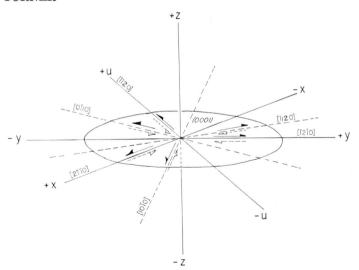

Fig. 5.   Convention for sense of slip on {0001} in directions ⟨11$\bar{2}$0⟩ and ⟨10$\bar{1}$0⟩. The arrows give the positive sense.

ideal kink band illustrated has the following geometric and other properties:

1.   Deformation is confined to the kink band; the host remains undeformed but not necessarily in its initial orientation with respect to σ.

2.   Deformation within the kink band takes place by distributed slip on the active glide plane, corresponding to a finite simple shear of magnitude 2 tan ω/2, where ω is the change in angle of the glide plane at the kink boundary (the external rotation).

3.   The interface between a kink band and the host is a microscopically sharp surface marked by no visible structural discontinuities. The material remains coherent across this boundary, which is thus an invariant plane in the shear transformation in the kink band. The other invariant plane is the glide plane.

4.   As a consequence of the coherence along the kink band–host interface, this boundary bisects the angle between the glide plane in the kink band and that in the host; i.e., in Figure 7, $\phi = \phi_K$.

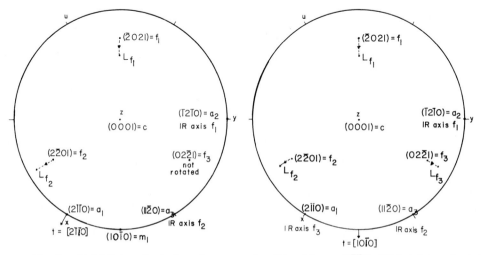

Fig. 6.   (Left) Internal rotation phenomena for glide with $t = [2\bar{1}\bar{1}0]$. (Right) Internal rotation phenomena for glide with $t = [10\bar{1}0]$.

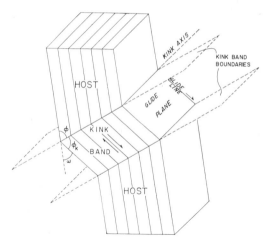

Fig. 7.   Geometry of an ideal kink band.

5.   The axis of kinking, defined by the line of intersection of the glide plane and the kink band boundary, is at right angles to the glide line.

The properties listed above are requirements for an ideal kink band in any material possessing one pervasive slip plane. Another possible property of kink bands in crystals has been emphasized by *Starkey* [1968], namely that the angles ϕ and ϕ_K of Figure 7 must be geometrically related to the Burgers vector for glide and the spacing of the glide planes according to the relationship

$$\tan \phi = \tan \phi_K = dT/[\tfrac{1}{2}(n \cdot V_t)]$$

where dT is the spacing of the glide planes (in the case of dolomite d_{0001}), V_t is the Burgers vector (primitive lattice vector) in direction t, and n is an integer. The most likely value for n in most kink bands is 1. This value means that each atom in the kink band has been displaced by one primitive lattice vector in the glide direction. The relationship can only be satisfied if slip is perfectly regular and no glide packets are present, so that displacements are uniformly distributed thoughout a deformed domain. Under these conditions, kinking would have to proceed in 'jumps,' the angles of external rotation corresponding to discrete values of $n = 1$, 2, and 3, and so on. Gradual growth and tightening of a kink band, such as has been observed in gypsum [*Turner and Weiss*, 1965], would be impossible. A more realistic interpretation of

the relationship is that it provides a series of limiting values toward which kink angles will tend as slip becomes uniformly distributed for each increment in n as deformation proceeds.

In ideal kinking the angle of external rotation provides a measure of the shear strain in the kink band. This determination can be used as a check on the values of shear strain (S) obtained from the internal rotations of passive markers.

RESULTS

In all experiments thin kink bands were formed by basal glide. In most of these, $t = [2\bar{1}\bar{1}0]$, as was expected from studies of translation gliding by previous workers [*Johnsen*, 1902; *Higgs and Handin*, 1959]. Examples of kink bands in two specimens are shown in Figure 8a and b. Because of high constraint the thin kink bands occurred in parallel swarms of variable thickness distributed generally throughout each specimen. In crystals compressed under low constraint as, e.g., in a triaxial apparatus using gas as a pressure medium, a single large kinklike band tends to form, generally with diffuse boundaries.

The detailed features of the kink bands varied widely. Their thickness ranged from <0.1 mm to >1.5 mm, and their spacing spanned a similar range. Some had very clear sharp boundaries visible only in polarized light; others had boundaries marked by thin dark laminas, which may be zones of cataclasis.

Confirmation of glide with $t = [2\bar{1}\bar{1}0]$ was based on the phenomena of internal and external rotation outlined above. Figure 9 is an example of the data from a typical kink band with $t = [2\bar{1}\bar{1}0]$. Shear strains in the kink band computed from rotation effects are as follows: (1) internal rotation ($7\tfrac{1}{2}°$) of f_1 to L_{f_1}, where $S = 0.24$; (2) internal rotation ($10\tfrac{1}{2}°$) of f_2 to L_{f_2}, where $S = 0.25$; and (3) external rotation ($14°$) of $\{0001\}$, where $S = 0.246$. These values are in close agreement. As far as could be determined (in the face of high tilts) f_3 was virtually unrotated, as is required for $[2\bar{1}\bar{1}0]$ glide (Figure 6, left).

Most kink bands conformed broadly to this example. Kink bands with $t = [2\bar{1}\bar{1}0]$ were formed in specimens with orientations both in areas 1A and 2A (slip negative) and in areas 1B and 2B (slip positive) of Figure 4.

In two specimens with orientations in areas

3A, where [2$\bar{1}$$\bar{1}$0] and [11$\bar{2}$0] are approximately equally stressed but the sense of slip is opposite (negative on [2$\bar{1}$$\bar{1}$0] and positive on [11$\bar{2}$0]), kink bands were formed in which glide with $t = [10\bar{1}0]$ could be unequivocally demonstrated. Data from one such kink band are given in Figure 10. Twin lamellas with orientations f_2 and f_3 are rotated through equal angles, whereas lamellas with orientation f_1 are also rotated but through a larger angle. The axis of external rotation of {0001} lies in the kink band boundary in direction [1$\bar{2}$10]. These observations are consistent with basal translation in the kink band with $t = [10\bar{1}0]$.

Shear strains in the kink band deduced from these rotation effects are as follows: (1) internal rotation (19°) of f_1 to Lf_1, where $S = 0.533$; (2) internal rotation (12°) of f_2 to Lf_2 and of f_3 to Lf_3, where $S = 0.468$; and (3) external rotation (23°) of {0001}, where $S = 0.408$.

Agreement among these values is only approximate. Errors may reflect the effects of high universal-stage tilt in measuring lamellas f_2 and f_3 and the possible presence of minor strain in the host domains. (The f lamellas in these regions depart from true rationality by a few degrees.)

Translation gliding with $t = [10\bar{1}0]$ of the sense shown in Figure 10 is designated as negative (Figure 5). Compression of specimens with orientations 3B (slip positive) failed to form kink bands with $t = [10\bar{1}0]$. Instead, very few thin kinks were formed with $t = [2\bar{1}\bar{1}0]$. This anomaly, which may reflect error in the orientation of cores, will be studied further.

Slip with $t = [10\bar{1}0]$ is best viewed as symmetric double slip involving glide in both [2$\bar{1}$$\bar{1}$0] and [11$\bar{2}$0]. Such double slip on adjacent directions in ⟨11$\bar{2}$0⟩ must involve both positive and negative senses, and thus the theory that the yield stresses for these two kinds of behavior cannot differ greatly is suggested.

Between kink bands most specimens appear to have rational or nearly rational twin lamellas, and plastic deformation seems to be minor or absent. But none of these domains retains its initial orientation, and an external rotation of the host domains always seems to accompany the development of kink bands. This 'back rotation' is in the opposite sense to that of external rotation in the kink band, and it tends to increase the initial angle between σ_1 and the pole of {0001} in the host domain. For example,

Fig. 8a. Kink bands in dolomite crystals compressed at small angles to {0001}. The whole crystal between crossed polarizers is shown. Swarms of kink bands appear as light stripes dipping very gently to the left. The other stripes are {02$\bar{2}$1} lamellas.

if the initial angle between [0001] and σ_1 is 80°, after about 10% shortening this angle approaches 90° in the host domains. In one specimen this angle has reached 90°, and thin swarms of lenticular second-generation kink bands have appeared between the main kink bands. In this second set of kink bands, translation is in the opposite sense (Figure 11). Very similar behavior has been observed in paper cards compressed initially at 15° to the plane of the cards [*Weiss*, 1969, p. 324].

Fig. 8b. Kink bands in dolomite crystals compressed at small angles to {0001}. A horizontal kink band (width of about 0.3 mm) can be seen in the center. Obliquely inclined {02$\bar{2}$1} lamellas (dipping about 45° right) are internally rotated as they cross the kink band.

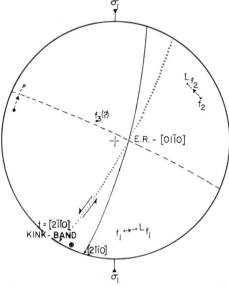

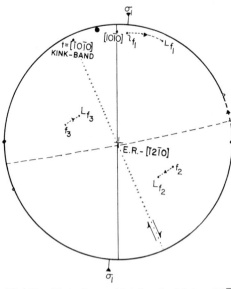

Fig. 9. Data from a kink band with $t = [2\bar{1}\bar{1}0]$. Solid circles, [0001] in the host band. Open circle, [0001] in the kink band. Cross within circle, the **pole of the** kink band boundary.

Fig. 10. Data from a kink band with $t = [10\bar{1}0]$. Solid circles, [0001] in the host band. Open circle, [0001] in the kink band. Cross within circle, the pole of the kink band boundary.

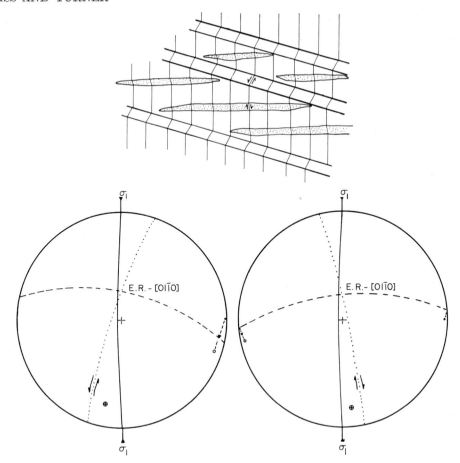

Fig. 11.   Conjugate senses of kinking in dolomite ($t = [2\bar{1}\bar{1}0]$). (Top) Diagrammatic sketch of specimen. Main kink bands are blank; kink bands with conjugate sense of slip are stippled. (Lower left) Data from main kink band. (Lower right) Data from conjugate kink band. Solid circles, [0001] in the host band. Open circle, [0001] in the kink band. Cross within circle, the pole of the kink band boundary.

This behavior seems to result from the strong lateral constraint exerted on the specimen. Ideally, the appearance of a single set of kink bands should be accompanied by a lateral offset of the specimen ends. Prohibition of this offset by the presence of a solid-medium and high end friction is accommodated in part by a backward rotation of the host domains toward the compression axis. This rotation seems to be a purely rigid body (external) rotation, perhaps because plastic deformation in the host domains involving the same glide system would require slip in the opposite sense from that of shear stress on the glide plane {0001}.

In a few specimens the host domains appeared to contain significantly irrational twin lamellas and to have been plastically deformed. These specimens were those with the smallest initial angle between σ_1 and [0001] (generally of $<65°$), and the most probable explanation is that some uniform translation gliding on {0001} preceded kinking. This view is supported by the observation that these specimens contain only very few thin kink bands.

DISCUSSION

Some of the experimentally induced kink bands conform closely to the model of ideal kink bands, as is described above. The host domains surrounding these kink bands are almost perfectly rational, and the kink band boundaries are generally sharp, clear, and unmarked by

TABLE 1. Data for Six Kink Bands

Kink Band	External Rotation, deg	ϕ, deg	ϕ_K, deg
Positive Glide			
ML − 1 − 1	20	79	81
ML − 1 − 3	25	75	78
ML − 1 − 4	25	80	75
Negative Glide			
MR − 2 − 1	10	84	86
MR − 2 − 2	17	82	81
MR − 2 − 3	14	84	82

visible structural discontinuities. The kink bands are generally visible only in polarized light by a change in the extinction angle at the boundary. Kink band boundaries are approximately symmetrically inclined to {0001} in the host and the kink band. Table 1, for example, presents measurements of ω, ϕ, and ϕ_K in six kink bands involving glide in $t = [2\bar{1}10]$. Three kink bands are for positive glide, and three for negative glide. Values of $\phi_K - \phi$ are in the range $-5°$ to $3°$, the mean value being $<1°$. Such a close approach to ideal symmetry has been observed in a large number of kinks. Examples of data from symmetric kink bands with $t = [2\bar{1}10]$ and $t = [10\bar{1}0]$ are given in Figure 12, left and right, respectively.

In symmetric kinking of this kind, coherence between the kink band and the host across the boundary surface can be maintained. This surface thus closely resembles the habit plane between an initial crystalline phase and a polymorphic equivalent formed from it by a martensitic transformation: both a kink band boundary and a martensitic habit plane are ideally invariant planes in the transformation from one state to another. Kink bands seem to resemble most closely martensitic transformations of the single interface type [e.g., *Wayman*, 1964, p. 81], in which a uniquely oriented interface propagates wholly or partly through a crystal from one or more martensitic nuclei. The strains accompanying such a phase transformation resemble those of kinking and mechanical twinning except that an actual change in the crystal structure occurs. But the resemblance is close enough to suggest that symmetric kink bands in crystals may have mobile interfaces with the host. Thus progressive deformation by kinking may proceed not only by an increase in the number of discrete kink bands but also by a widening and a coalescence of kink bands by the lateral migration of coherent boundaries.

Some of the kink bands produced in the experiments are markedly asymmetric. A few

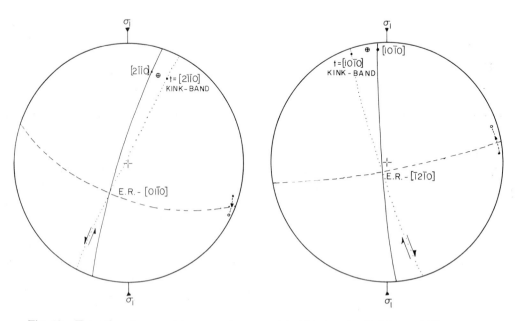

Fig. 12. Examples of symmetric or nearly symmetric kink bands: (Left) $t = [2\bar{1}10]$. (Right) $t = [10\bar{1}0]$. Solid circles, [0001] in the host band. Open circle, [0001] in the kink band. Cross within circle, the pole of the kink band boundary.

were observed in which the kink band boundary is approximately at right angles to $\{0001\}$ in the host. In these crystals, too, the host is almost unstrained, a fact suggested by the persistent near rationality of $\{02\bar{2}1\}$ lamellas; and no complete structural coherence along the kink interface is possible. The boundaries of such kink bands are commonly marked by thin dark layers representing structural discontinuity of some kind—probably cataclastic. It is likely that such boundaries are not mobile and cannot be propagated further into the host in the manner suggested for coherent symmetric boundaries.

For bulk shortening of the specimens of $\leq 10\%$ the experiments at high constraint demonstate that kink bands can remain symmetric and approximately ideal. But the host domains show a consistent sense of back rotation, in which the $\{0001\}$ glide plane approaches the direction of σ_1. If the back-rotated domain remains virtually undeformed, as seems to be true in most examples studied, the maintenance of a symmetric kink band boundary requires that slip must continue in the kink band and the kink band boundary must at the same time internally rotate in an opposite sense to that of the external rotation of the undeformed domain. Thus lateral growth of a symmetric kink band in a specimen with glide plane initially oblique to σ_1 requires a change in the state of strain in the kink and thus in the angles ϕ and ϕ_K.

Starkey's [1968] relationship for kinking in crystals predicts that only certain angles of external rotation should be observed in a given crystal, corresponding to first-order values ($n = 1$), second-order values ($n = 2$), etc., of the total slip vector. Ideal values of first- and second-order rotation angles for dolomite have

TABLE 2. Ideal Values of First- and Second-Order Rotation Angles for Dolomite

Glide Directions	ω	
	$n = 1$	$n = 2$
$[2\bar{1}\bar{1}0]$	17°8′	33°24′
$[10\bar{1}0]$	29°36′	55°

been computed for $t = [2\bar{1}\bar{1}0]$ and $t = [10\bar{1}0]$ and are reproduced in Table 2. The range of observed values for glide with $t = [2\bar{1}\bar{1}0]$ is 10°–25°, and for glide with $t = [10\bar{1}0]$ it is 17°–30°. The measurements are not, we feel, sufficient in number (about 20 kink bands were studied in detail) to permit any significant statistical treatment of these values. All that can be said on the data at hand is that they show no clear-cut correspondence with the predicted values. Study of large numbers of kink bands might yet show some correlation between predicted and observed values.

The present experiments have shown that kinking with glide line $[2\bar{1}\bar{1}0]$ is possible for both positive and negative slip. They have not shown that the yield stress for kinking is the same for the two senses. However, the demonstration of kinking with glide line $[10\bar{1}0]$, if it is interpreted as symmetric double slip on $[2\bar{1}\bar{1}0]$ and $[1\bar{1}20]$, suggests that these yield stresses must be approximately equal, since double slip involves both senses of glide. Failure to produce negative slip in $t = [10\bar{1}0]$ is not presently explicable.

Acknowledgments. We are indebted to Dr. M. S. Paterson for generous permission to use unpublished data.

We acknowledge support of this study by the National Science Foundation under grants GA-10636 and GA-29294.

Effect of Water on the Experimental Deformation of Olivine

JAMES D. BLACIC

Geophysics Program and Department of Geological Sciences, University of Washington
Seattle, Washington 98195

Experiments on single-crystal and polycrystalline samples of olivine have been carried out in a solid pressure medium deformation apparatus. Both 'wet' and 'dry' experiments were performed to test for the hydrolytic weakening mechanism in olivine. Drying a single-crystal sample at a high temperature under a vacuum increases the flow stress by a factor of about 2. This increase proves the operation of the hydrolytic weakening mechanism, since both the dried and the undried samples were deformed exclusively by glide. Recrystallization and shear induced by the presence of small amounts of water in dunite samples result in a plastic instability that may be important as a mechanism of intermediate and deep earthquakes.

It now seems certain from an extensive accumulation of geophysical, geochemical, and petrologic evidence that olivine, $(Mg, Fe)_2SiO_4$, is a major constituent of the upper mantle of the earth. Consequently, if we are to understand the nature of the large deformations that seem to be required in the upper mantle, we must determine the rheologic properties of this material. Recognizing the necessity of this information, several workers have carried out experimental studies of olivine deformation [*Griggs et al.*, 1960; *Raleigh*, 1965a, 1968; *Carter and Avé Lallemant*, 1970; *Post*, 1970]. The last two of these studies demonstrated that small amounts of water have a pronounced weakening effect on experimentally deformed olivine aggregates. This aspect of the problem is the subject of the work reported here.

The weakening effect of water in olivine was predicted by *Griggs and Blacic* [1965] and *Griggs* [1967], who showed that small amounts of dissolved water dramatically reduce the strength of quartz single crystals. These authors argued that water hydrolyzes the strong Si-O bonds in quartz, the hydrolysis allowing dislocations to move at a much reduced stress. Ultimately, the weakening is thought to be a reflection of the relatively weak hydrogen bonds that bridge the SiO_4 tetrahedra in the hydrolyzed regions. This phenomenon is called hydrolytic weakening. Griggs and Blacic predicted that, since all silicates contain Si-O or metal-O bonds

that are susceptible to hydrolyzation, hydrolytic weakening should be a general phenomenon in these materials. We present here some experimental observations that bear specifically on the effects of water on the flow stresses of olivine single crystals and aggregates.

EXPERIMENTS

Experimental apparatus. The experiments were carried out in two piston cylinder, solid pressure medium apparatuses designed by D. T. Griggs and his co-workers. The devices are basically similar, differing primarily in size. The small apparatus allows cylindrical samples approximately 3 mm in diameter by 8 mm in length to be deformed. The large apparatus accommodates samples approximately 6×19 mm. The sample is heated by an internal graphite tube furnace, and the temperature is measured by a Pt|Pt-10Rh thermocouple in contact with the sample. The temperatures listed here refer only to the central region of the samples, since in these apparatuses the samples are subject to large thermal gradients. The confining pressure is applied by compressing a relatively weak solid by means of a hydraulically actuated steel piston. Talc, pyrophyllite, AlSiMag 222 (a commercial ceramic material), and NaCl were used as confining media. An axial differential stress is applied by advancing a tungsten carbide deformation piston on to the sample at a constant rate. The deformation piston is driven by an electric motor through a gear train, a high-efficiency screw, a thrust bearing, and a load cell. Since the confining medium must be pushed aside as the sample deforms, the non-negligible strength of these materials introduces a certain amount of error into the differential stress measurements. The exact magnitude of this error is difficult to determine; however, *Blacic* [1971] estimated that the error in nominal flow stress is less than $\frac{1}{2}$ kb (the nominal flow stress is defined as the intersection of straight-line approximations to the stress-strain curves). This error is greatest when AlSiMag 222 is used as the confining medium and least when NaCl is used. Water may be introduced into the samples from dehydration of the talc or the pyrophyllite confining medium at high temperature or by jacketing the sample along with a small amount of water; the experiments are then termed wet experiments. Dry experiments are facilitated by using anhydrous confining media (AlSiMag 222 or NaCl) or by excluding dehydration water with a platinum jacket around the sample. More details of the apparatus and the experimental technique are given by *Griggs* [1967] and *Blacic* [1971].

Experimental materials. Experiments were carried out on samples of dunite and single-crystal olivine. The dunite is from Mount Burnett, Alaska, and consists almost entirely of medium to coarse grains of olivine ($\sim Fo_{90}$) with only minor amounts of pyroxene, chromite, and almost no serpentine. Analysis of the weight loss of samples heated in vacuum for 1–2 days at temperatures of up to 900°C indicates an average H_2O content of approximately 0.3% by weight. The olivine single crystal comes from Balsam Gap, North Carolina. This crystal (Fo_{95}–Fo_{100}) contains small amounts of serpentine and chromite. Weight loss analysis indicates a relatively high water content of approximately 1% by weight, which is confirmed by the presence of a sharp IR absorption band at 3680 cm^{-1}.

EXPERIMENTAL RESULTS

Dunite. The experiments in hydrous and anhydrous environments (Table 1) indicate that H_2O does have a significant effect on the flow stress of dunite. However, the experiments are complicated to some extent by the relatively large initial water content of the samples. Recrystallization along grain boundaries occurs in

the samples at temperatures of as low as 800°C in an anhydrous environment. This recrystallization is clearly due to the fluxing effect of water contained in the samples, since the onset of recrystallization in a very dry dunite is several hundred degrees higher (G. Dollinger, personal communication, 1971). (A similar effect is observed in quartz aggregates [*Green*, 1968; *Tullis*, 1971].) A peculiar mechanical effect is associated with the recrystallization in these experiments. This effect is illustrated in Figure 1, where the stress-strain curve for a dry experiment performed at a relatively high strain rate is shown. Note that, after about 13% plastic strain, the flow stress dropped by a factor of about 2 over a time span of about 2 min. A thin section of the sample (Figure 2) shows, in addition to the expected intragranular glide, an anastomosing set of narrow zones of fine-grained recrystallization. This shear zone appears to be associated with the stress drop in this experiment, since these zones are not observed in other experiments carried out within the ductile regime in which stress drops do not occur. Thus weakening promoted by the small amount of water contained in this dunite has resulted in the failure of a ductile material along a localized zone of recrystallization and shear. This case is not isolated. R. L. Post (personal communication, 1971) reports that a majority of his creep experiments on Mount Burnett dunite are terminated by these instabilities. This phenomenon is potentially important, since it seems to be akin to *Orowan*'s [1960] hot creep fracture mechanism of seismic faulting.

Olivine single crystals. To distinguish the mechanical effects of intragranular glide from those of grain boundary recrystallization, a number of experiments on olivine single crystals were carried out in hydrous and anhydrous environments (Table 1). The wet and dry experiments at first showed the same low strength, but it should be recalled that this crystal has a very large initial water content. *Griggs and Blacic* [1965] found that the relatively large amounts of water contained in some hydrothermally grown quartz crystals reduced the flow stresses of these crystals by approximately an order of magnitude in comparison with dry natural crystals tested at the same conditions of temperature, pressure, and strain rate. Hence

TABLE 1. Olivine Experiments

Strain Rate ($\dot{\epsilon}$), sec^{-1}	Temperature, °C	Confining Pressure (σ_3), kb	Nominal Flow Stress ($\sigma_1 - \sigma_3$), kb	Confining Medium	Strain* (ϵ), %	Deformation Features	Remarks	Experiment†
				Mount Burnett Dunite				
1.1 × 10^{-4}	1000	15.0	13.4	talc, Pt jacket	18.0	c bands, moderate recrystallization, ductile shear zone	dry	GB-20
7.7 × 10^{-6}	1000	15.5	2.7	talc	16.5	c bands, extensive recrystallization	wet	DT-107
7.7 × 10^{-6}	1000	15.0	8.1	AlSiMag 222	4.2	c bands, moderate recrystallization	dry	DT-108
7.9 × 10^{-6}	900	15.0	~1	pyrophyllite	2.5	c bands, slight recrystallization	wet	DT-116
1.0 × 10^{-6}	800	15.0	14.3	talc, Pt jacket	19.0		dry	GB-24
				Olivine Single Crystal				
7.2 × 10^{-6}	900	15.5	3.0	talc	10.0	slight recrystallization	wet	DT-640
7.5 × 10^{-6}	900	16.0	2.9	NaCl	13.0	(001) deformation lamellas	dry	DT-641
7.3 × 10^{-6}	900	15.5	4.8	NaCl	7.7	(001) and (100) deformation lamellas	dry‡	DT-648

* Strain is expressed in terms of a uniform shortening of the cylindrical samples. This shortening is calculated from micrometer measurements of the length of the sample before and after the experiment. In some cases where the strain is large, the diameter of the central region of the sample is measured in thin section and compared with the original diameter.

† The experiment numbers DT- indicate that the small sample size apparatus was used; the numbers GB- indicate that the large sample size apparatus was used.

‡ The sample was predried at 880°C under vacuum for 35 hours.

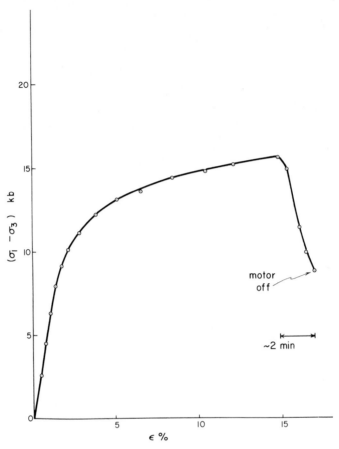

Fig. 1. Stress-strain curve for Mount Burnett dunite experiment GB-20, illustrating ductile shear failure behavior. Temperature, 1000°C; pressure, σ_3 = 15 kb; strain rate, $\dot{\epsilon}$ = 1 × 10^{-4} sec^{-1}; environment, anhydrous.

it was thought that the olivine crystal might be similar in its behavior to some of the synthetic quartz crystals with large OH contents. To test this theory, a sample from the olivine crystal was heated at about 880°C under vacuum for approximately 35 hours before deformation in an anhydrous confining medium. The fraction of the original water in this sample that was removed by the drying treatment is not known, but the amount that was removed apparently resulted in an increase in strength by a factor of about 2 (Figure 3). This increase is considered to be definite proof of the operation of hydrolytic weakening in olivine, since glide was the only mechanism of deformation evident in these samples. Also, note the apparent modulus reduction in the experiment on the undried sample. This behavior is characteristic of the

hydrolytic weakening phenomenon in quartz crystals [*Griggs,* 1967; J. D. Blacic and D. T. Griggs, unpublished manuscript, 1972], and it is cited as a likely explanation for the increase in strength of annealed single crystals of plagioclase by factors of 1.5–2.5 [*Borg and Heard,* 1970, p. 389].

DEFORMATION MECHANISM

The mechanisms of plastic deformation in olivine have recently been explored by a number of workers [*Raleigh,* 1965a, 1968; *Young,* 1969; *Carter and Avé Lallemant,* 1970]. The slip mechanisms deduced from the orientation of deformation lamellas and from the geometrical relations for kink bands in the deformed dunite and olivine single-crystal experiments reported here are in general agreement with the findings

0.5 mm

Fig. 2.   Photomicrograph of sample from experiment GB-20; crossed polarizers. The NW-SE
trending zone of fine-grained recrystallization and shear is shown. The compression direction
is approximately N-S.

of *Raleigh* [1968]. Namely, at temperatures of up to 900°C and a strain rate of about 1×10^{-5} sec^{-1} the predominant slip direction is [001]. The glide plane may be either (100) or {110}, the {110} planes being predominant at the higher temperatures. Inhomogeneous slip on these systems is commonly evidenced by kink bands parallel or nearly parallel to (001). The kink bands are often very numerous and closely spaced in the experimentally deformed samples.

Raleigh [1968] has identified a transition from the [001] to the [100] slip direction in experimentally deformed olivine at a temperature of

1000°C and a strain rate of about 10^{-4} sec^{-1}. Deformation lamellas are observed parallel to the most highly stressed planes in the zone of [100], this finding suggesting that a pencil glide mechanism may be operative at these conditions, i.e., {0kl}, [100]. However, *Carter and Avé Lallemant* [1970] show that at higher temperatures certain planes in the *a* zone may be preferred. Slip in the [100] direction was observed in the olivine single-crystal experiments in the present work (Figure 4). The single-crystal samples were compressed in a direction approximately normal to [010] such that high resolved

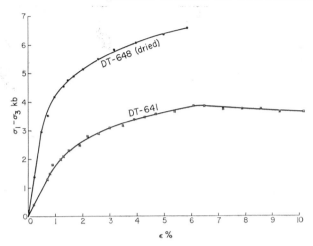

Fig. 3. Stress-strain curves for two olivine single-crystal experiments carried out under nearly the same experimental conditions, illustrating the influence of water on the flow stress of olivine. Temperature, 900°C; pressure, $\sigma_3 = 15.5$ kb; strain rate; $\dot\epsilon = 7.5 \times 10^{-6}$ sec^{-1}; confining medium, NaCl.

0.25mm

Fig. 4. Photomicrograph of sample from experiment DT-641; crossed polarizers. The NE-SW trending features are (001) deformation lamellas. The compression direction is approximately N-S.

shear stress was applied in the [100] and [001] directions. Well-developed deformation lamellas were observed nearly parallel to (001) in the deformed samples. External rotation associated with a slight bending of the lamellas indicates that [100] is the macroscopic slip direction. This finding is consistent with the high resolved shear stress on this system ($S_0 = 0.49$). The conditions of the experiments ($T = 900°C$, $\dot{\epsilon} = 10^{-5}$ sec^{-1}, $\sigma_3 = 15$ kb) would indicate that, according to the slip fields delineated by *Raleigh* [1968] and *Carter and Avé Lallemant* [1970], the system {110}, [001] ($S_0 = 0.43$) should have been operative. This system was not observed, but the conditions of the experiment are so close to the transition to the {0kl}, [100] system that this observation is not considered to be in serious conflict with the results of the authors cited above.

J. D. Blacic and J. M. Christie (unpublished manuscript, 1972) have examined these samples with the transmission electron microscope. They find a complex dislocation structure consisting of a high density of pure screw dislocations parallel to [100] and [001] and many other dislocations with Burgers vectors as yet undetermined. They did not find structures that could be definitely correlated with the (001) deformation lamellas in these samples; however, the orientation of the transmission specimens was unfavorable for observing these structures. The complex dislocation structure is very likely a reflection of the closeness to the slip system transition, as was discussed above in terms of macroscopic observations. (More detailed discussions of transmission electron microscopy of experimentally deformed olivine are given in papers by *Phakey et al.* [this volume] and *Green and Radcliffe* [this volume].)

CONCLUSIONS AND SPECULATIONS

The experiments reported here confirm the prediction that hydrolytic weakening is an important phenomenon that can explain the low stress deformation of geophysically relevant materials. In olivine the weakening is considered to be the result of the hydrolysis of Si-O-Mg bridges adjacent to slip dislocations. The small amount of water required for this mechanism (a few tenths of a per cent or less by weight) assures a wide applicability in the earth. The theory of the hydrolytic weakening mechanism (J. D. Blacic and D. T. Griggs, unpublished manuscript, 1972) predicts that the rate-controlling factor for the deformation of silicates in the earth is the diffusion rate of 'water' to dislocations. (The actual diffusing species, i.e., H_2O, H^+, OH^-, $O^=$, or H_3O^+, are unknown at present.) This prediction is important, since the activation parameters of this mechanism are expected to be substantially different from those associated with the self-diffusion-controlled mechanisms that have been suggested as being applicable to deformation in the earth [*Weertman*, 1970].

So far we have looked at the hydrolytic weakening of olivine in terms of the deformation of individual grains. However, the intergranular recrystallization and shear observed in the experimental deformation of dunite can also be analyzed in terms of this mechanism. Recrystallization without the presence of a free fluid phase requires the conservative and nonconservative motion of dislocations. Water can have an important effect on this process if, as is required by the hydrolytic weakening mechanism, the mobility of dislocations is greatly enhanced in the presence of water. This effect is in addition to that associated with the solvent character of the polar water molecule. Hence the experimental evidence reported here indicates that localized recrystallization and shear induced by the presence of small amounts of water can give rise to a type of plastic instability that may play a role in causing intermediate and deep earthquakes. This phenomenon is distinct from the thermal instability (shear melting) suggested by *Griggs and Baker* [1969] as a possible mechanism of deep earthquakes. Actually, water may also play a role in the shear melting mechanism, since it is well known that small amounts of water can severely depress the melting points of most silicates.

Acknowledgments. The phenomenon of hydrolytic weakening of silicates was codiscovered with David Griggs, whose enthusiasm and encouragement throughout this work was of inestimable value. Critical discussions with J. M. Christie, R. L. Post, and G. Dollinger proved most helpful. The experimental apparatus was fabricated and maintained by the UCLA Geophysics Machine Shop under the direction of W. Hoffman and H. Kappel. J. de Grosse prepared the excellent thin sections.

The research was supported by the National Science Foundation under grants GA-277 and GA-1394.

Transmission Electron Microscopy of Experimentally Deformed Olivine Crystals[1]

PREM PHAKEY,[2] GERALD DOLLINGER, AND JOHN CHRISTIE

Department of Geology and Institute of Geophysics and Planetary Physics
University of California, Los Angeles, California 90024

Forsterite crystals of very high perfection were compressed to small strains at strain rates of 10^{-4} and 10^{-5} sec^{-1} at temperatures from 600° to 1250°C under 10-kb confining pressure in a talc confining medium. They were examined by optical and transmission electron microscopy to determine the nature of the microscopic and submicroscopic deformation structures with a view to elucidating the deformation mechanisms. Slip dislocations with Burgers vectors $\mathbf{b} = c[001]$, $b[010]$, and $a[100]$ were observed, and the slip planes associated with these slip vectors were identified in most cases. The critical shear stress for slip parallel to $c[001]$ is less than that for the other directions between 600° and 1000°C and possibly at 1250°C. We confirm the operation of most of the slip mechanisms identified optically by previous workers, but there is evidence that they operate over ranges of temperature wider than those previously reported. Dense tangles and cell structures are developed at 800° and 1000°C. Cross slip occurs at 800°C and is extensive at 1000°C, where climb also takes place. Extensive recovery and recrystallization characterize the deformation at 1250°C.

Since olivine is generally accepted as one of the main constituents of the earth's upper mantle, there is considerable interest in the study of its mechanical properties. Several studies, reviewed below, have been made of the slip mechanisms operating in olivine by using (1) petrographic measurements of rotations in deformed crystals, (2) study of slip lines on polished surfaces, (3) optical examination of decorated dislocations or surface etch pits, and, more recently, (4) transmission electron microscopy. These investigations, however, have provided only limited information on the slip mechanisms: the petrographic and slip line studies provide only gross information regarding the predominant slip systems; etching and decoration techniques, although imaging individual dislocations if the density is low, do not provide information about slip vectors of individual dislocations; and transmission electron microscopy to date has been done on few specimens, mostly polycrystalline ones, and it has

yielded only limited information on the slip mechanisms. Hence a more detailed and systematic study of the slip mechanisms in olivine is necessary.

In the present study, gem quality peridot crystals, initially almost dislocation free, were deformed to small strains at temperatures from 600° to 1250°C. The dislocation configurations were examined by transmission electron microscopy, and the Burgers vectors were determined by diffraction contrast methods where possible.

The predominant slip systems in olivine aggregates experimentally deformed over a wide range of temperatures and strain rates were determined by *Raleigh* [1968] by observation of slip bands on polished surfaces combined with universal-stage examination of thin sections in transmitted light. *Young* [1969] reported several slip mechanisms observed in olivine single crystals, using optical methods of study of decorated dislocations and dislocation etch pits on polished surfaces. These observations were subsequently amplified by *Carter and Avé Lallemant* [1970], using decoration and universal-stage techniques, and by *Raleigh and Kirby* [1970], also using universal-stage meth-

---

[1] Publication 976, Institute of Geophysics and Planetary Physics, University of California, Los Angeles, California 90024.
[2] On leave from Department of Physics, Monash University, Clayton, Victoria, Australia.

TABLE 1. Slip Systems Previously Reported

Slip Direction	Slip Plane	Temperature Range, °C	Reference*
[001]	(100)	<300	1
		<350	2
		<500	3
[001]	{110}	300–1000	1
		<1000	4
		<950	2
		500–800	3
[100]	{0kl} (pencil)	>1000	1, 2, 4
		>1200	3
[100]	(010)	>1000	4, 3
		>1250	2
[100]	(001)	>1000	3
[100]	{032}	>1000	4
[010]	(100)	'low temperature'	1
		>1000	3

* 1, *Raleigh* [1968]; 2, *Carter and Avé Lallemant* [1970]; 3, *Young* [1969]; 4, *Raleigh and Kirby* [1970].

ods. The slip systems identified in these studies are summarized in Table 1, which lists the numerous slip systems and the approximate ranges of temperature over which they were reportedly observed (compared at a strain rate of about 10^{-4} sec^{-1}). Previous work on identifying the geometry and the Burgers vectors of dislocations in olivine with transmission electron microscopy is very limited. This work includes a study of one experimentally deformed peridotite sample by *Green and Radcliffe* [1972a], an olivine inclusion from a basalt [*Boland et al.*, 1971], and two experimentally deformed olivine samples, one single crystal and the other polycrystalline (J. D. Blacic and J. M. Christie, unpublished manuscript, 1972). The results of these studies, although confirming the operation of some of the slip systems identified optically (Table 1), are not sufficiently extensive to give a systematic picture of the details of the deformation process under various conditions of temperature and strain rate.

(*Green and Radcliffe* [this volume] describe the results of electron petrography of lherzolite peridotite aggregates deformed experimentally at 900°C at 10^{-6} sec^{-1} and at 1300°C at 10^{-4} sec^{-1}.)

DEFORMATION EXPERIMENTS

Materials. The starting material for the present set of deformation experiments consisted of gem quality peridot crystals from a single batch, the locality given only as the 'Red Sea.' Optical examination of several crystals showed that they have $2V_z = 88°$, corresponding to a composition Fo_{92}. No strain features, such as kink bands, deformation lamellas, or undulatory extinction, are seen in the thin sections. A sharp partly planar boundary separating regions of slightly different lattice orientation, however, was seen in the thin sections of several specimens. These boundaries, when planar, were found to be almost parallel to (010). It is thought that they are sub-boundaries produced by growth.

Most of the olivine crystals used were irregular in shape, having only a few planar growth faces. A few specimens, however, possessed good crystal form with well-developed (100), (010), {021}, and {110} faces. The crystals were elongated parallel to c [001] and flattened normal to a [100]. Striations parallel to c [001] on the (100) face were seen on all of them. The crystals ranged in size from about $1 \times 1.5 \times 1.5$ cm to approximately half that size.

Examination of crystals from the same batch by X-ray topographic methods (A. M. Aly and P. Phakey, unpublished manuscript, 1971) indicates that they are structurally very perfect, only a few dislocations appearing in a crystal plate with an area of 1 cm^2 and a thickness of ~1 mm; a typical example is illustrated in Figure 1. No information is available on the chemical homogeneity or variations in the batch of crystals, but consistency in the optic axial angle suggests uniformity among the crystals discussed here.

Experiments. All of the deformation experiments were conducted in a solid-pressure medium apparatus of the small type designed and described by *Griggs* [1967].

In the experiments, circular cylinders of ~3 mm in diameter and of variable length cored from the olivine crystals were deformed at strain rates of 10^{-4} or 10^{-5} sec^{-1} at temperatures of 600°, 800°, 1000°, and 1250°C. The confining pressure in all the experiments was 10 kb (± 1 kb), and the confining medium was talc. Along with variations in strain rate and temperature the orientation of the direction of compression σ_1 was also varied with respect to the crystallographic directions of the olivine (compressive stress positive $\sigma_1 > \sigma_2 = \sigma_3$) (Figure 2). The most commonly used orientation (A)

Fig. 1. X-ray topograph of an olivine crystal from the batch used in this study. The diffraction vector **g** = 101. The thickness of the crystal plate is ~0.1 cm. The dislocations labeled D are observed by diffraction contrast. (Courtesy of A. M. Aly).

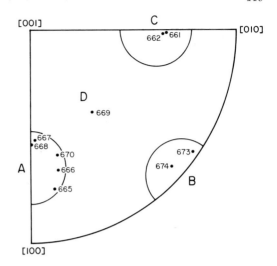

Fig. 2. Inverse pole figure showing compression directions in the experimentally deformed crystals. The groups of orientations are designated A, B, C, and D in the text.

was with the σ_1 direction approximately normal to b [010] and inclined to both a [100] and c [001] at angles of about 45°. This orientation was used because it gave high shear stress in the directions [100] and [001], which are the shortest axes of the unit cell and the commonest slip vectors previously reported (see the discussion above).

In the other experiments other pairs of crystal axes were oriented at ~45° to σ_1 (labeled B, C in Figure 2) to suppress slip on some possible slip systems; in one experiment the orientation was selected to give high shear stress parallel to all three crystal axes (orientation D, Figure 2). The experimental conditions are listed in Table 2, which also gives the average permanent strain and the orientation for each sample along with the stress difference ($\sigma_1 - \sigma_3$) at the termination of the experiment.

Stress-strain curves. Stress-strain curves for the experiments are shown in Figure 3. The precision of these curves is not high, as has been discussed elsewhere [*Tullis*, 1971]. Uncertainties are due to (1) the finite strength of the solid-pressure medium, (2) piston friction, (3) apparatus distortions, and (4) temperature errors

TABLE 2.   Conditions of the Experiments

Experiment	Orientation	Temperature, °C	Strain Rate, 10^{-5} sec^{-1}	Strain, %	Maximum ($\sigma_1 - \sigma_3$), kb
DT-665	A	600	1.3	8.5	5.6
DT-668	A	600	1.6	8.5	5.2
DT-661	C	800	14.0	12.0	10.0
DT-667	A	800	1.5	7.5	5.6
DT-670	A	800	1.8	10.0	7.9
DT-674	B	800	1.9	6.5	12.8
DT-662	C	1000	1.2	12.5	6.3
DT-666	A	1000	1.3	8.5	4.4
DT-673	B	1000	1.0	7.0	4.8
DT-669	D	1250	13.0	>10.0	~1.5

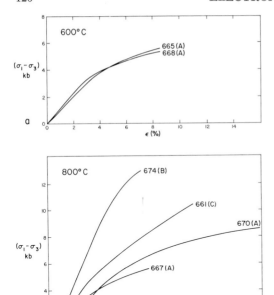

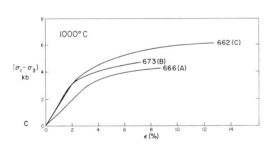

Fig. 3.   Stress-strain curves for the experiments at
(a) 600°C, (b) 800°C, and (c) 1000°C.

due to an inaccuracy in positioning the thermocouple with respect to the sample and the furnace and temperature gradients from the center
to the ends of the sample. Of these errors the
first is considered serious in this range of temperatures only when the samples are weak: for
this reason a stress-strain curve for DT-669
(1250°C) is not given in Figure 3; the initial
contact of the piston with the sample was inde-
terminate, and at no point did the axial stress
(σ_1) exceed the confining pressure by >1.5 kb
(Table 1). The second and third errors were
corrected by standard procedures. Strong alum-
ina end pieces were partly enclosed in the furnace in the experiments, so that the temperature

gradients were less than they are when metal or
carbide end pieces are used, but the temperature
uncertainty is still considered to be in the range
of ±50°C.

We note here the following features of the
stress-strain curves, discussed later in the text:

1. The initial portions of the curves for all
specimens are less steep than the slope indicated
by the elastic moduli (2–3 × 10^{12} dynes/cm^2),
this finding indicating that permanent deformation is taking place at low stresses. The bend in
the σ-ϵ curves is not, therefore, a yield phenomenon but a change in the hardening rates.

2. The curves for the two crystals deformed
at 600°C are consistent (Figure 3a), but they
are anomalously weak in comparison with
crystals of the same orientation deformed at
800° and 1000°C (Figure 3b, c).

3. At 800°C there is an appreciable effect of
orientation on stress, orientation B being considerably stronger than orientations A and C.

4. At 1000°C the effect of orientation on
flow stress is smaller and possibly insignificant;
the strengths and rates of work hardening are
slightly lower than those at 800°C.

5. There is no significant difference between the curves for crystals of orientation A
deformed at 600°, 800°, and 1000°C.

METHODS OF STUDY

Optical analysis. In the optical analysis of
experimental deformation structures it is usual
to cut thin sections parallel to the compression
axis (σ_1) and perpendicular to planar deformation features (if their orientation is known) for
the best observation of such features on a universal stage. However, our earlier experience of
the transmission electron microscopy of 'foils'
thinned from such sections revealed that planar
structures that are initially normal to the section are unsuitable for E.M. study, since dislocations in such steep planes commonly have short
projected lengths, are poorly resolved, and are
difficult to characterize by diffraction experiments. Arrays of dislocations inclined at ≤45°
to the plane of the foil are more suitable. For
this reason and also to maximize the number of
thin sections that could be cut from the small
sample cylinder, most samples were sectioned
perpendicular to the cylinder axis (σ_1). Several
30-μm-thick sections polished on both sides were

prepared from each sample and examined petrographically with the universal stage. One or more sections from the center (hottest) region of each sample was selected for thinning by ion bombardment and electron microscopy.

For a number of samples, sections of other orientations were prepared and examined optically, notably DT-673 and DT-674, which were sectioned parallel to the compression axis and perpendicular to c [001] for detailed observations of deformation lamellas. The orientations of kink boundaries, bending, and deformation lamellas observed in the samples are given below with the electron microscopic information.

Thinning and electron microscopy. Ultrathin electron transparent sections of the samples were made by using the ion bombardment technique, which has been described in considerable detail by *Barber* [1970] and has been used by various other investigators to study a number of minerals and lunar samples.

Areas for transmission electron microscopy were selected from the doubly polished petrographic sections and cored out as disks of about 3-mm diameter. The disks were cemented to copper grids (75 mesh) or rings to provide mechanical support to the specimen during subsequent handling. The cemented disks were floated off and sandwiched in a thin stainless-steel holder that was positioned at an angle of about 15° with respect to two opposed ion guns. The guns were operated with argon at a potential of 5-6 kv and beam currents of 50-60 μamp per gun. This treatment removed an average of 3 μm from the surface layer of the specimen per hour and yielded foils of good quality. A thin layer of carbon was evaporated on to the ultrathin specimen to avoid surface charging during examination in the JEM 120 kv electron microscope. Cementing of copper grids on the specimen not only provides a mechanical support to the specimen but is also advantageous, because the perforations in the specimen usually appear within a number of grid squares and provide many thin areas throughout the specimen.

EXPERIMENTAL RESULTS

600°C Experiments

Samples DT-665 and DT-668. The first sample (DT-665) was shortened ~8.5%. The compression axis made angles of 26° and 64° with [100] and [001], respectively (orientation A). The second sample (DT-668) was also shortened 8.5% at ~45° to [100] and [001]. For both these samples there was zero shear stress in the [010] direction on planes in this zone, but the (100) [001] and (001) [100] slip systems were highly stressed. Both experiments were performed at a uniform strain rate of 10^{-5} sec^{-1}.

Thin sections from sample DT-665 showed only slight undulatory extinction and no visible kink bands or deformation lamellas. The sample was locally fractured, but no systematic sets of fractures were observed. No optical strain features were observed in the thin section of DT-668, which was cut at ~45° to the cylinder axis and approximately parallel to [100], for reasons noted below.

From sample DT-665 a section suitable for transmission electron microscopy was prepared normal to the compression axis from a region close to the thermocouple. The electron micrographs revealed a homogeneous distribution of straight dislocations accurately aligned only along [001] (Figure 4a, b). The dislocation density was found to be ~2 $\times$ 10^9 cm^{-2}. The Burgers vector of these dislocations has been determined from their diffraction contrast images observed for different operating reflections. The dislocations seen in Figure 4b stayed in contrast for the $\bar{2}42$ and $\bar{2}11$ reflections, but they are out of contrast in the bright-field micrographs (Figure 4c, d), for which the operating reflections are $g = 140$ and 130, respectively. Thus the Burgers vector **b** for these dislocations is c [001]. These dislocations are, therefore, of pure screw character. Since the highest shear stress on (100) is in the [001] direction, the operating slip system is probably (100) [001], but the mixed or edge parts are scarcely evident, so that it is impossible to determine the slip plane more definitely. Since in this sample no dislocations were found to be aligned along [100] or other directions, a section was prepared from sample DT-668 so that [100] was nearly in the plane of the section and [001] was steeply inclined. For this specimen orientation any dislocations along [100] would have shown long projected images on the electron micrographs. However, Figure 5a shows that only short dislocations aligned along [001] are present. These dislocations are again of pure screw character with Burgers vector **b** = c

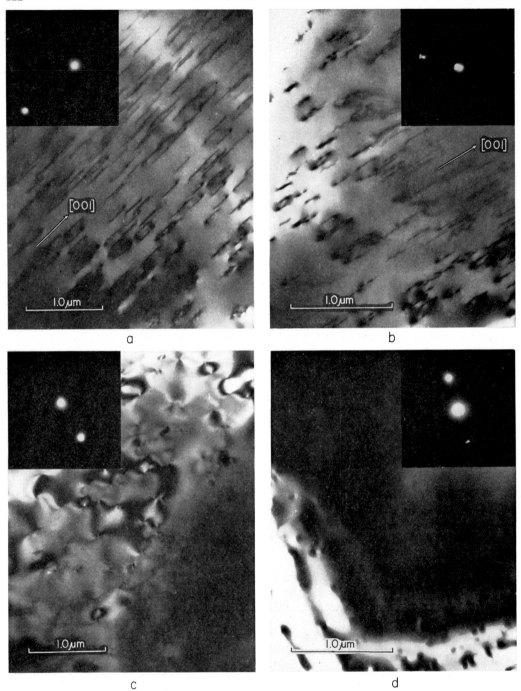

Fig. 4. (a) Bright-field (BF) micrograph for **g** = $\bar{2}42$ showing dislocations with **b** = $c[001]$. Note that the dislocations of opposite sign show a difference in contrast along their length (DT-665). (b) BF micrograph for **g** = $\bar{2}11$ (DT-665). (c) BF micrograph for **g** = $\bar{1}40$ for nearly the same area as that shown in (b). The strain contrast features mark the areas around the dislocations, which are out of contrast. (d) BF micrograph of nearly the same area as that shown in (b). The dislocations are out of contrast; **g** = 130. (All arrows denoting crystal directions in the micrographs are projections of these directions parallel to the electron beam.)

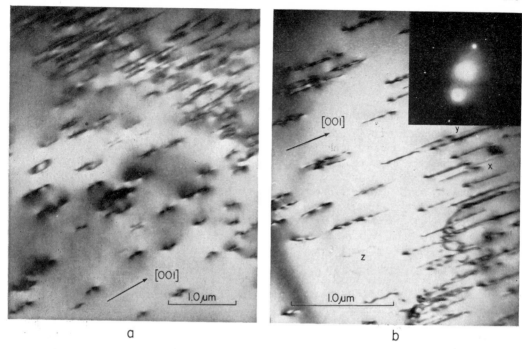

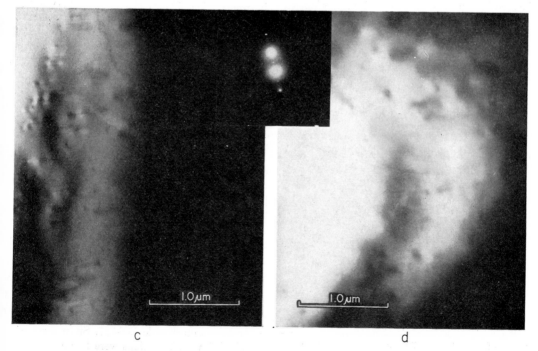

Fig. 5. (a) Bright-field (BF) micrograph for g = 021 (DT-668). (b) BF micrograph for g = 021. Note the dislocations labeled x, y, and z; they show different contrast from the other dislocations (see the text) (DT-667). (c and d) BF and dark-field micrographs, respectively, for g = 020. The area shown is nearly the same as that in (b). The dislocations are out of contrast.

[001], because they showed good contrast for the 333, 121, and 242 reflections but were out of contrast for the 120 reflection, for which **g·b** = 0. Once again, these conditions do not determine the slip plane uniquely.

800°C Experiments

Sample DT-667. This sample was compressed 7.5% in a direction inclined at 47°, 88°, and 43° to [100], [010], and [001], respectively (orientation A). The thin sections were cut perpendicular to the compression axis; the only strain features observed optically were fine fractures or cleavage, mostly parallel to the (010) plane.

Electron microscope examination of the section revealed a homogeneous distribution of predominantly straight dislocations of density of ~0.7 × 10⁹ cm⁻². Examples are shown in Figure 5*b*. When the specimen was tilted in the electron microscope, it was found that a majority of the dislocations were along [001]. The fact that dislocations parallel to both [100] and [001] project along the same line in Figure 5*b* is clear from the projection in Figure 6, which shows the specimen orientation for this case. It will also be seen in Figure 5*b* that the dislocations labeled

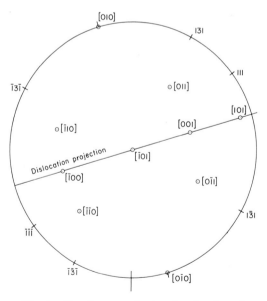

Fig. 6. Equal-area projection showing the orientation of crystal DT-667 with the important crystal zones and the poles of diffracting planes discussed in the text. The projection is oriented appropriately for micrographs *b*, *c*, and *d* of Figure 5; the compression axis is normal to the plane of projection.

X, Y, and Z show a contrast considerably different from that of the other dislocations present. The reason for the difference in contrast is that the labeled dislocations are along [100]. Since both kinds of straight dislocations remained in contrast for the operating diffraction vectors **g** = 111, 3̄13, 1̄11, 1̄41, 1̄31, and 202, they may be screw dislocations with Burgers vector **b** = *c* [001] and **b** = *a* [100]. Figure 6 shows that both types of screw dislocations should go out of contrast simultaneously for **g** = {0*l*0}; the validity of this observation is evident in Figures 5*c* and *d*, which are the bright-field and dark-field micrographs, respectively, for the 020 operating reflection.

Sample DT-670. This specimen was compressed in a direction inclined at 41°, 80°, and 50° to [100], [010], and [001], respectively (orientation A). Other experimental conditions were identical with those for DT-667 except that DT-670 was shortened 10%. As was done with the previous sample, sections normal to the compression axis were examined optically and in the electron microscope. The sections showed that the sample was highly fractured (probably in preparation, before the deformation), misorientations of ≤10° existing among the different regions. Two sets of faint planar features, resembling weak deformation lamellas, were observed subparallel to the {110} planes. Some undulatory extinction was present locally, but there were no kink boundaries.

Figure 7*a* illustrates the dislocation structure characteristic of this sample. The dislocation density of the sample was found to be ~2 × 10⁹ cm⁻². Although alignment of the dislocations along [001] is most evident, some of them have segments along the projections of [100] and [01̄1]. The diffraction contrast for the [001] dislocations was studied for a number of diffracting vectors, such as **g** = 3̄33, 2̄22, 031, and 130, from which they were found to be out of contrast for **g** = 130 only. From the invisibility criterion **g·b** = 0 it is quite clear that the [001] dislocations are of pure screw character with Burgers vector **b** = *c* [001]. This Burgers vector determination is consistent with our previous observations on sample DT-667. Since some of the [001] screw dislocations were found to have components along [100] and [01̄1] in Figure 7*a*, it appears that (010) [001] or (100) [001] slip systems might be operating. In other micro-

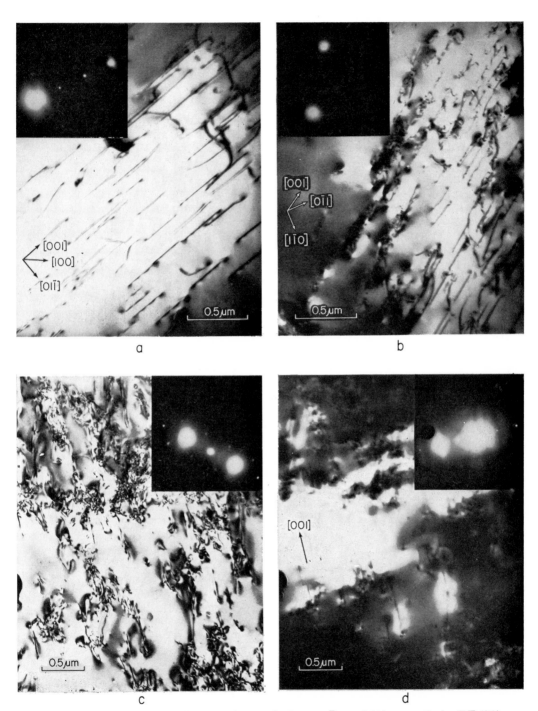

Fig. 7. (a and b) Bright-field (BF) micrographs for g = 3̄33 and 031, respectively (DT-670). (c and d) BF and dark-field micrographs of the same area for g = 222 and 200, respectively (DT-661).

graphs, e.g., Figure 7b, [001] dislocations curve into [1$\bar{1}$01] segments, which indicates that the slip system is (110) [001]. The Burgers vector for the dislocations oriented parallel to [100] was not determined.

In certain regions of the same specimen, tangling of dislocations was also observed (Figure 7b).

Sample DT-661. Since sample DT-670 showed that some slip on {hk0} planes may be taking place, sample DT-661 was compressed 12% at ~89° to [100], 35° to [010], and 55° to [001] to minimize the shear stress along [100] (orientation C). Optical examination of several sections cut perpendicular to the compression axis revealed some undulatory extinction and a substantial number of kink (tilt) boundaries subparallel to (001) planes. The orientation of these kink boundaries indicates that slip in the [001] direction was predominant. No deformation lamellas were observed, however, and the rotations across the kink boundaries were too small to reveal the slip plane. Numerous fractures subparallel to (001) resulted from the cracking of some of the kink boundaries.

Electron microscope observations show that this deformation has resulted in the formation of tangles of dislocations arranged in walls surrounding regions or cells almost free from dislocations. These cell walls tend to be oriented along the projections of the [001], [100], and ⟨110⟩ crystallographic directions. For example, see Figure 8a, where cell walls, most of which are parallel to [001] owing to tangling, are best developed. These figures also show the presence of straight dislocations along [001]. The dislocation density of the sample was found to be of the order of 5–10 × 10^9 cm^{-2}, which is higher than that of any other sample from the 800°C experiments. This high density of the dislocations prevented us from determining the Burgers vector of the short interacting dislocations. However, when the area in Figure 7c is compared with the same area in Figure 7d, it can be seen that in Figure 7d the dislocations along [001] and many other dislocations in the same general area are either completely or almost invisible when the diffracting vector g = 200 is operating. This observation suggests that dislocations with Burgers vector **b** = c[001] are present in the tangled regions and also shows that the straight dislocations are again of screw character. It can

also be seen in Figure 8a that some of the straight [001] dislocations have segments that lie along [010] and [$\bar{1}$10] projections, this fact indicating that (100) [001] and {110} [001] slip systems have probably operated.

Sample DT-674. In all the 800°C experiments discussed above, [001] was highly stressed, and the electron microscope observations demonstrate that, at this temperature, slip along [001] predominates over slip in other directions in the crystal. A few screw dislocations parallel to [100], indicating slip in this direction, are present (DT-670 and DT-667). Sample DT-674 was compressed in orientation B at 48° to [100], 43° to [010], and 82° to [001] (this orientation giving very low shear stress parallel to [001]) to determine the effect of suppressing slip along [001]. The crystal was shortened by 6.5%. Sections of two orientations from this crystal were examined optically, perpendicular to the cylinder axis and parallel to (001). The deformation was very heterogeneous, and the deformed regions showed undulatory extinction and two closely spaced sets of planar features subparallel to (100) and (010). The features parallel to (100) were sharply defined and associated with slight changes of orientation, resembling closely spaced kink boundaries of small tilt; the features subparallel to (010) were diffuse and did not sharpen clearly when the sections were tilted on the universal stage, resembling very weak closely spaced deformation lamellas.

The section examined in the electron microscope was cut parallel to (001) from a region near the thermocouple. Observations were made in a region of low strain and one of high strain. In the region of low strain the dislocation density was very low (3 × 10^8 cm^{-2}). Most of the dislocations were along [001]. However, in thicker parts of the specimen some of the dislocations had segments along [100]. Examples of the [001] dislocations can be seen in Figure 8b. The dislocations along [001] were found to show good contrast for diffracting vectors g = 130, 210, and 200, but they were out of contrast for g = 020 (Figure 8c). These observations show that the Burgers vector is **b** = a[100], so that the dislocations result from the (010) [100] slip mechanism. By contrast, the dislocation density of the highly strained region varied from 1 to ~5 × 10^9 cm^{-2}. A typical example of an area

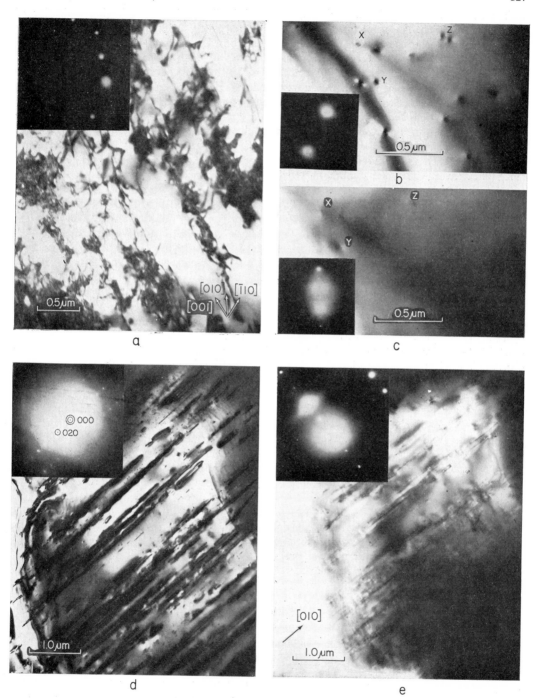

Fig. 8. (a) Bright-field (BF) micrograph for g = 021 showing extensive tangles along the projection of [001] (DT-661). (b) BF micrograph for g = 150 (DT-674). Some of the dislocations have been labeled for comparison with (c). (c) BF micrograph for the same area as that shown in (b). The dislocations are out of contrast; g = 020. (d) BF micrograph for g = 020 (DT-674). (e) BF micrograph for g = 200. This area is the same as that shown in (d). Note that the dislocations are in vanishingly weak contrast (see the text).

with moderately high dislocation density is depicted in Figure 8d. The characteristic features are some long and some short dislocations aligned along the projection of [010]; the heavy concentration of dislocations on certain (100) planes has the appearance of narrow walls or tangles that show extensive strain effects around them for the 020 reflection in Figure 8d. When the sample was tilted in the microscope, it was found that the shorter dislocations parallel to the [010] projection were curved into segments along [001]. Investigation of the diffraction contrast from both kinds of dislocations revealed that they remained in good contrast for $g = 020$ and 120; however, their contrast was weak for $g = 110$ and 210 and vanishingly weak for $g = 200$ (Figure 8e). The residual contrast is due to long-range strains associated with the dislocation arrays. These observations show that the dislocations along [010] are pure screw and their segments along [001] are pure edge. Consequently, the (100) [010] slip system is operating in this heavily deformed region of the crystal to the exclusion of the (010) [100] system observed in the less deformed region.

If we assume that the (100) [010] dislocations are undissociated, the shortest magnitude of their Burgers vector $b = b[010]$ is 10.21 A. Such high-energy ($\propto b^2$) dislocations should favor dissociation into partial dislocations. We made a careful search for evidence of stacking faults, but our diffraction contrast experiments showed no evidence of dissociation into partials of any kind.

1000°C Experiments

Sample DT-666. This sample was deformed 8.5% under compression and at a uniform strain rate of 10^{-5} sec^{-1}. The compression axis was inclined at angles of 35°, 55°, and 90° to [100], [001], and [010], respectively (orientation A).

Optical examination of sections cut normal to the compression axis showed slight undulatory extinction, but no deformation lamellas or kink boundaries were observed when the sections were tilted on the universal stage. The sections contained some systematic fractures, most of which were parallel to (100).

The electron microscope observations were made on one of the sections cut normal to the compression axis. Examples of the dislocation content of this sample are shown in Figure 9a

and *b*. The dislocations, having a density of $\sim 0.8 \times 10^9$ cm^{-2}, are fairly homogeneously distributed. As can be seen in this figure, a majority of the dislocations are straight and parallel to [001]. The diffraction contrast experiments showed that their Burgers vector is $b = c[001]$. The dislocations labeled A in Figure 9a are aligned along [100] and probably have Burgers vector $b = a[100]$.

Some of the $b = c[001]$ screw dislocations were found to have composite jogs along their length. For examples, see points B and C in Figure 9a. Also note that the jogs labeled B and C are aligned in [010] and [100] edge orientations, respectively, as was expected. At points D in Figure 9b, dislocation dipoles aligned in the [010] edge orientation are also present. An example of a single-ended source occurs at point E (Figure 9a); its dipoles are also in the [010] edge orientation. It appears that at some stage of the deformation process this dipole has broken to lower its energy by forming the row of loops seen near E. The presence of both small and elongated loops in Figure 9a and b suggests that this debris was left by the movement of jogged dislocations. Loops originating from this process would be of edge character throughout (prismatic loops) and therefore would tend to be sessile. Although the elementary jogs along these dislocations may be due to their intersections with edge and screw dislocations on other slip planes, a likelier explanation for the composite jogs (on the [001] or [100] screw dislocations) is that they have originated by double cross slip, which is quite common in alloys [*Low and Turkalo*, 1962]. It seems reasonable, therefore, to suggest that in this experiment the screw dislocations have experienced a considerable amount of double cross slip. If it is further assumed that some of these [001] dislocations originally resulted from slip on {110}-type planes, it is probable that later during the deformation process they may have tended to move on the (010) plane, on which the resolved shear stress is maximum.

Sample DT-662. This crystal was compressed 12.5% in orientation C, high shear stress being parallel to the [010] and [001] directions. The thin sections were cut perpendicular to the compression axis. Optically they show undulatory extinction, and the many measurable planar boundaries are subparallel to (001) ; slight orien-

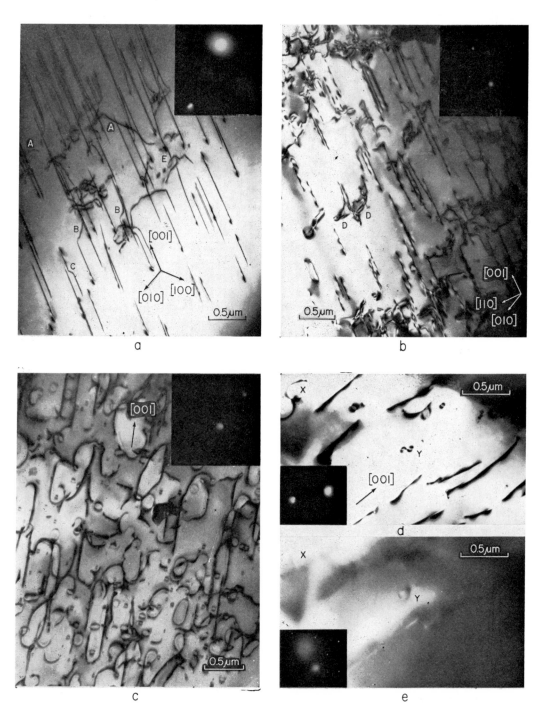

Fig. 9. (a) Bright-field (BF) micrograph for **g** = 06$\bar{2}$. The labeled features are discussed in the text (DT-666). (b) BF micrograph for **g** = 1$\bar{1}$2. D marks dislocation dipoles (DT-666). (c) BF micrograph for **g** = 122 (DT-662). (d) BF micrograph for **g** = 112. Two dislocations are labeled for comparison with (e) (DT-662). (e) BF micrograph of the same area as that shown in (d). All dislocations are out of contrast; **g** = 1$\bar{1}$0.

tation changes across these boundaries indicate that they are kink boundaries with small tilts.

Figure 9c shows examples of the dislocation content of a relatively thick area of the specimen. The dislocation density varies between 1 and 3×10^9 cm^{-2}. The features seen in these micrographs are very similar to those described for sample DT-666. For example, dislocation loops of different sizes are present, the nearly straight dislocation segments are aligned along [001], and some of them are jogged. The Burgers vector determination of the straight dislocations was made in a thinner region near the edge of a hole in the sample. Figure 9d shows that the dislocations are slightly wavy but they are aligned approximately with [001]. Some small dislocation loops are also present. For this micrograph the operating reflection $g = 112$ makes an angle of $\sim$20° with the trace of the [001] dislocations. Although these dislocations remained in contrast for 231, $12\bar{1}$, and $\bar{2}42$ reflections, their contrast was poor when the angle between g and [001] approached 90°. The dislocations were out of contrast for the $1\bar{1}0$ reflection (Figure 9e), for which the invisibility criterion $g \cdot b = 0$ is satisfied. These observations show quite clearly that the Burgers vector of the dislocations is $b = c$ [001]. The diffraction contrast from the dislocation loops was the same as that for the [001] dislocations; it can be seen that the loops are also out of contrast in Figure 9e. Hence, their Burgers vector must also be $b = c$ [001]. It seems plausible, therefore, that the dislocation loops represent the debris left by movement of the jogged [001] screw dislocations, as was discussed with specimen DT-666.

Sample DT-673. The observations on samples DT-662 and DT-666 demonstrate that, at $\sim$1000°C, slip along [001] still predominates over slip along [100]. To investigate what other slip mechanisms could be operating, sample DT-673 was stressed in a direction such that there was no shear stress along [001] (orientation B).

Sections perpendicular to the compression axis and sections parallel to the compression axis in the (001) plane were prepared and examined optically. Those perpendicular to the axis contained a sharp planar boundary parallel to (010) with 2°–3° of misorientation across it, similar to the sub-boundaries seen in some of the undeformed crystals; we think that it pre-

dated the deformation. The longitudinal (001) section showed well-developed deformation lamellas parallel to (010) with slight undulatory extinction. Two sets of planar fractures were also present: long fractures parallel to the (010) lamellas and sometimes coinciding with them and short fractures subparallel to (100), commonly restricted between pairs of (010) lamellas.

Electron microscopy was performed on the section parallel to (001). The dislocation distribution was found to be heterogeneous, varying from 0.6 to 2×10^9 cm^{-2}. The electron micrographs revealed two crystallographic orientations for the dislocations. First, some long and some considerably shorter dislocations were aligned along the projection of [100] (Figure 10a). Second, a considerable number of dislocations were also found to be aligned along the projection of [001], which is nearly normal to the plane of the specimen (Figure 10a). When the sample was tilted through a large angle in the microscope, it was found that the shorter dislocations parallel to the [100] projection were in fact curved into segments along [001]; this finding is illustrated in Figure 10b. The diffraction contrast shown by both [001] and [100] dislocations was investigated for a number of operating reflections, and it was found that dislocations of both orientations remained in contrast for $\bar{1}20$, $\bar{1}30$, $\bar{1}40$, 210, and 200 reflections, the contrast being the best for the 200 reflection. However, they were out of contrast, or very nearly out of contrast, for the 020 reflection (Figure 10c). Since the invisibility criterion $g \cdot b = 0$ is fulfilled only by the 020 reflection, it is quite clear that the Burgers vector must be $b = a$[100]; the dislocations along [100] are, therefore, pure screw, and those along [001] are pure edge. It can be seen in Figure 10c that the screw dislocations are completely out of contrast, but the edge dislocations still show a small residual contrast. This observation can be explained, because we know that for the edge dislocations the invisibility criteria are $g \cdot b = 0$ and also $g \cdot (b \times l) = 0$, where l is a unit vector along the dislocation line. It can now be seen that, although $g \cdot b = 0$, for the 020 reflection, $g \times l \neq 0$, and this gives rise to the residual contrast. The operating slip system is (010) [100].

Another characteristic property of this sample is that single rows of edge dislocations of Burgers vector $b = a$ [100] are aligned, forming closely

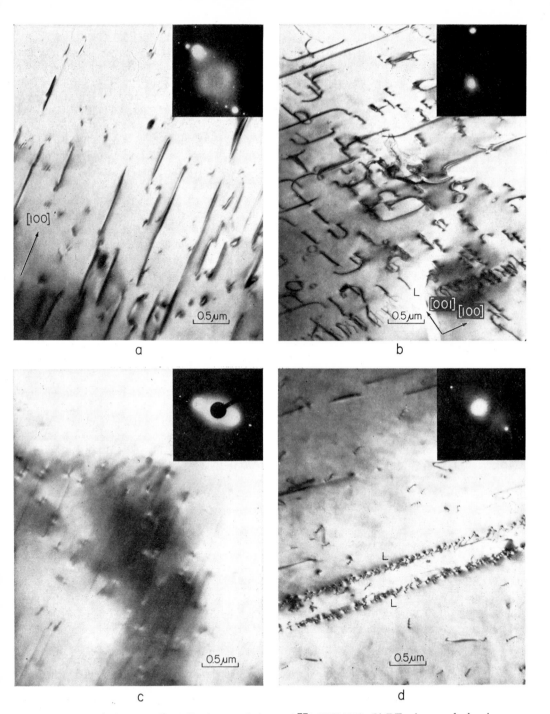

Fig. 10. (a) Bright-field (BF) micrograph for **g** = $\bar{1}$30 (DT-673). (b) BF micrograph showing parts of loops with [100] screw segments and [001] edge segments; note the array (L) of edge dislocations in a (010) plane. Among the isolated dislocations, those of opposite sign can be distinguished from the nature of the contrast; **g** = 231 (DT-673). (c) BF micrograph of the same area as that shown in (a), showing dislocations that are out of contrast for **g** = 020 (DT-673). (d) BF micrograph showing a closely spaced pair of arrays (L) like that shown in (b). The nature of the beaded contrast shows that these edge dislocations (**b** = a[100]) are predominantly of one sign in one array and of the opposite sign in the other; **g** = 130 (DT-673).

spaced arrays in the (010) slip plane. Good examples of these arrays are labeled L in Figure 10*b* and *d*. The average spacing between these arrays was found to be of the order of 5 μm in a heavily deformed part of the specimen. A few pairs of more closely spaced arrays of this type were also observed, as in Figure 10*d*. These arrays of edge dislocations correspond in spacing and orientation to the optical deformation lamellas observed in the thin section and therefore conform exactly with the model proposed for such features [*Christie et al.*, 1964]. *Green and Radcliffe* [1972a] have also observed arrays of edge dislocations in olivine, corresponding to deformation lamellas. A more detailed discussion of these features will be given in another paper.

1250° C Experiment

Sample DT-669. The sample for this experiment was compressed ~10% in a direction designed to give high shear stress simultaneously along [100], [010], and [001]. The angles between the compression axis and [100], [010], and [001] were 58°, 66°, and 40°, respectively. The sections for optical and transmission electron microscope investigation were cut at right angles to the compression axis.

Optical examination showed that parts of this specimen were recrystallized and some enstatite, produced by reaction with SiO_2 released from the talc sleeve, was locally present. Slight undulatory extinction was observed in some of the olivine, but no other strain features were visible optically. The electron microscope observations reveal a much lower average density of dislocations, estimated to be of the order of 10^8 cm^{-2}, than that in the other specimens. The dislocation configurations in the specimen are illustrated in Figures 11 and 12.

One characteristic feature is that the dislocation distribution is no longer homogeneous. For example, the dislocations seem to form a narrow array in Figure 11*a*. The individual dislocations in this array are aligned along [100] or [001]. Figure 11*b* shows an area adjacent to that of Figure 11*a*. The dislocations seen in Figure 11*b* are out of contrast in Figure 11*c* for $g = 040$, and thus these dislocations definitely are of pure screw character. Since the exact alignment of the dislocations can be parallel to [001] or [100], the Burgers vector may be either $\mathbf{b} = a$ [100] or c [001].

In the region of the sample shown in Figure 11*d* and *e* the arrays demonstrably consist of [001] screw dislocations, and segments of other orientations are also present. The micrographs show that some of the [001] dislocations have segments that are curved into an orientation along [110], and thus the {110} [001] slip mechanism operated. Other dislocation orientations are represented in these arrays, but it was impossible to analyze their characteristics.

Another characteristic feature of dislocations in a different region of this sample is shown in Figure 12*a*. The dislocations are helical, the axes of the helices approximately aligned along [001]. The presence of these helical dislocations indicates that extensive climb has occurred.

A few comparatively large dislocation loops are also present. Two mechanisms of forming loops are possible. First, the movement of elementary jogs leaves a single trail of point defects (vacancies or interstitials); if the temperature is sufficiently high to allow diffusion of the defects, they will collect to form a vacancy or an interstitial platelet bounded by a dislocation loop. Second, the dislocation loops are pinched off from dipoles and single-ended sources. From our experiments we know that such dipoles and single-ended sources are present at 1000°C. These pinched-off loops can also grow by the redistribution of material close to the loops. It is probable that both processes may have contributed to the formation of the loops in this sample. From examination of Figure 12*a* it is not clear whether the loops enclose typical low-energy stacking fault fringes. There are indications that a few broad fringes may be present in the loops labeled M (Figure 12*a*); the number of fringes enclosed by the loops may be limited, because the plane of the loops is very nearly parallel to the plane of the specimen.

Yet another characteristic structure of sample DT-669 is illustrated in Figure 12*b*, where the dislocation arrangements show what appear to be large- and small-angle grain boundaries. The regions between the boundaries are relatively free from dislocations. These boundaries or subboundaries are consistent with subgrain development due to the recovery of the original crystal material or with the recrystallization observed optically in parts of the sample.

The low dislocation density, heterogeneous

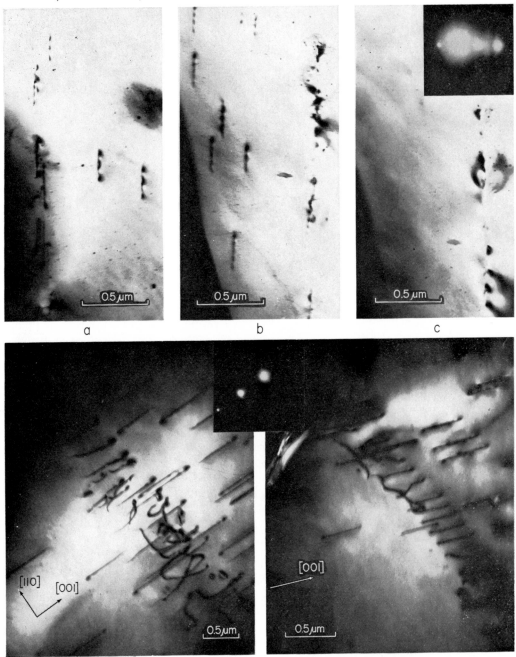

Fig. 11. (a) Bright-field (BF) micrograph of a region with straight parallel dislocations forming a vertical array on the left; g = 224 (DT-669). (b) BF micrograph of a region close to that of (a), again with g = 224. (c) BF micrograph of same region as that shown in (b) showing dislocations that are out of contrast for g = 040. Residual contrast on the right is due to strain. (d) BF micrograph of dislocations forming an array; g = $\bar{1}$11. The dislocations are mostly screw, parallel to [001]; some can be seen curving into segments parallel to [110] and other orientations (DT-669). (e) BF micrograph of region close to that of (d), showing another part of the same array; g = $\bar{1}$11. The diffraction pattern is not correctly oriented with respect to this micrograph.

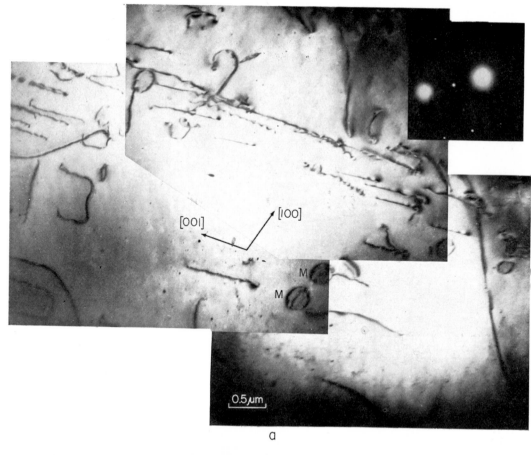

a

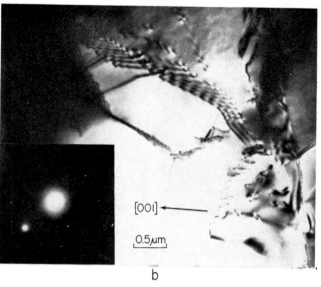

b

Fig. 12. (a) Composite of bright-field (BF) micrographs showing helical dislocations, the axes of the helices being parallel to [001], and loops; g = 222. The loops marked M may contain stacking fault fringes (see the text) (DT-669). (b) BF micrograph from another area showing subboundaries separating relatively dislocation-free subgrains; g = 242 (DT-669).

distribution of dislocations, and presence of helices, loops, and sub-boundaries are consistent with the optical evidence of extensive recovery and recrystallization in this sample.

DISCUSSION AND CONCLUSIONS

The types of dislocations observed at different temperatures and the slip systems determined from them are summarized in Table 3. Slip parallel to c[001] is predominant in all experiments between 600° and 1000°C where there was high shear stress in this direction, even when other slip vectors were also highly stressed. This finding indicates that the critical shear stress for slip parallel to [001] is lower than that for all other systems over this temperature range. Many of the isolated helical screw dislocations observed at 1250°C also had $b = c[001]$, this fact suggesting that it is also the easiest slip direction at this temperature. There is evidence for some slip parallel to [100] at 800° and 1000°C, but it is minor. The vectors $a[100] = 4.76$ A and $c[100] = 5.98$ A are the shortest Burgers vectors for unit dislocations in olivine, so that slip in these directions is predictable. Slip parallel to [010] is unexpected, since the magnitude of the Burgers vector $b[010] = 10.21$ A for complete dislocations is large, but it appeared to predominate over slip with $b = a[100]$ in the crystal of orientation B deformed at 800°C; the high flow stress of this crystal reflects the high energy of these dislocations. In the crystal of orientation B (where slip parallel to c[001] is suppressed) deformed at 1000°C, only dislocations with Burgers vector $b = a[100]$ were observed. Thus it is demonstrated that the motion of these dislocations is easier than that of dislocations with

$b = b[010]$ at this temperature. The flow stress of this crystal (DT-673) is almost as low as that of the crystal of orientation A (DT-666), and thus the stress required to form or move dislocations with $b = a[100]$ is nearly equal to that required to form or move dislocations with $b = c[001]$ at 1000°C.

The dislocations observed in all the crystals deformed between 600° and 1000° generally show a very strong crystallographic control of their orientation, as was also observed by J. D. Blacic and J. M. Christie (unpublished manuscript, 1972) in their two samples deformed within this range. The mixed segments of the dislocation loops are short or absent at 600° and 800°, this situation giving rise to nearly rectangular loops, and, where they are present at 1000°C, the mixed segments are relatively straight (except in the cases of small loops). These observations indicate that the Peierls force is strong and the dislocations are constrained to lie along directions of low energy, notably the crystallographic axes.

In crystals deformed at 600°C and, to a lesser extent, at 800°C the edge segments of the dislocations with $b = c[001]$ are very short in comparison with the screw segments, so that the loops must be greatly elongated parallel to the slip direction. This fact probably indicates that the edge components of an expanding loop travel with much greater velocity than the screw components, as has been observed in some ionic crystals [Gilman and Johnston, 1957] and other materials [e.g., Stein and Low, 1960]. For dislocations of the system (010) [100] produced at 1000°C the densities of edge and screw segments were found to be approximately equal, this

TABLE 3. Slip Systems Identified in This Work

Temperature, °C	Types of Dislocations	Slip Systems
600	screw, $b = c[001]$ (edges very short)	(?) [001]
800	screw and edge, $b = c[001]$	(100) [001], {110} [001], and (010) [001]
800	screw and edge, $b = b[010]$	(100) [010]
800	screw and edge, $b = a[100]$	(010) [100]
1000	screw and edge, $b = c[001]$, and loops	{110} [001] and probably (100) [001]
1000	screw and edge, $b = a[100]$, and some loops	(010) [100]
1250	helices, loops, and networks	not determined

finding indicating that the configuration of the average loop is square. It has not been possible to confirm that the dislocations with $b = c[001]$ also behave in this way, but there appear to be more edge segments of these dislocations also in the 1000° crystals. Thus it appears that the disparity in the velocities of edge and screw components is reduced with increasing temperature.

The slip systems identified here are in agreement with those determined in the earlier optical studies (Table 1), but a comparison of Table 3 with Table 1 indicates some apparent inconsistencies.

1. We observe slip parallel to c [001] at 1000°C and possibly at 1250°C, as was noted above, extending the range of this mechanism. There is evidence of slip on the (010) [100] system as low as 800°C at least. Thus the actual ranges of these slip mechanisms are more extensive than was previously reported.

2. We note extensive operation of the (100) [010] system at 800°C in the crystal of orientation B; the operation of this system was observed before at 'low temperature' and above 1000°C (Table 1). This system probably operates over a wider range of temperatures in suitably stressed crystals, but it requires shear stresses higher than those for other systems.

3. We have not observed definite evidence of (001) [100] slip or the more general pencil glide on {0kl} [100] (see *Green and Radcliffe* [this volume] for a description of this system in natural olivines), but this lack of observed evidence may be due to the fact that the edge and mixed components of the dislocations with $b = a$ [100] are short and obscured by the commoner dislocations with $b = c$ [001] in most of our samples, even at 1000°C. However, there is extensive cross slip (see discussion presented below) of screw dislocations at 1000°C in our experiments, so that there is little tendency for the dislocations to remain in a single slip plane; thus cross slip of [100] screw dislocations is equivalent to pencil glide on the {0kl} [100] system reported microscopically above 1000°C by all the earlier authors.

Boland et al. [1971] identified complete screw dislocations with $b = [1\bar{1}2]$ forming arrays in naturally deformed olivine. We have observed no dislocations of this type and think that they are unlikely to develop by combination of the types we have observed (with $b = [100]$, [010], and [001]), in view of the large magnitude of the $[1\bar{1}2]$ Burgers vector ($|b| = 16.43$ A) and the high energy of such dislocations.

Inasmuch as the earlier observations by *Raleigh* [1968], *Carter and Avé Lallemant* [1970], and *Raleigh and Kirby* [1970] were based on optical examination of deformed polycrystals, they yielded evidence of only the predominant slip mechanisms. It is therefore not unexpected that our submicroscopic observations indicate minor dislocation motion in most systems over a wider range of temperatures. The single-crystal studies [*Young*, 1969; this paper] are based on relatively few crystals in a limited range of orientations and temperatures, so that the most favorably stressed systems will tend to operate at the expense of the others. In polycrystalline aggregates it is likely that the constraints of neighboring grains will force a grain to deform (at least locally) on several slip systems, including those requiring high shear stresses, so that we expect an even wider range of operation of the observed slip systems in polycrystals and possibly operation of some systems not yet observed.

In this connection it is worth noting that all the slip mechanisms observed so far in olivine (Table 1) fail to provide more than three independent slip systems [*Groves and Kelly*, 1963; *Kelly and Groves*, 1970]. Since all the slip vectors are parallel to the crystal axes a, b, and c, elongations parallel to these axes are not permitted by the observed mechanisms alone. In a polycrystal the condition of 'material continuity' at grain boundaries requires that five independent systems operate in the grains for a general homogeneous strain to occur by slip (the Von Mises condition). The lack of five independent systems in olivine is presumably responsible for the brittleness of dunite at low to moderate pressures and temperatures [*Griggs et al.*, 1960]. The impossibility of elongating or shortening a crystal parallel to the crystal axes probably explains why there is invariably kinking subparallel to the pinacoids in dunite and peridotite deformed in compression at higher pressures and temperatures (below those at which extensive recovery occurs). The kinking produces shortening parallel to the crystal axes by a concertinalike buckling of the slip planes and is accomplished by motion of dislocations in the slip systems noted above.

Dislocation densities were estimated in most areas of the crystals sampled by transmission electron microscopy and are noted with the other data given above. These areas generally included the most heavily deformed regions of the crystals. The densities range from 3×10^8 cm^{-2} in less deformed parts of DT-674 to 3×10^{10} cm^{-2} in the most deformed parts of DT-661. In the cooler relatively undeformed end regions the densities are too small to be detected by transmission microscopy ($<10^7$ cm^{-2}), and they may be as dislocation free as the starting material. In the crystal deformed at 1250°C (DT-669) the dislocation density is very low, and the distribution is heterogeneous. It is not possible to estimate the density reliably, but it is probably between 10^7 and 10^8 cm^{-2}. We have attempted to correlate the dislocation densities with the maximum stress ($\sigma_1 - \sigma_3$) and the resolved shear stresses on the operating slip systems in each experiment. The scatter of the data, however, is such that we could detect no systematic relationship between the stresses and the dislocation densities. This lack of correlation may be due partly to uncertainties in the experimental data and possibly partly to our inability to sample the most heavily deformed parts of some crystals. However, the dislocation densities are relatively consistent: they range between 10^8 and 10^{10} cm^{-2} in the range of temperatures, strain rates, and stresses represented by the experiments.

Tangling of the dislocations is present in many areas of the crystals deformed at 800°C and ubiquitous in those deformed at 1000°C. The tangles are best exemplified in DT-661 (Figures 7c and 8a). In extreme cases like this one, dense tangles form in zones both parallel and perpendicular to the dominant slip vector ($\mathbf{b} = c$ [001] in DT-661), their formation giving rise to a cell structure of relatively unstrained regions surrounded by dense tangles. The dense tangles perpendicular to the [001] slip direction correspond in orientation to the numerous tilt boundaries in the sample and therefore contain many edge dislocations. There are two possible mechanisms that could cause the tangled structure. (1) The presence of dislocations with different slip vectors in the denser tangled zones suggests that they are formed by intersection of

the common dislocations (with $\mathbf{b} = c$ [001] in DT-661) with other dislocations of different Burgers vectors moving in intersecting slip planes. This intersection would give rise to jogs that would obstruct the motion of the common slip dislocations in the absence of appreciable climb. This mechanism is probably the more important one at 800°C. (2) The tangles may be caused by cross slip of the long screw dislocations, which gives rise to jogs and dipoles. There is slight evidence of cross slip in the experiments at 800°C, and it is widespread in those at 1000°C. Large jogs are present in otherwise long straight screw dislocations (Figure 9a, b, and c), and good examples of narrow edge dislocation dipoles trailing from screw dislocations (Figure 9a, b) are also present. These dipoles are formed by cross slip and double cross slip of the screw dislocations. In some cases at 1000°C (Figure 9a) and 1250°C (Figure 12a) the dipoles degenerate into rows of loops by climb. In view of the widespread evidence of cross slip at 1000°C, this process may be the dominant cause of tangling at this temperature. Thus there is clear evidence of both intersection and cross slip at 800°C and 1000°C, and both processes are probably contributing to the formation of the tangled zones.

There is no evidence of climb at 600°C, and the only indications of climb at 800°C are the large jogs on cross-slipped screw dislocations noted above; their motion requires only enough climb of the edge components of the dislocations to accommodate cross slip of the screws. At 1000°C, however, enhanced climb is indicated by the presence of curved dislocations and the formation of small loops and other debris from dipoles, as was previously discussed. The dislocation densities are still relatively high, however, and extensive recovery does not accompany the deformation at 1000°C. In the experiment at 1250°C the low dislocation density, the helical dislocations, the loops, and the extensive stabilized networks (forming subgrain and recrystallized grain boundaries) indicate extensive climb and recovery during deformation.

In summary, at these strain rates (10^{-4}–10^{-5} sec^{-1}), conservative motion of the dislocations appears to be general at 600°C, cross slip occurs at 800°C and is marked at 1000°C, and climb of the dislocations begins somewhat below

1000°C and leads to extensive recovery and recrystallization at 1250°C.

We believe that further transmission electron microscopy of single crystals is necessary to determine the full ranges of temperature over which the slip systems operate in olivine and to elucidate the early stages of the mechanisms of tangling and interaction of the dislocations. Similar work on polycrystals deformed at these conditions may reveal further slip mechanisms. Transmission microscopy of dunite specimens deformed under conditions of steady-state creep should also be useful in characterizing the flow processes and providing a basis for comparison with naturally deformed olivine rocks.

Acknowledgments. We are most grateful to our colleague David Griggs, who gave liberal advice and constructive criticisms in the course of this study, which was carried out in part in his laboratory. We also thank Dr. Alan Ardell, who arranged access to the electron microscope facility of the Materials Department, School of Engineering, University of California at Los Angeles, and offered advice on the manuscript. S. Kirby, H. Green, and S. V. Radcliffe gave valuable criticisms of the manuscript. The difficult task of preparing specially oriented petrographic sections was carried out with skill by E. Gonzales. We are indebted to L. Weymouth, who provided a prodigious amount of photographic work of excellent quality. Mrs. J. Schachter and Mrs. V. Jones did the typing, and Mrs. J. Martinez and Miss J. Guenther the drafting.

The work was supported by National Science Foundation grant GA-26027.

Deformation Processes in the Upper Mantle

H. W. GREEN II

Department of Geology, University of California
Davis, California   95616

S. V. RADCLIFFE

Division of Metallurgy and Materials Science
Case Western Reserve University, Cleveland, Ohio   44106

High-voltage (≤ 1 Mev) electron petrography of naturally and experimentally deformed peridotites has been conducted to elucidate the dislocation flow mechanisms in two of the major phases existing in the upper mantle, olivine $(Mg, Fe)_2SiO_4$ and orthorhombic pyroxene $(Mg, Fe)_2Si_2O_6$. Some observations have also been made on a minor phase, spinel $(Mg, Fe)(Cr, Fe, Al)_2O_4$. The specific materials examined include metamorphic-tectonite xenoliths from alkalic basalts exhibiting optical textures indicative of low, moderate, and high deformation, a naturally deformed alpine peridotite, and specimens of a previously undeformed peridotite strained under compression in the laboratory at high temperatures and pressures. The xenoliths, fragments of the upper mantle incorporated in magma during passage to the surface, exhibit dislocation substructures characteristic of deformation and partial recrystallization, the second-generation crystals being deformed also. Slip in olivine is by movement of unit dislocations on {0kl} [100] accompanied by development of abundant (100) kink bands and subgrains in the form of rectangular prisms elongated parallel to [010]. Orthopyroxene (enstatite) contains submicroscopic diopside $(CaMgSi_2O_6)$ exsolution lamellas and exhibits irregular subgrains. The alpine peridotite shows similar deformation of olivine, but the dislocation densities are greater. The orthopyroxene (bronzite) crystals are also heavily deformed, their deformation occurring by slip of partial dislocations primarily on (100). The experimental specimens examined show still higher densities of dislocations. Olivine contains dislocations of the {110} [001] system at lower temperatures as well as the more abundant pencil glide on [100]. At the higher temperatures the former system disappears and rectangular-parallelepiped subgrains are abundant. Enstatite deformation is extremely complex, abundant partial dislocations and stacking faults, together with less common subgrains and, possibly, inversion to clinoenstatite, occurring. The peridotites deformed in nature are shown to flow by the same mechanisms as those in the deformation experiments at the highest temperatures. The substructure of the alpine peridotite indicates that it has undergone deformation at a lower temperature subsequent to the high-temperature flow responsible for the features visible optically, the colder deformation corresponding to strains imposed during or after emplacement of this material in the crust. In general, the xenoliths are highly recovered. However, the lack of appreciable recovery in the Lunar crater xenolith examined suggests that mantle material can be transported to the surface by volcanism without undergoing significant modification. Finally, it is concluded from the substructures observed that intracrystalline flow in the upper mantle is dominated by dislocation movement and that such flow processes are controlled by dislocation climb. The results support the extension of the empirical flow rules from laboratory experiments to the flow conditions within the mantle.

The concepts of sea floor spreading [*Hess,* 1962; *Vine and Matthews,* 1963] and plate tectonics [*Morgan,* 1968; *Le Pichon,* 1968; *Isacks et al.,* 1968] have recently produced a marked advance in the understanding of differential movements of large sections of the earth's outermost layer, the lithosphere. The current models of plate motions establish the localization of the large majority of earthquakes and volcanic activity along plate margins. An important component in these models is the asthenosphere, considered to be a weak low-density layer of low seismic velocities and high seismic attenuation, in which the flow responsible for plate movements is believed to be concentrated. The possible mechanisms of flow in the material of the mantle, in particular in the asthenosphere, have been the subject of several theoretical analyses, which were reviewed recently by *Weertman* [1970].

As is evidenced by ophiolite suites on continental margins and xenoliths from basalts and kimberlite pipes, the rocks of the upper mantle appear to consist predominantly of various types of magnesium-rich peridotite. In laboratory experiments on peridotites a power law relationship between stress and strain rate has been determined by Griggs and his co-workers, by *Carter and Avé Lallemant* [1970], and by *Raleigh and Kirby* [1970]. However, the uncertainties involved in the extension of such empirical flow laws to the much lower (6–8 orders of magnitude) strain rates involved in mantle flow require a definitive link between the deformation mechanisms in the experimental and natural environments. Determination of such flow laws for the asthenosphere should assist in determining the driving force responsible for plate tectonics.

Optical analysis by *Raleigh* [1963, 1965a, 1967, 1968] indicates that olivine in peridotite deformed in the laboratory at temperatures below approximately 1000°C at a strain rate of 10^{-4} sec^{-1} slips primarily on {110} [001], but at higher temperatures or lower strain rates changes to pencil glide about [100]. The latter system is responsible for the deformation structures found in olivine in naturally deformed peridotites. *Carter and Avé Lallemant* [1970] confirmed and expanded Raleigh's results by systematically delineating the regimes of temperature, pressure, and strain rate dominated by the slip systems {110} [001] and {0kl} [100]. They also reported that, at temperatures above about 1300°C, slip was largely restricted to (010) [100]. *Avé Lallemant and Carter* [1970] showed that syntectonic recrystallization of olivine in the pencil glide regime results in patterns of preferred crystallographic orientation similar to those commonly found in alpine peridotites and peridotite xenoliths from basalts. Deformation of orthorhombic pyroxene (enstatite) in the laboratory results in an apparently martensitic transformation to clinoensta-

tite [*Griggs et al.*, 1960; *Turner et al.*, 1960; *Borg and Handin*, 1966; *Raleigh et al.*, 1971] and/or slip and associated kinking on (100) [001] [*Raleigh et al.*, 1971]. The latter mechanism is responsible for virtually all optically visible deformation features in naturally deformed orthopyroxene; clinoenstatite has only been well documented from one tectonite [*Trommsdorf and Wenk*, 1968].

It is apparent that these optical studies provide part of the necessary link between natural and laboratory deformation processes. This paper presents the results of a high-voltage electron petrographic study of naturally and experimentally deformed peridotites. This high-resolution technique allows the direct observation of dislocations and other fine-scale features and hence an analysis of deformation processes that is more detailed and complete than that possible by other means.

SPECIMENS AND EXPERIMENTAL PROCEDURES

Peridotite xenoliths from basalts showing features in the petrographic microscope indicative of a spectrum of imposed strains [*Raleigh*, 1968; *Raleigh and Kirby*, 1970; *Carter and Avé Lallemant*, 1970; *Avé Lallemant and Carter*, 1970] were examined and compared with a highly deformed alpine peridotite and specimens strained in the laboratory under high temperatures and pressures. The xenolith specimens selected for study were lherzolites from palagonitized tuff of Salt Lake crater, Hawaii (low strain), dunites from the 1800–1801 Kaupulehu basalt flow of Hualalai volcano, Hawaii (moderate strain), and lherzolites from basalts at Lunar crater, Nevada (high strain). The specimen of alpine peridotite investigated was a harzburgite mylonite from the Vourinos ophiolite complex, Greece (specimen V-645A of *Moores* [1969]). The experimental specimens consisted of previously unstrained (optical examination) lherzolite cylinders deformed 30% in axial compression under 15-kb confining pressure, one specimen at 900°C and a strain rate of 10^{-6} sec^{-1} and the other at 1300°C and a strain rate of 10^{-4} sec^{-1} (specimens N-25 and N-41, respectively, of *Raleigh and Kirby* [1970]). (Detailed discussions of electron petrography of olivine single crystals deformed experimentally under comparable conditions are given by *Phakey et al.* [this volume].) These experiments were performed in a solid-pressure-medium device. Although unavoidable temperature gradients exist along the lengths of the cylinders [*Green et al.*, 1970, p. 281], these gradients do permit several different temperatures of deformation to be examined from each specimen in addition to the maximum temperature noted above.

Electron microscope foils of the order of 1 μm thick and 3 mm in diameter were prepared by coring disks from standard petrographic thin sections and thinning to electron transparency by low-angle ion bombardment [*Paulus and Reverchon*, 1961; *Tighe*, 1970; *Radcliffe et al.*, 1970; *Heuer et al.*, 1971]. Optical microscopy studies were conducted at various stages of thinning, and thus detailed correlation could be established for several features observable by both electron and optical petrography.

The prepared foils were examined in the U.S. Steel million-volt electron microscope. The high accelerating voltages used in this study (800–1000 kv) offer several distinct advantages over conventional (100 kv) transmission electron microscopy. The foil thickness of less than a few thousand angstroms required for 100-kv microscopy results in extremely fragile specimens that are easily broken and destroyed by handling or air flow during the pumping out of the microscope column. The several-fold thicker

specimens usable at high voltages are much more resistant to accidental destruction during handling, and thus they facilitate reliable correlation with optical studies by minimizing losses during foil preparation. In addition, the greater transparency at high voltages allows the study of much larger areas of individual foils and greatly facilitates the visualization of structures in three dimensions. Furthermore, the high voltage permits increased resolution by a reduction of chromatic aberration, an increased precision of selected-area electron diffraction, and a reduced possibility of specimen damage in the microscope from ionization and beam-heating effects.

RESULTS

Naturally Deformed Peridotites

Xenoliths. The three sets of xenoliths studied include low (Salt Lake), moderate (Hualalai), and high (Lunar crater) deformation, as was indicated by textures and structures observed in thin sections in the petrographic microscope. Salt Lake crater spinel lherzolite specimens display igneous textures: rare euhedral olivine, common subhedral olivine and orthopyroxene, orthopyroxene-spinel intergrowths, and late crystallizing interstitial crystals of clinopyroxene and spinel (Figure 1a). Broad deformation

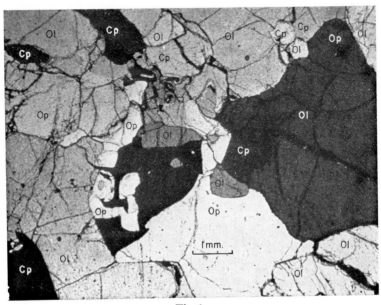

Fig. 1a.

Fig. 1b.

Fig. 1c.

Fig. 1. Photomicrographs of peridotite xenolith thin sections viewed between crossed polarizers. (a) Spinel lherzolite from Salt Lake crater, Hawaii. Ol, olivine; Op, orthopyroxene; Cp, clinopyroxene; black area, spinel. Note the intricate intergrowths of spinel and orthopyroxene which may reflect growth from a melt. No grain boundary recrystallization is present, and deformation is slight. (b) Dunite from the 1800–1801 Kaupulehu lava flow, Hualalai volcano, Hawaii. Any igneous textures were greatly modified by moderate deformation and extensive recrystallization. (c) Lherzolite from Lunar crater, Nevada. All pre-existing textures were obliterated by extensive deformation and recrystallization. Relict olivine crystals show abundant deformation bands and subgrains of rectangular cross section.

(kink) bands occur in olivine as a result of {0kl} [100] slip, but no undulatory extinction indicative of lattice bending is present, and no subgrains are evident. Pyroxene crystals display exsolution lamellas and rarely contain subgrains. Recrystallization is lacking except for low-temperature alteration products on grain boundaries. Hualalai dunite specimens are metamorphic tectonites (Figure 1b); grain boundary recrystallization is abundant, as are kink bands and long prismatic subgrains with long axes parallel to [010] and boundaries parallel to (100) and (001). Subgrains differ in orientation from their neighbors by small rotations about axes lying in (100). The new grains (on grain boundaries and in kink bands) show evidence of deformation subsequent to their origin by recrystallization. However, they are less strained than the host crystals. Lunar crater lherzolite specimens exhibit the same general optical textures as Hualalai xenoliths but to a greater degree (Figure 1c); these nodules are highly foliated tectonites indistinguishable optically from many mylonites common in alpine peridotite terranes and similar to the Vourinos specimen discussed below. Relict olivine ribbons are

extensively deformed and show abundant recrystallization along grain boundaries and in deformation bands. Pyroxenes are highly contorted and show large irregular subgrains, partial recrystallization, and numerous exsolution lamellas.

Electron microscopy of the thin foils from Salt Lake specimens shows that they contain low dislocation densities. In olivine, dislocations are rare except for those in kink band boundaries, which are seen in Figure 2 to be simple tilt boundaries composed of a single set of edge dislocations. All the dislocations observed were {0kl} edges with Burgers (slip) vector $\mathbf{b} = [100]$ as determined by diffraction contrast experiments involving specimen tilting in the microscope; neither screws nor dislocations belonging to other slip systems were found. Enstatite crystals are virtually devoid of dislocations except for those in irregular sharply stepped low-angle (subgrain) boundaries (Figure 3). The few isolated dislocations observed were partials connected by stacking faults (See *Barrett and Massalski* [1966, p. 387] for a discussion of this structural feature in metals.) Numerous thin planar features parallel to (100) were seen in

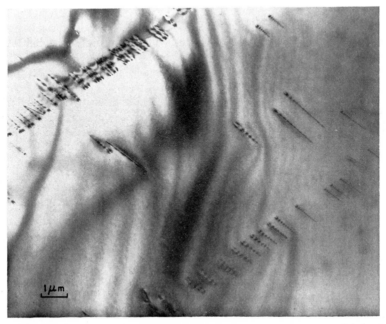

Fig. 2. Kink band in olivine from Salt Lake lherzolite. Walls are parallel to (100) and composed of edge dislocations with Burgers vector $\mathbf{b} = [100]$. The isolated dislocations shown here parallel to those in the band boundaries are rare. Long curved grey lines are bend contours.

Fig. 3. Subgrain boundary in enstatite from Salt Lake lherzolite. The boundary consists of steps in low-index directions. On an optical scale the boundary would appear curved. The steps end at dislocation lines. The dark and light bands in the boundary are extinction fringes due to diffraction contrast phenomena. The curved fringes are thickness extinction contours. No dislocations are visible outside the boundary.

enstatite (Figure 4), and it is presumed that they are diopside exsolution lamellas, in keeping with the optical visibility of the coarser lamellas. Large elastic strains were always associated with these lamellas and probably produced during travel to the surface by the different compressibilities and thermal expansions of the two phases. In the spinel crystals, greater numbers of dislocations are displayed. They are generally straight, but occasional curved dislocation lines, triple nodes (Figure 5), and intersecting arrays were visible. No stacking faults were visible at nodes, and thus a high stacking fault energy for spinel is indicated [Barrett and Massalski, 1966, p. 390].

In electron transmission, olivine in Hualalai specimens also contains a low density of dislocations, but kink bands are more abundant, and subgrains (Figure 6) and small voids (Figure 7) are also observed. Larger, optically visible bubbles are abundant in these crystals. In an extensive optical study of xenoliths from a variety of sources, including Hualalai and Salt Lake crater, Roedder [1965] observed bubble densities as great as 3% by volume. Although

many bubbles appeared empty, he found that large numbers of them were filled with liquid CO_2 under pressure. The bubbles were interpreted by Roedder [1965] as droplets of an immiscible fluid incorporated during growth from a melt. More recently Green [1972] has suggested that many of the bubbles may be intracrystalline and grain boundary precipitates of volatiles outgassed from the deep interior.

A one-to-one correspondence was established between optical and electron optical observations on an olivine kink band (Figure 8). The band boundaries consist of parallel equally spaced edge dislocations; the dislocation lines are parallel to the optically determined kink axis [001] (the axis of external rotation), as they must be for true kinking. The electron microscope images show that the band is composite, an additional wall appearing within the band. The inclination across this interior boundary was too small to be seen optically.

In the electron microscope, Lunar crater specimens show much more evidence of plastic strain than the other xenoliths and allow a more detailed analysis of deformation. In olivine, kink bands

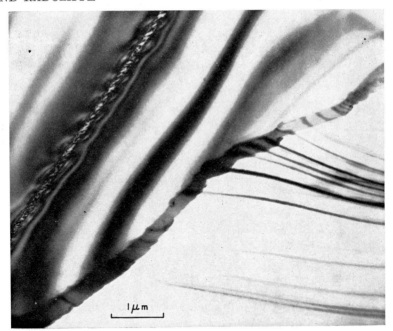

Fig. 4. Diopside exsolution lamella (upper left) in enstatite from Salt Lake lherzolite. Lattice bending due to elastic strains at the boundaries of the lamella causes the peculiar pattern of the extinction contours. The subgrain boundary (diagonal) separates the areas of distinctly different extinction band trends.

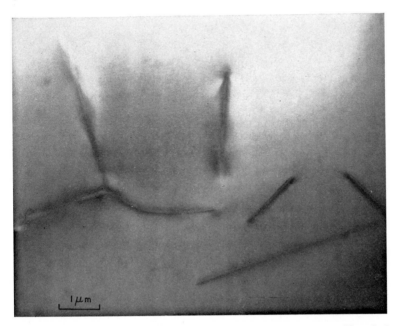

Fig. 5. Dislocations in spinel from Salt Lake lherzolite. Lack of a stacking fault at the triple-point node indicates that spinel has a high stacking-fault energy.

Fig. 6. Subgrain boundaries in olivine from Hualalai dunite. The boundaries appear to be of three types. They can contain one set of parallel dislocations, as in lower boundary; they can be predominantly of one type with occasional lines of different orientation, as in boundary at right; or they can consist of an intersecting array of lines, as at the left. Occasional dislocations outside the boundaries are rare. (The 'wispy' lines trending NE-SW are surface contamination.)

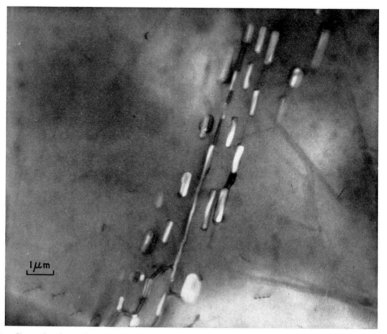

Fig. 7. Capsule-shaped voids in olivine from Hualalai dunite. These bubbles, sometimes interconnected by long tubes, are extremely abundant locally, and many optically visible ones have been found to be filled with liquid CO_2 under pressure [*Roedder,* 1965].

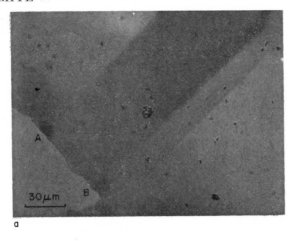

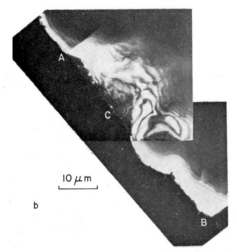

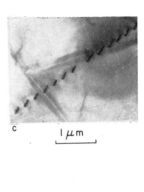

Fig. 8. Kink band in olivine from Hualalai dunite. (*a*) Optical micrograph (approximately crossed polarizers) showing a kink band at the edge of the electron transparent foil. The boundaries are labeled A and B. (The surface mottling and the circular lumps are artifacts introduced during ion bombardment thinning.) (*b*) Dark-field electron micrographs of the same area showing an additional wall (indicated at C) within the optically visible band. (*c*) Bright-field electron micrograph showing the interior wall. It contains dislocations parallel to [001], as do the exterior band walls, but the dislocation spacing is greater.

on {0kl} [100] are narrower, more numerous, and more closely spaced than those in the Hawaiian samples. Dislocations approximately parallel to [100] are common as lines extending across kink bands and aligned in walls parallel to (001) (Figure 9). Diffraction contrast experiments established that these dislocations are screws with **b** = [100]. The (001) walls, in combination with the (100) kink boundaries, comprise subgrains with rectangular cross sections identical to those visible optically but on a

much finer scale. The newer recrystallized olivine grains are also deformed and show the same strain features as those shown by the primary crystals. Substructures in the two generations of crystals were so similar that it was possible to determine the observed generation only by correlation with low magnification (optical) photographs. Bubbles of ≤ 2-μm diameter were observed distributed on grain boundaries and along dislocation arrays comprising subgrain boundaries. The pyroxenes from Lunar crater have not yet been examined.

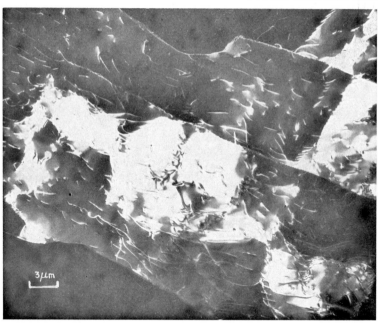

Fig. 9. Dark-field electron micrograph of Lunar crater lherzolite. NW-trending boundaries are walls of [100] screws. Note the distribution of abundant dislocations outside the boundaries, indicating that post-deformation recovery is not great.

Alpine peridotite. Optical study of the Vourinos specimen shows it to be a highly foliated harzburgite consisting of large extensively deformed relict crystals of olivine and bronzite in a groundmass of much smaller recrystallized grains (principally olivine), which also show abundant evidence of deformation. The bronzite crystals display exsolution lamellas of diopside. Undulatory extinction, deformation bands, and subgrains are common in both olivine and pyroxene (Figure 10). The optically visible structures thus indicate extensive high-temperature deformation accompanied by recrystallization [*Raleigh and Kirby*, 1970]. As in the xenoliths, deformation features in olivine indicate slip on {0kl} planes in the [100] direction.

Electron petrography of this harzburgite reveals a density of dislocations in both generations of crystals that is much higher than that found in any of the xenoliths. The major structures visible in olivine (Figure 11a, b) are the same as those seen in the xenoliths; kink band boundaries parallel to (100) are common and consist of {0kl} [100] edge dislocations, and less common (001) walls contain [100] screws. However, the average width of and spacing between kink bands are

smaller in the alpine peridotite. Also, the abundant dislocations not in walls are much straighter than those in xenoliths and form rectangular loops of pure edge and screw components (Figure 11c).

Bronzite crystals are extensively faulted, partial dislocations lying in and terminating the (stacking) faults (Figure 12). The majority of these faults are parallel to (100), their bounding dislocations being commonly curved and assuming many directions in the (100) plane. Partial dislocations and stacking faults on some other plane at a high angle to (100) but parallel to [001] are also present and probably are parallel to (010). Rarely, dislocations are seen between (100) faults. No stacking faults were detected in association with these latter dislocations. The Burgers vectors have not yet been determined for any of these dislocations; from optical measurements it is known that the dominant slip system is (100) [001] [*Raleigh et al.*, 1971], but the magnitude of the Burgers vector is still unknown.

Experimentally deformed peridotites. The two experimental specimens studied, because of their considerable temperature gradient, provided material representative of the three olivine

Fig. 10. Photomicrograph of Vourinos alpine peridotite (harzburgite) thin section viewed between crossed polarizers. The specimen closely resembles the Lunar crater specimen but is somewhat more recrystallized (Figure 1c). This mylonite exhibits extensive deformation features in new recrystallized crystals as well as in relict olivine and pyroxene augen.

'slip fields' in temperature–strain rate space delineated by *Carter and Avé Lallemant* [1970]. For low temperatures *Blacic* [1971] recently has confirmed by electron microscopy the {110} [001] slip mechanism discovered by the earlier optical studies, and our observations are similar. In the present paper, discussion will be restricted to deformation conditions that induce plastic flow in olivine by pencil glide about [100], because this slip mechanism is the one exhibited by rocks

Fig. 11a.

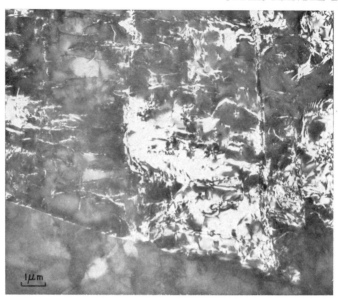

Fig. 11b.

Fig. 11c.

Fig. 11. Typical substructures developed in olivine in Vourinos harzburgite. (a) Kink bands trending NW are parallel to (100). The edge dislocations comprising band boundaries are parallel to several different 0kl directions. Within the bands, dislocations are abundant and both edges and [100] screws (trending NE) are common. (b) Walls parallel to (001) consisting of [100] screw dislocations (N-S in the micrograph) combine with (100) edge walls (WNW in the micrograph) to yield prismatic subgrains with the long axis parallel to [010]. (c) Straight dislocation segments that are parts of rectangular loops with pure edge components (in this case parallel to [001]) trending NW and pure screw components parallel to [100] trending NE.

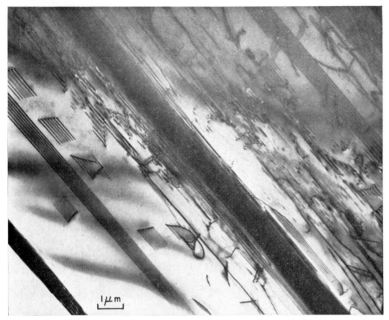

Fig. 12. Typical substructure of bronzite from Vourinos harzburgite. Partial dislocations and associated stacking faults on (100) are abundant. Less commonly, faults and partials are seen at a high angle to (100), possibly on (010). Note also the dislocations of apparently unit strength that intersect (100) faults.

deformed in the mantle, as was discussed in the previous section. The features visible optically in olivine in the experimental specimens deformed in this flow regime are the same as those exhibited by the naturally deformed specimens [*Raleigh*, 1968; *Carter and Avé Lallemant*, 1970]. Orthopyroxene, however, tends to transform to clinoenstatite in the experiments but to slip on {100} [001] in the natural specimens [*Raleigh et al.*, 1971].

At the lower temperatures of the pencil glide field, olivine displays unit dislocations of the low-temperature system, {110} [001], as well as those of the dominant system, {0k1} [100]. Slip bands consisting of planar arrays of closely spaced parallel edge dislocations are common [*Green and Radcliffe*, 1972a]. The dislocations in these arrays apparently are predominantly of one sign, for they produce long-range elastic strains visible optically as deformation lamellas [*Christie et al.*, 1964; *Christie and Green*, 1964; *Raleigh*, 1968; *Carter and Avé Lallemant*, 1970]. Despite the 'cold-worked' appearance of the substructure within olivine grains, characterized by high to extreme dislocation densities (locally of $>10^{11}$ cm^{-2}), recrystallization occurs at grain

boundaries. Figure 13 shows some new grains growing into a region with slip bands. These new crystals have distinctly fewer dislocations than the material they are replacing, but they too are deformed. Elsewhere in this specimen, new grains were seen with dislocation densities approaching the density of the host crystal.

For deformation of olivine at a higher temperature or a lower strain rate *Green and Radcliffe* [1972b] have shown that slip bands (deformation lamellas) do not develop; edge dislocations climb into (100) microkink band boundaries. Screw dislocations, after breaking away from pinning points in kink walls (Figure 14), climb and/or cross slip into (001) walls. The result is long prismatic subgrains parallel to [010] identical to those found in Lunar crater xenoliths (Figures 9 and 15). No dislocations with a Burgers vector other than [100] were found. Recrystallization, more abundant than at lower temperatures, yields equiaxed aggregates with moderate to high dislocation densities and microkinks [*Green and Radcliffe*, 1972b].

Enstatite crystals from both experimental specimens are so extensively faulted on (100)

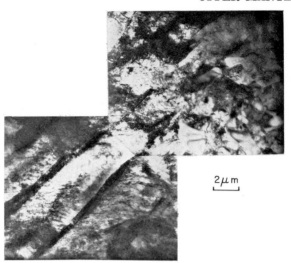

Fig. 13.   Recrystallization of slip bands in olivine from experimental specimen N-25. Slip bands (NE-trending features) consist of arrays of edge dislocations with **b** = [100] lying in their slip plane and spaced about 150 A apart. Generally distributed background dislocations have **b** = [100] or [001]. Note that new recrystallized grains are also deformed. (The deformation conditions are T = 900°C, P = 15 kb, $\dot{\epsilon}$ = 10^{-4} sec^{-1}, and $\epsilon \simeq 30\%$.)

that other features are usually obscured (Figure 16); faults are commonly separated by <100 A. However, in some regions, tilting in the microscope reveals abundant partial dislocations lying in the (stacking) faults as well as numerous other dislocations inclined to the faults and

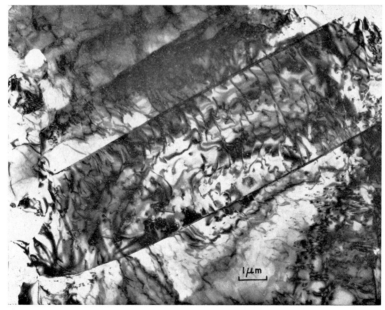

Fig. 14.   Olivine kink bands in experimental specimen N-41. Band boundaries are (100) walls of edge dislocations that have their lines parallel to the normal to (0$\bar{1}$1). Dislocations trending across the bands are [100] screws and can be seen to bow out where they are pinned at band boundaries. All the dislocations have **b** = [100]. The new recrystallized grains at the left are also deformed. (The deformation conditions are T = 1300°C, P = 15 kb, $\dot{\epsilon}$ = 10^{-4} sec^{-1}, and $\epsilon \simeq 30\%$.)

apparently of unit Burgers vector (Figure 16). All the regions investigated show electron diffraction patterns characteristic of the orthorhombic polymorph, but small regions may be transformed to clinoenstatite. Subgrains (Figure 16a) are common in enstatite, although recrystallization (Figure 16b) is much less abundant than it is in olivine.

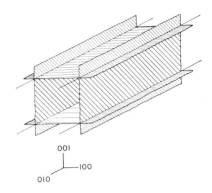

Fig. 15. Idealized drawing showing the rearrangement of the slip dislocations into walls defining rectangular-parallelepiped subgrains. Walls parallel to (100) contain edge dislocations of the system {0kl} [100], and walls parallel to (001) contain screws of the same system.

DISCUSSION

The experimental specimens of this study display all the major features observed in the corresponding phases from mantle-derived rocks plus many transitional structures. These electron petrographic observations provide a framework to discuss the development of optically visible deformation features in olivine. However, for the orthorhombic pyroxene, further substructural information is necessary before detailed characterization of flow can be developed.

At the lower temperatures of the pencil glide deformation regime, flow in olivine is predominantly by conservative slip of rectangular loops of unit dislocations on a wide variety of planes in the [100] zone, the Burgers vector always being [100]. Dense arrays of edge dislocations lying in the slip planes (slip bands) develop (Figure 15 and *Green and Radcliffe* [1972a]), presumably owing to pile-ups at obstacles [*Christie et al.*, 1964; *Hirsch et al.*, 1965, p. 52]. Under conditions slightly more conducive to climb of dislocations around such obstacles, slip bands are less numerous, and (100) microkink bands are abundant. *Green and Radcliffe* [1972b] have shown that during deformation under climb-dominated conditions, such as that at the highest-temperature experimental conditions reported here and in the xenolith specimens, the [100] screw dislocations remaining after kink band formation climb into (001) walls and lead to the development of long prismatic subgrains with the long axis parallel to [010] (Figures 9, 14, and 15). Dislocation theory and observations on metals suggest that such simple screw dislocation arrays should be unstable and the actual wall could also contain additional accommodation dislocations. We have not yet been able to obtain unequivocal evidence for such arrays.

The discovery of small optically invisible subgrains in Lunar crater specimens does not support the suggestion of *Raleigh and Kirby* [1970]

that subgrain size may vary inversely with differential stress and hence provide an indirect measure of stress difference in the mantle. However, our observations do not rule out the possibility that subgrains of large enough size and misorientation to be visible optically may vary inversely with shear stress, as they suggest. More optical measurements on experimental specimens are necessary to resolve this point.

In the alpine peridotite we have observed the same suite of substructures as that in the xenoliths and the experiments, and we believe they originated similarly during flow in the mantle. The Vourinos harzburgite specimen, however, shows the additional kinking and isolated rectangular dislocation loops discussed above. The structures are characteristic of lower temperature or more rapid deformation and were probably 'overprinted' during or after 'cold' emplacement in the crust. Thus it is possible that only these minor optically invisible structures were developed during the alpine orogeny proper, the major features having been formed earlier, during mantle flow. The mantle deformation may have occurred in association with the formation of a mid-oceanic ridge [*Moores and Vine*, 1971] or during the lithospheric plate convergence that culminated in the formation of the Alps.

In the case of the orthorhombic pyroxenes, marked similarity also exists between the experimentally and the naturally deformed material. Both show dominant slip by partial dis-

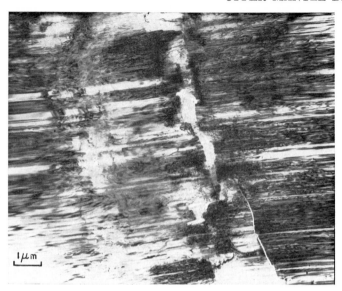

Fig. 16a.

Fig. 16b.

Fig. 16. Substructure of experimentally deformed orthorhombic pyroxene. (a) Extensive deformation by slip of partial dislocations on (100) produces a very high density of stacking faults parallel to (100); specimen N-25. Dislocations parallel to the faults are abundant but difficult to resolve. Irregular subgrains are common. (b) At higher temperatures, stacking faults are less abundant. Additional dislocations, apparently with unit Burgers vector, are seen inclined to (100). Recrystallization, shown at the lower right, is much less advanced than in olivine.

locations on (100) and evidence of other (as yet unidentified) mechanisms, some of which appear to consist of propagation of unit dislocations. Climb effects are noticeably less im-portant than they are in olivine, a finding that is in keeping with the optical evidence of a lesser tendency to polygonize and recrystallize. Clinoenstatite was not identified by us in the lower-temperature regions of these experimental specimens, but optical identification [e.g., *Raleigh et al.*, 1971] indicates that this transformation marks an important difference between most naturally deformed orthopyroxenes and those strained in the laboratory. However, our observations on the highest-temperature experiments in the region determined by *Raleigh et al.* [1971] to be characterized by slip confirms the operation of deformation mechanisms in the laboratory that are strikingly similar to those in nature.

A distinct difference between the xenolith specimens and those deformed in the laboratory is the much lower general dislocation density in many of the xenoliths, even though the major substructures are indistinguishable from those produced in the laboratory at the highest temperatures studied. Even the Lunar crater xenoliths, although they have considerable numbers of isolated dislocations within bands and subgrains, have distinctly fewer than the experimental specimens. We attribute this difference to post-deformational static recovery that probably occurred during passage to the surface in the high-temperature ($\simeq 1200°C$) basaltic magma. The major features are not significantly affected by static recovery, because the processes by which the dislocations rearrange themselves involve dislocation climb and migration of point defects, the same processes active in the dynamic recovery responsible for development of kink bands and subgrains during deformation. Static recovery, therefore, has the primary effect of clearing out isolated dislocations into the nearest walls and promoting recrystallization. The alpine peridotite from Anita Bay, South Island, New Zealand, examined by *Boland et al.* [1971] has apparently undergone a similar high-temperature static recovery rather than the later low-temperature deformation experienced by the Vourinos specimen described here.

The presence in rocks deformed in the mantle of appreciable densities of dislocations and of kink band and subgrain structures resulting from such deformation mechanisms reflects the presence of stresses sufficient to generate and move dislocations. Under such conditions the strain produced by crystal slip must predominate over any contribution from point-defect migration (diffusion creep). Thus the Newtonian viscosity frequently presumed is unlikely to be characteristic of flow in the mantle.

SUMMARY AND CONCLUSIONS

With high-voltage transmission electron microscopy, comparison has been made of the substructures of peridotites deformed naturally in the upper mantle and similar rocks deformed in the laboratory. The principal results and conclusions are:

1. Olivine and orthorhombic pyroxene deformed in the upper mantle flow by the same mechanisms as they do in very high-temperature experimental deformation. Olivine plasticity is accomplished by slip of unit dislocations on {0kl} planes with **b** = [100] (pencil glide along [100]). Simultaneous dynamic recovery results in (100) kink bands and prismatic subgrains of rectangular cross sections composed of (100) walls of edge dislocations and (001) walls of screw dislocations [*Green and Radcliffe*, 1972b]. Orthorhombic pyroxene deforms primarily by slip on (100) of partial dislocations with **b** parallel to [001]. Subsidiary strain is accomplished by slip on other systems. The apparent difficulty of dislocation climb in pyroxene under these conditions may necessitate activation of the additional slip systems. Recrystallization occurs in and around crystals of both phases during deformation but is considerably more advanced in olivine.

2. The substructures in the Salt Lake and Hualalai xenoliths indicate that annealing in a hot environment after deformation promotes static recovery and produces olivine and enstatite crystals that retain major deformation features but lose the generally distributed dislocations. Consequently, the lack of static recovery in the Lunar crater specimens indicates that it is possible for rocks to be extracted from the mantle and transported to the surface in magma with little alteration. Such xenoliths, then, are almost as useful for chemical and structural study as specimens from a deep core would be.

3. There is no direct evidence from the electron microscopy for a significant contribution in over-all plastic strain from grain boundary sliding. However, in view of the presence of the CO_2-filled bubbles as intracrystalline and grain boundary precipitates in the mantle specimens, it cannot be precluded that several volume

per-cent fluid present in this manner may affect grain boundary sliding rates and hence the relation between stress and macroscopic strain rate. These observations suggest that experiments to simulate flow in the mantle should be conducted under conditions of $P_{CO_2} \simeq P_{total}$ to examine the extent of this possible contribution.

The presence of extensive dislocation substructures points to control of the over-all plastic strain in the mantle peridotites by the generation and the propagation of dislocations. This conclusion should be true even if grain boundary sliding contributes to deformation; the rate-limiting step will still probably be the dislocation processes so abundantly evident in the mantle rocks represented by the xenoliths examined here. Furthermore, the observed densities of dislocations preclude any major contribution to the over-all strain by diffusion creep (migration of point defects unrelated to dislocation climb). The crystal slip appears from our observations to be rate limited by dynamic recovery mechanisms controlled by self-diffusion. Accordingly, it is unlikely that flow in the upper mantle can be characterized by a Newtonian viscosity; the power law relation established experimentally between stress and strain rate is probably closer to the true relation.

Acknowledgments. This study has drawn heavily on the experimental studies of rock deformation conducted by David T. Griggs and his co-workers and students over many years. These previous studies of a number of minerals established that it is possible to duplicate experimentally many of the optically visible features in tectonites. It is only with this background that high-resolution electron petrography can contribute to further refinement of laboratory-nature correlations. We thank E. D. Jackson for Hawaiian xenolith specimens, C. B. Raleigh and S. H. Kirby for the experimentally deformed and Lunar crater specimens, and E. M. Moores for the Vourinos specimen. We also thank R. M. Fisher and J. S. Lally and associates at the 1-Mev microscope facility of the U.S. Steel Corporation for their interest and cooperation.

This work was supported by the National Science Foundation under grant GA-13409.

Seismic Velocity Anisotropy Calculated for Ultramafic Minerals and Aggregates

DAVID W. BAKER

Department of Geological Sciences, University of Illinois at Chicago Circle
Chicago, Illinois 60680

NEVILLE L. CARTER

Department of Earth and Space Sciences, State University of New York
Stony Brook, New York 11790

Velocity surfaces for compressional and shear wave velocities have been calculated for single crystals and aggregates of mafic and ultramafic minerals and rocks from the elastic stiffness coefficients and the Christoffel equation. The elastic stiffnesses for monomineralic aggregates are obtained by calculating the Voigt-Reuss-Hill average from single-crystal constants and their temperature and pressure derivatives, where available, and universal-stage data that give the complete orientations of individual grains. The same procedure is employed for each major mineral species in a polymineralic aggregate, and the composite average is obtained by weighting according to the volume fraction occupied by each mineral. The velocity surfaces so calculated are presented on the lower hemisphere of equal-area projections for direct comparison with pole figures. The extent to which each mineral in a polymineral aggregate controls the velocity variation in the aggregate can be assessed by comparing individual velocity surfaces with those of the aggregate. The difference in the velocities of the two shear waves in a polymineral aggregate and the plane of polarization for each shear wave are shown on the lower hemisphere.

The important discovery by *Raitt* [1963] and *Shor and Pollard* [1964] that the velocity of seismic compressional waves varies with azimuth in the northeast Pacific has stimulated much discussion concerning the origin of the anisotropy and its significance. More recent studies have been made of the Flora area (off northern California) and the Quartet area (off Baja, California) by *Raitt et al.* [1969], where the anisotropy was found to be about 0.3 km/sec, the maximum velocity being in a nearly east-west direction. *Morris et al.* [1969] also found the maximum velocity to be in an east-west direction across the Hawaiian arch, and the anisotropy was found to be 0.6 km/sec. *Keen and Barrett* [1971] observed an anisotropy of 0.7 km/sec in the mantle off the coast of British Columbia, the maximum velocity also being in roughly an east-west direction. Since this pronounced anisotropy was first discovered, several papers, beginning with *Hess* [1964], have appeared that

treat the most probable explanation that the anisotropy results from preferred crystal orientations in rocks composing the upper mantle. Recent research to evaluate this hypothesis has taken two directions: (1) the velocity of the ultrasonic waves has been measured in specimens cut in several orientations from naturally deformed material [*Christensen,* 1966, 1971; *Christensen and Ramananantoandro,* 1971; *Kasahara et al.,* 1968; *Kumazawa et al.,* 1971] and (2) the velocity variation has been calculated (or inferred) by use of the elastic constants of single crystals and the preferred orientations of the crystals [*Kumazawa,* 1964; *Christensen and Crosson,* 1968; *Crosson and Christensen,* 1969; *Francis,* 1969; *Avé Lallemant and Carter,* 1970; *Crosson and Lin,* 1971; *Kumazawa et al.,* 1971].

In this paper we extend the second method to calculate velocity surfaces for both compressional and shear waves for monomineralic and polymineralic aggregates composed, in various proportions, of olivine, orthopyroxene, clinopyroxene, garnet, and amphibole. The velocity anisotropy of these crystals and the effect on the anisotropy of aggregates will be considered in detail and applied to Twin Sisters dunite (calculated as monomineralic) and a Galician crustal garnet clinopyroxenite. The seismic refraction studies sample only the top few kilometers of the upper mantle, and we assume that the physical conditions are about 225°C and 3 kb; these considerations affect only the olivine and the orthopyroxene for which temperature and pressure derivatives of the elastic constants are known. In the following paper [Carter et al., this volume] we apply the results of this study and others to questions concerning the mechanisms of flow and the constitution of the upper mantle.

ELASTIC PROPERTIES OF AN AGGREGATE

The elastic stiffness coefficients of an individual crystal in an aggregate with respect to the specimen coordinate frame c'_{ijkl} are given in terms of the stiffness coefficients with respect to the crystal axes c_{ijkl} and the transformation matrix (a_{ij}), which relates specimen axes and crystal axes [Nye, 1957]:

$$c'_{ijkl} = a_{im}a_{jn}a_{ko}a_{lp}c_{mnop} \qquad (1)$$

(The Einstein summation convention is used throughout this paper.) The transformation matrix for the crystal can be determined by universal-stage measurements of the orientation of the optical indicatrix and crystallographic axes with respect to the specimen coordinate axes as described, for example, by Crosson and Lin [1971]. By averaging the elastic stiffness coefficients c'_{ijkl} for a population of crystals with different orientations, one obtains the Voigt average $\bar{c}'_{ijkl}{}^m$ for the elastic stiffnesses of that mineral species. If the aggregate contains M different minerals, the Voigt average for each species can be calculated separately and the averages combined by weighting the volume fraction v_m occupied by each mineral (obtained from modal analyses):

$$\bar{c}'_{ijkl}{}^M = \sum_{m=1}^{M} v_m c'_{ijkl}{}^m \qquad (2)$$

This Voigt average assumes that each crystal in the aggregate is subject to exactly the same strain, that the grains are not interconnected, and thus that discontinuities in stress arise at grain boundaries. These conditions clearly are not fulfilled in a real aggregate, but the Voigt average is nevertheless useful, because it provides an upper bound to the stiffness coefficients of the aggregate. The Reuss scheme of averaging makes the equally unlikely assumption that all crystals in the aggregate are subject to the same stress and amounts to averaging the elastic compliances s'_{ijkl}. By converting the $\bar{s}'_{ijkl}$ to matrix notation $\bar{s}'_{ij}$ and inverting the matrix $(\bar{s}'_{ij})$ [Nye, 1957, p. 156], one obtains the Reuss average for the elastic stiffness coefficients, which provides a lower limit to the average stiffness coefficients of the aggregate. Hill [1952], on the basis of energy density considerations, showed that the actual stiffness coefficients of the aggregate should lie between those provided by the Voigt and Reuss schemes and suggested the use of the arithmetic mean:

$$(\bar{c}'_{ij}{}^M)_{\mathrm{VRH}} = [(\bar{c}'_{ij}{}^M)_V + (\bar{s}'_{ij}{}^M)_R{}^{-1}]/2 \qquad (3)$$

We have elected to use the Voigt-Reuss-Hill (VRH) average for determining the elastic properties of aggregates in this paper and the companion paper [Carter et al., this volume]. Several studies of metal aggregates and alloys [e.g., Bunge and Roberts, 1969] have shown that the Hill average compares well with the values measured experimentally. Crosson and Lin [1971] compared VRH values for two dunite specimens with values determined from compressional wave velocity measurements and found the experimentally determined values to be slightly closer to the Voigt (upper) values than to the Reuss values. The more refined and elaborate theoretical approaches, such as those based on inclusion theory [Kröner, 1958; Eshelby, 1961; Morris, 1969a, 1970] and the approach based on variational techniques [Hashin and Shtrikman, 1962], require far greater computational effort but lead to better approximations of the elastic properties of aggregates. However, knowledge of the elastic constants for silicate single crystals, especially as functions of composition and also of temperature, pressure, and imperfection state, does not warrant use of the more exact methods at this time, and so we have used the simpler VRH approximations exclusively.

VELOCITY OF PLANE WAVES IN ANISOTROPIC MEDIA

The phase velocity of infinite plane waves traveling in an anisotropic medium can be obtained by solving the Christoffel equation [*McSkimin*, 1964, p. 325; *Musgrave*, 1970, p. 84]:

$$\det |c_{ijkl}l_j l_l - \rho V^2 \delta_{ik}| = 0 \quad (4)$$

In this equation det indicates the determinant with respect to subscripts i and k, V is the phase velocity, ρ is the aggregate density, δ_{ik} is the Kronecker delta, and the l_j are the direction cosines of the wave normal. For any given wave normal, (4) represents a cubic equation in ρV^2. For the ultramafic materials considered here and in the companion paper there are always three real and unequal roots to this equation that correspond to one quasi-longitudinal and two quasi-transverse waves. Although the displacement vectors for these waves are orthogonal, they are, in general, neither parallel nor perpendicular to the wave normal. An algebraic solution for V from (4) is cumbersome; numerical solutions, however, were easily obtained with a Fortran 4 program and a high-speed computer that solved (4) in a stepwise fashion.

The displacement vector $p_k{}^M$ was obtained in the following manner [*Musgrave*, 1970, p. 85]. Let

$$c_{ijkl}l_j l_l = \Gamma_{ik} = \alpha_i \alpha_k \qquad i \neq k$$
$$c_{ijkl}l_j l_l = A_k \qquad i = k$$

Then

$$\alpha_1 = (\Gamma_{12}\Gamma_{13}/\Gamma_{23})^{1/2}$$
$$\alpha_2 = \Gamma_{12}/\alpha_1$$
$$\alpha_3 = \Gamma_{13}/\alpha_1$$

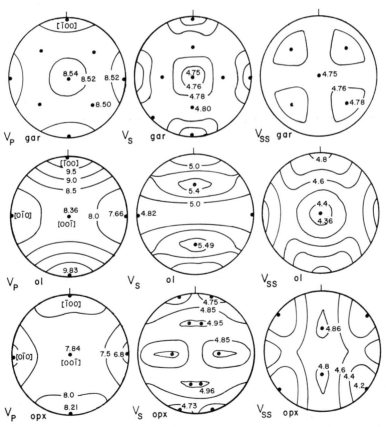

Fig. 1. Seismic velocity surfaces for garnet, olivine, and orthopyroxene single crystals for compressional (P) and shear (S and SS) waves. Equal-area projection is shown on the lower hemisphere. The units are kilometers per second.

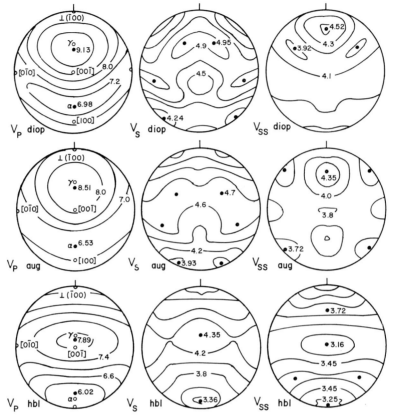

Fig. 2. Seismic velocity surfaces for diopside, augite, and hornblende for compressional (P)
and shear (S and SS) waves. Equal-area projection is shown on the lower hemisphere.

and

$$p_k{}^M = \alpha_k N / [\rho(V^M)^2 - A_k + \alpha_k{}^2] \qquad (5)$$

where $N = \alpha_k p_k$ is a normalization constant and
$M = 1, 2,$ and 3 refers to SS, P, and S waves,
respectively.

VELOCITY SURFACES FOR MINERALS

Velocity surfaces for several important minerals in mafic and ultramafic rocks are shown
in Figures 1 and 2 on the lower hemisphere of
equal-area projections. The diagrams labeled
V_p are for quasi-longitudinal waves, those
labeled V_s are for the faster quasi-transverse
waves, and those labeled V_{ss} are for the slower
shear waves. The velocity at each point on a
diagram is the velocity of a wave having its
normal given by that point. The three velocities
for each wave normal on a grid of 241 points
were obtained by solving (4); a linear interpolation between points was used.

The elastic constants used to obtain the velocity surfaces of Figures 1 and 2 are given in
Table 1. The constants for garnet are those
for almandine measured at STP by *Verma*
[1960]. Those for a natural olivine crystal
(Fo_{93}) were calculated at 225°C and 3 kb from
the elastic constants and temperature and pressure derivatives given by *Kumazawa and Anderson* [1969]. The elastic constants for orthopyroxene (bronzite), also evaluated at 225°C
and 3 kb, were calculated from the unpublished
constants and their derivatives obtained by
Frisillo and Barsch [1971] and A. L. Frisillo
(personal communication, 1971).

The only elastic constants for clinopyroxenes
currently available are those for augite and
diopside measured at STP by *Aleksandrov et
al.* [1964]. Their constants $c_{15}, c_{25}, c_{35},$ and c_{46}
for diopside are given as negative but should
be positive, as listed in Table 1, because the
crystal was misoriented by 180° about [001]

TABLE 1.   Elastic Stiffness Coefficients c_{ij} for Single Crystals*

Conditions	Garnet	Garnet (isotropic)	Olivine	Ortho-pyroxene	Diopside	Augite	Hornblende
Temperature, °C	20	20	225	225	25	25	25
Pressure, bars	1	1	3000	3000	1	1	1
Density, g/cm³	4.183	4.183	3.3027	3.3475	3.31	3.32	3.12
c_{ij}, mb							
11	3.048	3.033	3.1926	2.249	2.04	1.816	1.160
22	3.048	3.033	1.9378	1.566	1.75	1.507	1.597
33	3.048	3.033	2.3132	2.051	2.38	2.178	1.916
44	0.944	0.951	0.6271	0.7987	0.675	0.697	0.574
55	0.944	0.951	0.7654	0.7360	0.588	0.511	0.318
66	0.944	0.951	0.7659	0.7558	0.705	0.558	0.368
12	1.123	1.131	0.6572	0.6880	0.844	0.734	0.449
13	1.123	1.131	0.7106	0.5080	0.883	0.724	0.614
23	1.123	1.131	0.7571	0.4086	0.482	0.339	0.655
15					0.193	0.199	0.043
25					0.196	0.166	−0.025
35					0.336	0.246	0.100
46					0.113	0.043	−0.062

* See text for the sources of the data.

(M. Kumazawa, personal communication, 1971). Chemical analyses of the augite and the diopside were not given by *Aleksandrov et al.* [1964], although the diopside was listed as baikalite, a variety of salite. The elastic constants for hornblende are those for crystal 1 measured at STP by *Aleksandrov and Ryzhova* [1961].

The velocity surfaces projected in Figures 1 and 2 have the same point group symmetry as the crystals: garnet, $m3m$; olivine and bronzite, mmm; and diopside, augite, and hornblende, $2/m$. For the orthorhombic minerals the maximums and minimums in the P-wave velocity coincide exactly with principal axes of the optical indicatrix. For the monoclinic crystals the maximums are subparallel to the slow ray γ, whereas the minimums are parallel or subparallel to the fast ray α. Thus a qualitative estimate of the type and magnitude of the velocity anisotropy can be made simply by determining preferred orientations of the principal axes of the optical indicatrices of crystals in the aggregate. However, the velocity surfaces of shear waves in these crystals are more complicated, and thus estimates of the shear velocities are much less certain. The P-wave velocity anisotropies of these crystals, in descending order, are (1) olivine, 2.17 km/sec at 225°C and 3 kb; (2) diopside, 2.15 km/sec at STP; (3) augite, 1.98 km/sec at STP; (4) hornblende, 1.87 km/sec at STP; and (5) bronzite,

1.41 km/sec at 225°C and 3 kb. The P-wave velocity anisotropy of garnet (0.04 km/sec) is so low that it can be neglected in the analyses that follow.

VELOCITY VARIATION IN A MONOMINERALIC AGGREGATE

A dunite specimen from the Twin Sisters area, Washington (specimen TW-3 of *Christensen* [1971] and dunite A of *Christensen and Ramananantoandro* [1971]) was used to illustrate the method as applied to monomineralic aggregates. The specimen, although containing 12% serpentine and 5% other phases (Table

TABLE 2.   Modal Analyses of Twin Sisters Dunite and Galicia Eclogite

Mineral	Twin Sisters Dunite Mode	Twin Sisters Dunite VRH	Galicia Eclogite Mode	Galicia Eclogite VRH
Olivine	83.1	100.0		
Clinopyroxene			49.1	53.6
Orthopyroxene	2.1			
Garnet			36.6	40.0
Hornblende			5.9	6.4
Quartz			6.1	
Serpentine	12.1			
Zoisite			1.2	
Sphene			1.1	
Magnetite	2.7			
Total	100.0	100.0	100.0	100.0

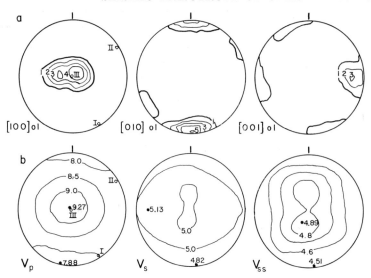

Fig. 3. (*a*) Pole figures and (*b*) calculated seismic velocities for the Twin Sisters dunite specimen. Equal-area projections are shown on the lower hemisphere. One hundred grains are represented in (*a*) with contours of multiples of the expected number for a random distribution; the counting area is 3.8%. Numerals I, II, and III indicate the orientations of the cores taken for velocity measurements.

2) was treated as a pure olivine rock. The transformation matrix (a_{ij}) was obtained for each crystal from the five-axis universal-stage angles of the inner vertical axis, the inner east-west axis, and the north-south axis (data provided by N. I. Christensen, private communication, 1971) by using an algorithm similar to that given by *Crosson and Lin* [1971]. Pole figures (lower hemisphere) for the [100], [010], and [001] crystal axes are given in Figure 3*a*. The fabric is that of a quasi-single crystal with three well-developed point maximums.

The maximum in the diagram showing calculated compressional wave velocities (V_p, Figure 3*b*) coincides with the [100] maximum, and the minimum velocity coincides with the [010] maximum, as was expected from single-crystal values (Figure 1). The velocities calculated at points 1, 2, and 3 in the diagram are 8.0, 8.1, and 9.2 km/sec, respectively, and can be compared with the values measured at 4 kb and 25°C (N. I. Christensen, personal communication, 1971) of 7.96, 7.91, and 8.60 km/sec [*Carter et al.*, this volume, Table 3]. The maximum *P*-wave velocity anisotropy ΔV_p calculated in these directions is therefore 1.2 km/sec as compared with the anisotropy of 0.69-km/sec obtained from the measurements. The maximum

anisotropy calculated for the specimen as a whole is 1.39 km/sec as compared with a maximum of 0.79 km/sec (at 4 kb and 25°C) measured by *Christensen and Ramananantoandro* [1971]. Thus the measured velocity anisotropies are lower than those calculated by about 40%, a result that can be accounted for, at least partly, by the presence of serpentine and other phases in the specimen.

VELOCITY VARIATION IN A POLYMINERALIC AGGREGATE

A specimen of garnet clinopyroxenite (GAL-143, provided by D. E. Vogel) from Cabo Ortegal in the Galicia district of northwest Spain [*Vogel*, 1967] will be used to illustrate the analysis of a specimen with several major phases (Table 2). According to *Vogel* [1967], this specimen is crustal in origin, having been transformed to eclogite from gabbro during catazonal metamorphism and then further altered during retrograde metamorphism in the hornblende-clinopyroxene-almandine subfacies. The clinopyroxene contains about 30% jadeite, and the garnet is composed of about 33% pyrope, 43% almandine, and 24% grossularite (Table 3). The rock has a very strong foliation defined by tabular or flattened (aspect ratio of about 3:1) pyroxene and amphibole crystals.

Pole figures for the various optical and crystallographic elements of 50 grains each of the clinopyroxene and amphibole are shown in Figure 4. The clinopyroxene fabric (Figure 4a and b) is very strong and dominated by a 7% maximum of [010] axes normal to the foliation s. The remaining subfabric elements are arranged in girdles with maximums, or partial girdles, parallel to the foliation. Because of the strong fabric and because virtually identical fabrics have been obtained for 10 additional specimens (200 measurements each) of eclogite from this region by J. P. Engels (personal communication, 1971), we felt that a 50-grain sample was sufficient. The fabric for the amphiboles (Figure 4c and d) is similar to the clinopyroxene fabric in that it appears to be controlled by a [010] concentration normal to s, but apparently the fabric is considerably weaker.

To calculate the velocity surfaces for this specimen, we have considered only the effects of clinopyroxene, amphibole, and garnet. The clinopyroxene was treated elastically as augite, and the amphibole as hornblende (Table 1). The orientation of the optical indicatrices and two crystal directions, such as [001] and the pole to (100) (Figure 4), was obtained from universal-stage data. The garnet is treated as almandine (Table 1), but, because the very low anisotropy ($\Delta V_p = 0.04$ km/sec) of the single crystal would be reduced still further by any spread in the orientation of crystals in the aggregate, we have used the VRH isotropic elastic constants listed in Table 1 [*Anderson and Liebermann*, 1966] as a sufficient approximation.

Separate velocity surfaces for the clinopyroxene and amphibole have been determined and are presented in Figure 5a and b. The minimum in the V_p diagram for the clinopyroxene coincides with the maximum [010] (equal to β, which equals the intermediate velocity) concentration normal to the foliation. However, it can be seen from Figure 2 that the V_p surface for augite is nearly axially symmetric, so that the lowest and intermediate velocities are very close, and both are much lower than the maximum velocity. The elongate maximum in V_p in Figure 5a can be correlated to a maximum in the girdle of γ (within a few degrees of the highest V_p, Figure 2) axes parallel to the foliation. The

TABLE 3. Electron Probe Analyses of Eclogite GAL-143*

	Pyroxene	Garnet	Amphibole
SiO_2	53.5	38.3	46.03
Al_2O_3	8.2	19.8	13.17
TiO_2	0.15	0.8	0.48
FeO	4.9	22.3	9.11
MnO		0.54	
MgO	10.5	8.5	15.68
CaO	16.5	8.5	10.11
Na_2O	4.3		2.76
Cr_2O_3			
Total	98.05	98.74	97.34

Cations Based on Oxygens†

	Pyroxene	Garnet	Amphibole
Si	1.97	6.03	6.56
AlIV	0.03		1.41
AlVI	0.33	3.68	0.82
Ti	0.004	0.01	0.05
Fe	0.15	2.95	1.09
Mn		0.07	
Mg	0.57	1.99	3.35
Ca	0.65	1.44	1.55
Na	0.30		0.77

* Averages of three analyses by Louis A. Fernandez.

† The number of oxygens for pyroxene is 6; for garnet, 24; and for amphibole, 23.

maximum V_p for the amphibole (Figure 5b) is also correlative with a maximum in the pole figure for γ (Figure 5d; about the highest V_p, Figure 2), and the elongate minimum in V_p correlates with a cleft girdle of α and [100] axes (about the slowest seismic waves, Figure 2).

The velocity surfaces for the composite aggregate, shown in Figure 5c, were calculated according to the scheme of 0.536 clinopyroxene + 0.40 garnet + 0.064 amphibole. The resulting values for $(\bar{c}'_{ijkl}{}^M)_{VRH}$ were substituted in (4) and solved for eigenvalues in the 241 directions. The density of the aggregate was calculated by weighting the density of each mineral by the corresponding volume fraction. The velocity surfaces for the aggregate are, of course, similar to those for the clinopyroxene. The amphiboles, although representing only a small volume fraction, are oriented in such a way as to add to the anisotropy of the aggregate. The effect of the garnet is to increase the overall velocity but to decrease velocity anisotropy. (The ΔV_p for the composite is 0.47 km/sec as compared with 0.79 km/sec for the clinopy-

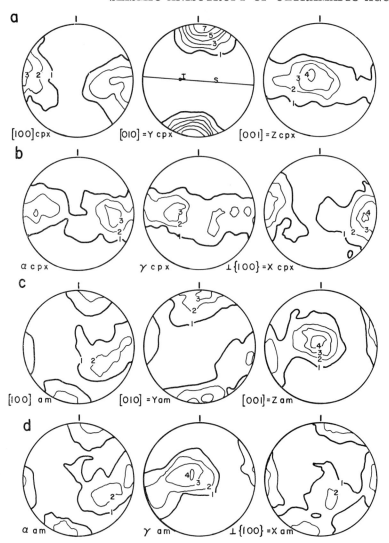

Fig. 4. Pole figures for clinopyroxene and amphibole in garnet pyroxenite (GAL-143) from Galicia in northwest Spain. Equal-area projections are shown on the lower hemisphere. The contours are in multiples of the expected number in a random distribution; there are 46 clinopyroxene grains with 7.8% counting area and 48 amphibole grains with 7.7% counting area. The plane s in (a) is the foliation plane, and the point I shows the orientation of the core taken for velocity measurements.

roxene alone.) The single core (marked by I in Figures 4a and 5c) from GAL-143 yielded a P-wave velocity at 4 kb and 25°C of 7.88 km/ sec (N. I. Christensen, personal communication, 1971), which is in excellent accord with the calculated value of 7.85 km/sec.

In Figure 5d the behavior of shear waves is considered in greater detail. The distinction between the two shear waves is twofold: there is a difference in velocity, and, because the displacement vectors for the two waves are perpendicular, the planes of polarization are also perpendicular. For the eclogite specimen analyzed here there are no directions in which $V_s = V_{ss}$. This finding is shown in the diagram on the left (Figure 5d) by the fact that the minimum of $V_s - V_{ss}$ is still positive. The other two diagrams show the plane of polarization for

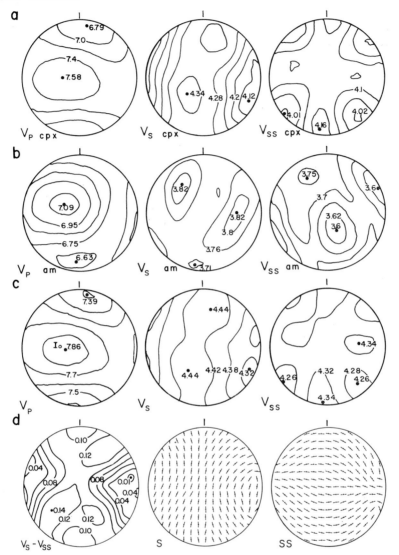

Fig. 5. Seismic velocities for compressional (*P*) and shear (*S* and *SS*) waves in garnet pyroxenite specimen (GAL-143) for clinopyroxene (*a*), amphibole (*b*), and the aggregate (*c*) calculated as 53.6% clinopyroxene, 40% garnet, and 6.4% amphibole. Equal-area projections are shown on the lower hemisphere. Shear waves in the aggregate are shown (*d*). The difference in velocities of *S* and *SS* shear waves is given ($V_s - V_{ss}$). The polarization plane of the shear wave is indicated by the great circle extending from wave normal (dot) toward the displacement vector.

each shear wave by plotting part of the great circle that extends from the wave normal shown by the dot toward the displacement vector, obtained from (5).

SUMMARY AND DISCUSSION

Universal-stage data, which give the complete orientation of individual grains in an ag-gregate, have been combined with the elastic constants for single crystals to calculate seis-mic velocity surfaces for aggregates. We have employed the Voigt-Reuss-Hill scheme for averaging the elastic constants for the aggre-gate, substituted these $(\bar{c}'_{ijkl}{}^{M})_{\mathrm{VRH}}$ values into the Christoffel equation, and solved for the eigenvalues as a function of direction in the specimen coordinate frame. The quasi-longitudinal and quasi-transverse velocity surfaces so obtained are presented as equal-area projections on the lower hemisphere for direct comparison with the pole figures for major phases of the aggregates. Thus the mineral or minerals that to the greatest extent control the anisotropy can be identified immediately.

The analytical method presented has the advantage of omitting the effects of alteration products of the mafic minerals (particularly serpentinization of olivine), which must encumber experimental measurements. Furthermore, the effects of both temperature and pressure can be taken into account insofar as their derivatives of the elastic constants are known. Suitable data for the elastic constants and their temperature and pressure derivatives are now available for olivine and orthopyroxenes, but a great deal more work is needed on clinopyroxenes and amphiboles of various chemical compositions. In the absence of such data the simple VRH averaging scheme used in this and the following paper [Carter et al., this volume]

seems justified, although it is recognized that more exact methods are currently available.

Acknowledgments. We are very pleased to dedicate this and the companion paper [*Carter et al.*, this volume] as a tribute to David Griggs. As former graduate students of Dave, with our silhouettes firmly embedded in his blackboard, we have been inspired by his brilliance, incisiveness, and devotion to his science, students, and colleagues.

Wayne Jarke wrote the program to solve (4) and, together with Andrea Krivz, did most of the data processing. We are grateful to N. I. Christensen for supplying universal-stage data and measured velocities for the Twin Sisters specimen and the velocity for GAL-143 and to A. Frisillo for the use of his unpublished elastic constants and their derivatives for orthopyroxene. We wish to acknowledge D. E. Vogel, who supplied GAL-143, and J. P. Engels, who sent us his unpublished fabric diagrams for Galician eclogites. L. A. Fernandez kindly supplied us with his unpublished microprobe data for the minerals in GAL-143. Drs. I. Borg, H. Bunge, N. Christensen, R. Crosson, M. Kumazawa, R. Lieberman, and M. Paterson made many helpful suggestions.

This research was supported by National Science Foundation grant GA-15412 and by the Research Board of the University of Illinois at Chicago Circle.

Seismic Anisotropy, Flow, and Constitution of the Upper Mantle

Neville L. Carter

Department of Earth and Space Sciences, State University of New York
Stony Brook, New York 11790

David W. Baker

Department of Geological Sciences, University of Illinois at Chicago Circle
Chicago, Illinois 60680

Richard P. George, Jr.

Department of Earth and Space Sciences, State University of New York
Stony Brook, New York 11790

Experiments on syntectonic recrystallization of compacted powders of enstatite and diopside were conducted in axial compression at a confining pressure of 15 kb, temperatures of 900°– 1300°C, and strain rates from 7.8×10^{-4} to 7.8×10^{-7} sec^{-1}. Both recrystallized enstatite and diopside aggregates have preferred orientations with [010] oriented parallel to σ_1, and the remaining principal axes and crystallographic axes measured are arranged in girdles in the $\sigma_2 \simeq \sigma_3$ plane. A similar fabric was obtained previously for experimental syntectonic recrystallization of olivine, and such fabrics have common counterparts in naturally deformed peridotites, dunites, and some garnet clinopyroxenites. Variations in seismic body wave velocities have been calculated for the experimentally recrystallized aggregates and eight naturally deformed peridotites, dunites, and garnet clinopyroxenites by using the method described in the previous paper (Baker and Carter, this volume). The calculated compressional wave velocities and anisotropies compare well with measured values when allowance is made primarily for alteration products. The average mean P-wave velocity for these and certain other peridotites and dunites is 8.3 km/sec, and the average maximum anisotropy is 0.76 km/sec, as compared with average mean velocities and maximum anisotropies for garnet clinopyroxenites of 7.8 and 0.22 km/sec, respectively. The average mean P-wave velocity in the upper mantle of the northeast Pacific is 8.1 km/sec, and the anisotropy ranges from 0.3 to 0.7 km/sec. Thus the observed anisotropies can not be produced by a dominantly garnet pyroxenite upper mantle but are easily accounted for by peridotite and/or dunite with preferred crystal orientations produced primarily by syntectonic recrystallization. If syntectonic recrystallization, accompanied by plastic deformation and recovery, is the dominant mechanism governing power law creep in the upper mantle, our results support the hypothesis that thermal convection is the driving force for the sea floor spreading process.

The constitution of the earth's upper mantle and its flow properties and processes are important questions that have been debated extensively, especially since the discovery of the sea floor spreading process. Indirect information concerning possible compositions comes from experimental and natural propagation of seismic waves and free oscillation data combined with the known mass and moment of inertia and inferences drawn from high pressure–temperature experimental petrology and also from meteoritic and cosmic elemental abundances. This information severely limits permissible upper-mantle compositions, but to date it has not been sufficiently sensitive to allow a definitive decision between a dominantly eclogitic [e.g., *Ito and Kennedy*, 1970] or a dominantly peridotitic [e.g., *Ringwood*, 1969] upper mantle. More direct methods include systematic petrological and chemical investigations of inclusions in basalts known to have emanated from the upper mantle [e.g., *Jackson and Wright*, 1970] or in diamond-bearing kimberlite pipes [e.g., *MacGregor and Carter*, 1970]. The generally great preponderance of inclusions of peridotite over those of eclogite, combined with the indirect information mentioned above, has convinced most workers that the upper mantle is dominantly peridotitic.

Equally important to determinations of the constitution of the upper mantle are determinations of the mechanical equation of state governing the high-temperature creep and of the atomic processes responsible for it. The stress-strain rate relationship generally has been assumed to be linear (for mathematical simplicity), but recent theoretical [*Weertman*, 1970] and empirical [*Carter and Avé Lallemant*, 1970; *Raleigh and Kirby*, 1970; *Post*, 1970] studies have indicated that flow over most of the upper mantle is probably governed by power law creep. This conclusion, which is of great importance to estimates of the depth dependence of viscosity, is supported by similarities in the deformational processes that have operated in materials derived from the upper mantle and those observed in the high-temperature steady-state experiments. These similarities led to the suggestion [*Avé Lallemant and Carter*, 1970] that the dominant mode of flow during steady-state creep in the upper mantle is probably syntectonic recrystallization (hot working).

An independent test of this hypothesis is now possible because of the careful seismic refraction studies of the upper mantle in the northeast Pacific by *Raitt et al.* [1969], *Morris et al.* [1969], and *Keen and Barrett* [1971]. Their results have revealed a systematic variation of compressional wave velocity with azimuth, the greatest velocity being nearly normal to the ridge axis and the velocity anisotropy being in the range 0.3–0.7 km/sec. Any hypothesis concerning the constitution of the upper mantle and/or its flow properties must also account for these most important results. Using the methods discussed in the previous paper [*Baker and Carter*, this volume] and other arguments, we show that a dominantly peridotitic upper mantle alone, having crystals with preferred orientations induced primarily by syntectonic recrystallization, is most consistent with these as well as other observations.

FLOW PROPERTIES AND PROCESSES IN THE UPPER MANTLE

There can no longer be any serious question concerning the importance of high-temperature creep in at least the upper mantle in being primarily responsible for the distributions of continents with time, the origin of their first-order structures, and the evolution and destruction of island arcs and ocean basins. Some important remaining questions pertain to the driving mechanisms and their vertical extent and to the flow laws governing the creep and the atomic processes responsible for it. Most theoretical attempts to model motions in the mantle, when various initial boundary conditions are assumed (for a review, see *Knopoff* [1969]), have also assumed a Newtonian viscous behavior of the material, mainly for mathematical convenience. Such an approach has been justified frequently by citing certain data from creep studies on some very fine-grained metals and ceramics near melting [*Garafalo*, 1965; *Weertman*, 1968; *Sherby and Burke*, 1967; *McKenzie*, 1968] for which the atomic deformational process is supposed to be stress-induced vacancy migration (diffusion or Nabarro-Herring creep). Assuming that Nabarro-Herring creep dominates in the mantle, *Gordon* [1965] found the increase in viscosity with depth below the low velocity zone to be so rapid as to preclude the possibility of mantle-wide convection. Other arguments against mantle-wide convection are based mainly on possible chemical stratification [e.g., *Anderson et al.*, 1971] and phase changes [*Knopoff*, 1967], although these arguments are by no means incontrovertible [*Wang*, 1970; *Ringwood*, 1969; *Knopoff*, 1969].

More recently, *Weertman* [1970], *Raleigh and Kirby* [1970], *Carter and Avé Lallemant* [1970], *Green* [1970], and *Green and Radcliffe* [1972b] have pointed out that Nabarro-Herring creep must be of limited extent and that the law governing flow over most of the upper mantle is probably nonlinear. The steady-state creep data available for dunite and lherzolite [*Carter and Avé Lallemant*, 1970; *Raleigh and Kirby*, 1970; *Post*, 1970] are best fit by a power creep equation of the form

$$\dot{\epsilon} = A \exp\left(-Q/RT\right)\sigma^n \qquad (1)$$

where, in the absence of externally released H_2O, a material constant A is near 10^{10} when σ

is expressed in kilobars, the creep activation energy Q is near 110 kcal/mole, and the stress exponent n is near 5. The values of these constants are generally somewhat lower for dunite and lherzolite deformed in the presence of H_2O, which is released during the alteration of the talc confining medium [Carter and Avé Lallemant, 1970; Post, 1970].

There is a considerable theoretical [e.g., Weertman, 1968] and empirical [e.g., Sherby and Burke, 1967] basis for a flow law of the form of (1) in the high-temperature creep regime, and the constants determined for the dry dunite and lherzolite are in reasonable accord with those found in extensive studies on high-temperature creep of metals and metal compounds. In such experiments the activation energy for creep is near that for the self-diffusion of the least mobile atomic species, and generally it can be shown that diffusion is rate controlling, although a similar result can be obtained if the creep rate is controlled by the glide of jog-dragging screw dislocations [Weertman, 1968, 1970]. An equation of the form of (1) is also best fit by high-temperature steady-state creep data for ice [Kamb, 1964], halite single crystals [Carter and Heard, 1970], halite aggregates [Heard, this volume], and marble [Heard and Raleigh, 1972]. (Goetze [1971] obtained a different result in his wet creep experiments on Westerly granite, but, because of the low strain amplitudes used (10^{-3}–10^{-5}), it is unlikely that steady-state flow was achieved.)

In Figure 1a we have estimated the variation with depth of stress and viscosity (solid lines) for dry dunite deformed at a constant rate of 10^{-14} sec^{-1}, using (1) and the constants determined by Carter and Avé Lallemant [1970]. We have also used the empirical relation employed by Weertman [1970; see also Sherby and Simnad, 1961].

$$\dot{\epsilon} \simeq f(\sigma)D = f(\sigma)D_0 \exp(-aTm/T) \quad (2)$$

where D is the diffusion coefficient and $f(\sigma)$ is a constant that depends primarily on stress. This equation assumes, with some experimental justification [Weertman, 1970], that the pressure effect on D is given approximately by its effect on the melting temperature Tm or on the ratio of Tm/T in (2). The constant $a = Q/RT$ at melting is 28.7, as was determined by using our activation energy (120 kcal/mole) and the

melting temperature of forsterite at 15 kb (about 2100°K [Davis and England, 1964]). The coefficient of oxygen diffusion in olivine (expected to be rate limiting) is not known, but, for most materials near melting, $D \simeq 10^{-8}$ cm^2 sec^{-1} [Shewmon, 1963]. With this value and our constant a, $D_0 = 2.9 \times 10^4$ cm^2/sec.

The stress estimate $(\sigma_1 - \sigma_3)$ in Figure 1a, calculated at a representative geological strain rate of 10^{-14} sec^{-1}, is based on the equation

$$\sigma^{4.8} = \frac{\dot{\epsilon}}{A \exp(-28.7Tm/T)}$$

$$= \frac{8.38 \times 10^{-25}}{\exp(-28.7Tm/T)} \quad (3)$$

where Tm/T (Figure 1a) is taken as the ratio of the melting temperature of forsterite with depth [Davis and England, 1964] to the ambient temperature as given by the oceanic geotherm of Ringwood [1969]. Over the depth interval of 100–400 km the shearing stresses range from 7 to 16 bars. The effective viscosity ($\eta = \sigma/3\dot{\epsilon}$; [Griggs, 1939]) in the same interval ranges from 5×10^{20} to 5×10^{21} poises and increases only slowly with depth below 200 km. Both the stress and the viscosity estimate are in excellent accord with estimates based on other geophysical considerations [e.g., Crittenden, 1967; McConnell, 1968].

For comparison we have calculated the effective viscosity from a Nabarro-Herring relationship of the form

$$\eta = \frac{\sigma}{\dot{\epsilon}} = \frac{L^2 kT}{5V_a D_0 \exp(-28.7Tm/T)}$$

$$= \frac{95L^2 T}{\exp(-28.7Tm/T)} \quad (4)$$

where L is the grain diameter, V_a is the atomic volume (about 10^{-23} cm^3), $D_0 = 2.9 \times 10^4$ cm^2/sec, and k and T have their usual meaning. The viscosities shown (dashed lines in Figure 1a) were calculated for average grain diameters of 0.5 and 5 cm, which encompass the range in grain size observed in peridotites and dunites derived from the upper mantle [Raleigh and Kirby, 1970; Avé Lallemant and Carter, 1970]. In the depth interval 100–400 km the viscosities range from 7×10^{19} to 3×10^{23} poises, depending critically on grain size and increasing rapidly at depths below 200 km for both grain sizes.

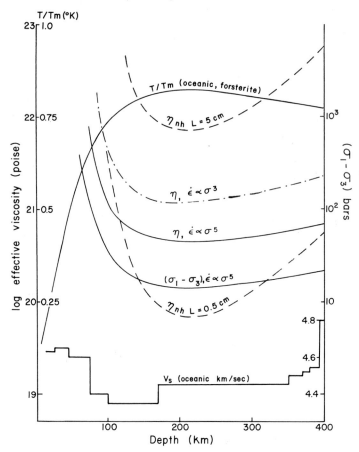

Fig. 1a. Stresses and effective viscosities of dunite in the upper mantle under the oceans, calculated at a strain rate of 10^{-14} sec^{-1}. The ratio of T/Tm in degrees Kelvin is calculated from the oceanic geotherm [*Ringwood*, 1969] and the melting of forsterite [*Davis and England*, 1964]. Shear wave velocity data showing the low-velocity channel are from *Anderson* [1967]. The solid lines are calculated from (3), the dashed-dotted line from (6), and the dashed lines from (4).

Steady-state creep cannot be achieved by the Nabarro-Herring mechanism, nor can large creep strains, because, as the diffusion path becomes longer, the strain rate decreases and viscosity increases [*Weertman*, 1970; *Green*, 1970]. The elongate grains must ultimately break down to subgrains, which may continue to creep by the vacancy diffusion mechanism provided that the subgrain boundaries are good sources and sinks for vacancies and that the subgrain size is independent of stress. However, for metals and alloys [*Sherby and Burke*, 1967; *Weertman*, 1968] the subgrain size is a function of stress, the dependence being given by

$$L = L_0(\mu/\sigma) \qquad (5)$$

where L_0 is a constant and μ is the shear modulus. *Raleigh and Kirby* [1970] also found that the subgrain size developed during the experimentally produced polygonization of olivine in lherzolite was dependent on stress, the constant being $L_0\mu = 2.4 \times 10^7$ dynes cm^{-1} (incorrectly printed as 2.4×10^5 in their paper). Substituting (5) into (4) and using $L_0\mu$ gives the creep equation

$$\sigma^3 = \frac{\dot{\epsilon}kT(L_0\mu)^2}{5V_aD_0 \exp\left(-28.7Tm/T\right)}$$

$$= \frac{5.47 \times 10^2 T}{\exp\left(-28.7Tm/T\right)} \qquad (6)$$

where now the creep rate is proportional to σ^3

and independent of grain size. The effective viscosity calculated for subgrain creep ranges from 1.2×10^{21} to 4.3×10^{21} poises in the depth interval 100–400 km.

Thus all three creep equations give reasonable effective viscosities near 10^{21} poises in the low-velocity zone, if we assume an intermediate grain size in the range 1–3 cm for the Nabarro-Herring mechanism. The values of the viscosity calculated depend rather critically, of course, on the temperature gradient chosen (through the Tm/T terms in (3), (4), and (6)). For such independent processes the process that gives the fastest creep rate at constant stress and T/Tm should be the dominant one [*Sherby and Burke*, 1967; *Weertman*, 1970]. In Figure 1b we plot the creep rate versus stress at $T/Tm = 0.75$; for intermediate grain size (≥ 1 cm), σ^5 creep should dominate at shearing stresses of ≥ 10 bars and at strain rates between 10^{-15} and $\geq 10^{-14}$ sec^{-1}.

The relationships at higher T/Tm values can be visualized by shifting the abscissa in Figure 1b to the right; thus for $T/Tm = 0.825$ (maximum in Figure 1a) the position of $\dot{\epsilon} = 10^{-14}$ sec^{-1} would be approximately that shown by the vertical bar. Under these conditions it is not certain which of the processes would dominate. At still higher T/Tm values, Nabarro-Herring creep should dominate at all reasonable strain rates, stresses, and grain sizes, but, as was mentioned above, the crystals must ultimately break down to give subgrain or power law creep. Thus it would appear that power law creep should govern flow over most upper-mantle conditions, but for confirmation we must compare the atomic processes in the steady-state experiments and naturally deformed upper-mantle materials available.

Steady-state creep in the experiments is accomplished by plastic deformation (dislocation

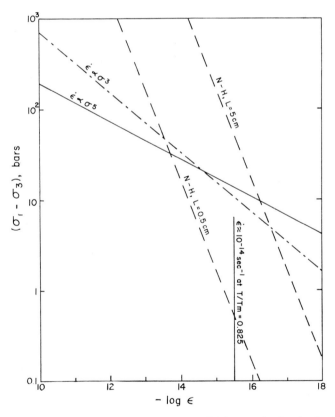

Fig. 1b.   Calculations of creep rate versus stress at $T/Tm = 0.75$ for dunite, based on (3), (4), and (6). The process giving the fastest creep rate at constant stress and T/Tm should dominate. The vertical bar indicates the shift of the abscissa to the right at the higher T/Tm of 0.825.

glide), polygonization (dislocation climb), and recrystallization, any process of which can dominate, depending on the physical conditions. In general, all three processes are observed, and in most instances there is evidence that diffusion is important and probably rate controlling. The plastic-deformation mechanisms in olivine change with increasing temperature and pressure and decreasing strain rate (Figure 2) from T (slip plane) = {110} and t (slip direction) = [001] through {0kl}[100] to (010)[100] [*Raleigh*, 1968; *Carter and Avé Lallemant*, 1970]. Those systems were determined by optical microscopy but have since been confirmed by transmission electron microscopy [*Phakey et al.*, this volume; *Green and Radcliffe*, 1972b]. The pencil glide system {0kl}[100] is most commonly observed in naturally deformed olivine [*Raleigh and Kirby*, 1970] and apparently occurs by the conservative motion of rectangular loops of unit dislocations [*Green and Radcliffe*, 1972b]. There is some evidence of slip on the high-temperature system (010) [100], but diffusion-controlled processes probably dominate at moderate depths in the upper

mantle because of the rapid increase in temperature along the geotherm (Figure 2).

At the higher temperatures and lower strain rates and in natural deformations, plastic flow of orthopyroxenes occurs by kinking and slip on the system {100}[001] [*Raleigh et al.*, 1971]. Under similar conditions, clinopyroxenes deform primarily by twin gliding and translation gliding on {100}[001] [*Raleigh and Talbot*, 1967], although slip may also occur on other systems (S. H. Kirby, personal communication, 1972).

At temperatures above about 1000°C at a strain rate of 10^{-3} sec^{-1} in the experiments the diffusion-controlled process of polygonization becomes increasingly important (Figure 2) [*Raleigh and Kirby*, 1970; *Carter and Avé Lallemant*, 1970]. Excellent examples of the polygonization process, whereby subgrains form by dislocation climb and by cross slip during recovery, have been described for both experimentally and naturally deformed olivine by *Raleigh and Kirby* [1970]. The subgrains are rectangular with their sides parallel to (100) and (001). According to *Green and Radcliffe*

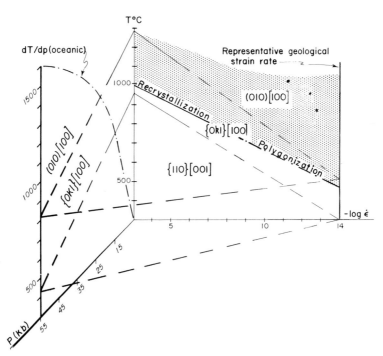

Fig. 2. Deformational processes in olivine as functions of temperature, pressure, and strain rate. The data are from *Carter and Avé Lallemant* [1970] and *Avé Lallemant and Carter* [1970]. The geotherm is from *Clark and Ringwood* [1964].

[1972b], who studied the same specimens by electron petrography, edge components of the {0kl}[100] dislocation loops climb into walls parallel to (100), and the screw components cross-slip into walls parallel to (001). Polygonization is also observed in the pyroxenes but is much less common and requires higher temperatures and/or lower strain rates.

At temperatures and strain rates near those required for polygonization, syntectonic recrystallization (hot working) also becomes important and in many instances clearly dominates. This diffusion-controlled process, whereby new, relatively strain-free grains grow at the expense of old, strained ones, leads ultimately to strong preferred crystal orientations that are controlled by the orientations of the principal stress axes. For coarse-grained starting material the new grains are first observed at grain boundaries, and at higher temperatures (or lower strain rates) new grains appear within the host crystal with orientations related to the host [Avé Lallemant and Carter, 1970]. Finally, grain boundary recrystallization advances until the host and the new grains oriented unfavorably with respect to the stress are consumed. For fine-grained starting material only the grains oriented favorably (statistically) with respect to the stress grow at the expense of the others. The final preferred orientations produced by using both fine-grained and coarse-grained aggregates are similar to each other and to fabrics commonly observed in olivine of peridotites from the upper mantle (with allowance for the fact that σ_2 need not equal σ_3 in natural deformations, as is required in the axially symmetric tests), a topic discussed more fully later.

Similarities of orientations of olivine crystals induced experimentally by syntectonic recrystallization and those observed in upper-mantle peridotites led Avé Lallemant and Carter [1970] to suggest that this process might dominate in the upper mantle, a hypothesis that we still favor and one that is tested in this paper. We emphasize, however, that syntectonic recrystallization is generally accompanied by plastic deformation and polygonization [Raleigh and Kirby, 1970; Green and Radcliffe, 1972b; Nicolas et al., 1972], any process of which might dominate under certain physical conditions. The similarity in these flow processes observed in

the steady-state experiments and those observed in peridotites presumably derived from the upper mantle, as well as other considerations outlined above, lead us to suspect that power law creep prevails over most of the upper mantle (or at least over those portions available to us). It should be pointed out, however, that processes such as grain boundary sliding (ultimately controlled by internal deformation) and Nabarro-Herring creep, if they have operated, would be very difficult to detect in naturally deformed peridotites and dunites.

EXPERIMENTAL SYNTECTONIC
RECRYSTALLIZATION

Syntectonic-recrystallization experiments were conducted on compacted pellets of powdered ($<37 \mu$) Mount Addie dunite and ground enstatite and diopside single crystals in the manner described by Avé Lallemant and Carter [1970]. All experiments were done in Griggs's solid-pressure-medium apparatus [Griggs, 1967; Green et al., 1970] at constant strain rate, and the physical conditions and extent of grain growth are indicated in Figure 3. The specimens were deformed in compression and shortened by 10–20% after compaction. The results for olivine (Figure 3a) are from Avé Lallemant and Carter [1970], and those for diopside and enstatite (Figure 3b) are from the present study. Under comparable conditions, olivine recrystallizes and grows more readily than the pyroxenes. This difference is more pronounced for the deformation of coarse-grained aggregates of dunite and lherzolite in which the olivine can recrystallize totally, whereas the pyroxenes can only slip and polygonize. In all experimentally recrystallized specimens a foliation develops normal to σ_1 and is defined by the shapes of the new grains that are shortest parallel to σ_1.

In Figure 4a we present an inverse pole figure for syntectonically recrystallized olivine powder, a pattern that is typical of that observed for olivine over the entire range of pressure-temperature-strain rate examined [Avé Lallemant and Carter, 1970]. Pole figures for crystallographic axes of grains in this specimen (N-154) are given in Figure 5a. The dominant features are statistical alignment of [010] parallel to σ_1 (the axis of maximum principal com-

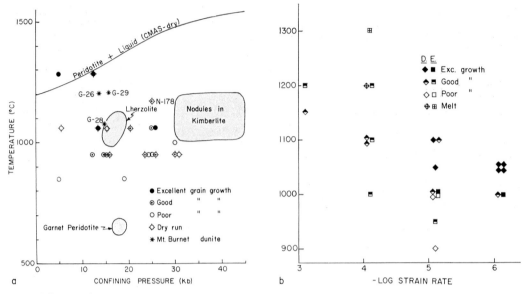

a

b

Fig. 3. Physical conditions of experiments on syntectonic recrystallization and the extent of grain growth for compacted powders of (a) olivine, deformed at a strain rate of about 8×10^{-6} sec^{-1} [Avé Lallemant and Carter, 1970], and (b) pyroxenes, enstatite E and diopside D, deformed at 15-kb pressure. Fields of lherzolite, garnet peridotite, and nodules in kimberlite and peridotite solidus in (a) are from O'Hara [1957].

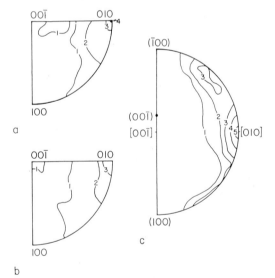

Fig. 4. Inverse pole figures for experimentally recrystallized powders: (a) olivine, N-154, (b) enstatite, N-352, and (c) diopside, N-367, showing variation in concentration of σ_1 axes in relation to crystal axes. The contours are in multiples of the expected number for a random distribution. The aggregates were deformed at 1000°C, 15 kb, and 7.8×10^{-7} sec^{-1}. Equal-area projections are shown in the lower hemisphere.

pressive stress; north-south in Figures 5 and 6) with [001] and [100] girdles in the $\sigma_2 = \sigma_3$ plane (Figure 5a). Underpopulated areas in the girdles of Figures 5 and 6, at about 45° from the periphery, are ascribed to sampling errors inherent in universal-stage measurements of principal axes of optical indicatrices of crystals in fine-grained aggregates. The fabrics shown in Figures 4a and 5a were observed both for powders and for recrystallized coarse-grained starting material [Avé Lallemant and Carter, 1970].

Fewer experiments have been done on pyroxenes, and even fewer have been analyzed for fabric; the results should therefore be regarded as being preliminary. The results for a specimen of powdered enstatite compressed at 1000°C at a strain rate of 8×10^{-7} sec^{-1} are presented in Figures 4b and 5c; another specimen deformed under similar conditions showed the same fabric. As was observed for olivine, [010] axes of recrystallized enstatite grains are oriented statistically parallel to σ_1 with poorly defined partial girdles of [100] and [001] axes in the $\sigma_2 = \sigma_3$ plane. The [010] axes of both olivine and bronzite are most compliant elastically [Baker and Carter, this volume, Figure 1],

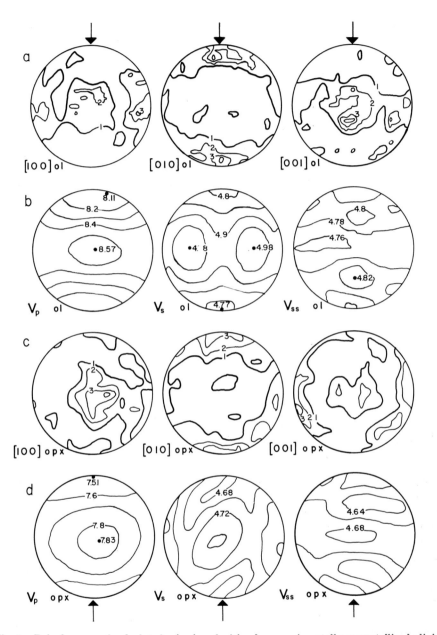

Fig. 5. Pole figures and calculated seismic velocities for experimentally recrystallized olivine and enstatite; equal-area projections are shown in the lower hemisphere. Axis σ_1 is oriented north-south. (*a*) Pole figures for 338 olivine grains in specimen N-154 deformed at 1000°C, 13.4 kb, and 7.8 × 10⁻⁷ sec⁻¹ [*Avé Lallemant and Carter*, 1970]. Contours are in multiples of the expected number for a random distribution; the counting area is 1.2%. (*b*) Seismic velocities calculated for specimen N-154. (*c*) Pole figures for 200 enstatite grains in specimen N-352 deformed at 1000°C, 15 kb, and 7.8 × 10⁻⁷ sec⁻¹. The contours are in multiples of the expected number for a random distribution; the counting area is 2%. (*d*) Seismic velocities calculated for specimen N-352.

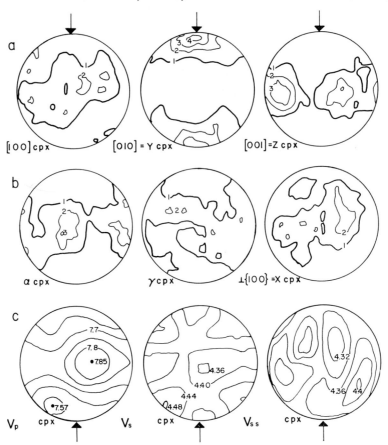

Fig. 6. (a and b) Pole figures and (c) calculated seismic velocities for 76 diopside grains in specimen N-367 deformed at 1000°C, 15 kb, and 7.8×10^{-7} sec^{-1}. Equal-area projections are shown in the lower hemisphere. Axis σ_1 is oriented north-south. The contours in (a) and (b) are in multiples of the expected number for a random distribution; the counting area is 4.9%.

and in the presence of an interstitial fluid they should orient parallel to σ_1 in accord with *Kamb*'s [1959a] thermodynamic theory as it is applied to ultramafic minerals by *Hartman and den Tex* [1964].

Only one recrystallized diopside specimen has been fully analyzed to date, and the results are given in Figures 4c, 6a, and 6b. Again, [010] axes of the new grains align parallel to σ_1, although a distinct concentration is observed at about 45° to σ_1 (Figure 4c). The remaining subfabric elements determined (Figure 6a, b) form girdles with maximums in the $\sigma_2 = \sigma_3$ plane. For diopside, however, [010] is not the most compliant direction [*Aleksandrov et al.*, 1964; *Kumazawa*, 1969; *Baker and Carter*, this volume, Figure 2] at STP, but its magnitude is near that of the most compliant direction

(nearly parallel to the indicatrix axis α). Temperature and pressure derivatives of elastic constants of clinopyroxenes have not yet been determined, and it is possible that [010] is the most compliant direction at the elevated temperatures of our experiments.

Variations in compressional and shear wave velocities for these experimentally recrystallized specimens (Figures 5b, d and 6c) have been calculated in the manner described in the previous paper [*Baker and Carter*, this volume]. For the olivine (Figure 5b) the compressional wave velocity anisotropy ΔV_p is 0.46 km/sec, the minimum velocity (8.11 km/sec) being near σ_1 and the maximum (8.57 km/sec) being in the $\sigma_2 = \sigma_3$ plane. The shear wave velocity anisotropies ΔV_s and ΔV_{ss} are 0.21 and 0.06 km/sec, respectively. The enstatite fabric produces a maximum ΔV_p of 0.32 km/sec (Figure 5d), maximum ΔV_s and ΔV_{ss} both being 0.04 km/sec. Experimentally recrystallized diopside yields a maximum ΔV_p of 0.28 km/sec (Figure 6c), the maximum ΔV_s and ΔV_{ss} being 0.12 and 0.08 km/sec, respectively.

Thus syntectonic recrystallization of olivines and pyroxenes in these axially symmetric tests leads to strong preferred crystal orientations that are related to the orientations of principal stress axes. These preferred orientations give rise to appreciable anisotropies in velocitites of seismic compressional waves. In the next section we compare these fabrics and velocity anisotropies with those observed for naturally deformed ultramafic aggregates, most of which may have been derived from the upper mantle.

FABRICS AND SEISMIC ANISOTROPY OF NATURAL PERIDOTITES AND GARNET CLINOPYROXENITES

Peridotite

A great deal of fabric work has been done on olivine in peridotites and dunites from many sources and localities since the early work of *Andreatta* [1934] (for recent reviews, see *den Tex* [1969], *Avé Lallemant and Carter* [1970], and *Lappin* [1971]). With the exception of some recent studies, pyroxenes in these rocks have received considerably less attention, presumably because they form relatively minor volume percentages. In particular, clinopyroxenes form only a few per cent of the most common peridotites, and hence they will be ignored in the discussions and the analyses that follow.

We have selected two lherzolites, a garnet peridotite, a chlorite peridotite, and a dunite (discussed in the previous paper) to represent the range of olivine and orthopyroxene fabrics and seismic anisotropies expected in upper-mantle aggregates. Of these fabrics, only those of the dunite and the two lherzolites are typical of fabrics observed in olivine-rich tectonites. Modal analyses of these specimens are presented in Table 1 with the recalculated modes used in the Voigt-Reuss-Hill (VRH) averages of the elastic constants [*Baker and Carter*, this volume]. All the specimens are tectonites having fabrics, on the basis of textural and structural evidence and comparisons with the experiments, that are believed to have originated by syntectonic recrystallization. There is evidence of

TABLE 1. Modal Analyses of the Specimens

Mineral	Peridotite											Garnet Pyroxenite			
	L-62		John Day		Ar-26		Ar-18(17)		OFS-7		OFS-8	
	Mode	Hill	Mode*	Hill	Mode	Hill	Mode	Hill	Mode	Hill	Mode	Hill
Olivine	60.0	73.2	35.6	93.9	61.0	81.2	60.0	82.2				
Orthopyroxene	22.0	26.8	6.1	6.1	14.0	18.8	13.0	17.8				
Clinopyroxene	10.3				5.0				49.1	64.4	47.5	54.8
Garnet					11.0				27.1	35.6	39.2	45.2
Amphibole					9.0		13.0					
Spinel	4.8		0.3				8.0					
Serpentine			57.8				6.0					
Chlorite												
Micas and alteration	2.8		0.2						23.3		13.2	
Miscellaneous									0.5			

* A more recent modal analysis of the specimen by H. G. Avé Lallemant, who has distinguished between host crystals of the serpentine, gives 76.8% olivine, 14.6% orthopyroxene, 8.2% clinopyroxene, and 0.4% spinel.

mild to moderate plastic deformation according to the system {0kl} [100] in the specimens, but this deformation has been superimposed on a pre-existing fabric.

Alpine lherzolite L-62 from the French Pyrenees. An olivine fabric commonly observed in peridotites is that shown by specimen L-62, a type spinel lherzolite from Etang de Lers in the French Pyrenees [*Avé Lallemant*, 1967]. The fabric (Figure 7a) has a [010] maximum, spreading into a girdle, normal to a foliation *s* that contains more or less well-developed [100] and [001] partial girdles. The foliation is defined by the shape of olivine grains whose shortest semi-axis is normal to the foliation plane, as is typical of many olivine tectonites. This fabric is similar to that obtained for 200 crystals from the same specimen by *Avé Lallemant* [1967, Figure 27], although the girdles are developed somewhat better in his diagrams. The fabric for the enstatite crystals (Figure 7b) is stronger than but similar to that of the olivine. The orientations of both olivine and enstatite are remarkably similar to those produced experimentally (Figure 5a, c) and are consistent with σ_1 having been oriented normal to *s* with $\sigma_2 \simeq \sigma_3$.

In Figure 7c we show the longitudinal and transverse wave velocities for this specimen by using the recalculated mode (Table 1) and the averaging scheme and the procedure discussed in the previous paper [*Baker and Carter*, this

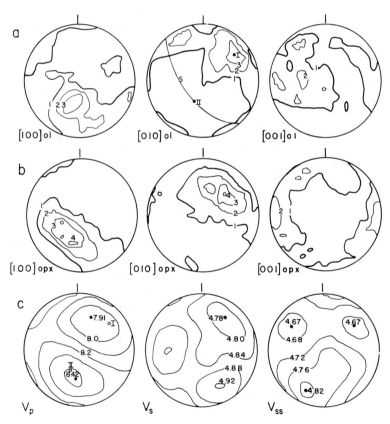

Fig. 7. Pole figures and seismic velocities for lherzolite specimen L-62 from Etang de Lers, French Pyrenees (specimen supplied by E. den Tex). Equal-area projections are shown in the lower hemisphere. (a) Pole figures for 98 olivine grains with contours in multiples of the expected number for a random distribution; the counting area is 3.9%. The great circle *s* shows the orientation of foliation, and points I and II show the orientations of specimens cored for seismic measurements (Table 3). (b) Pole figures for 49 enstatite crystals with contours in multiples of the expected number for a random distribution; the counting area is 7.5%. (c) Seismic velocities calculated from the data in (a) and (b) and the VRH mode (Table 1).

volume]. The minimum P-wave velocity, 7.91 km/sec, is subparallel to the [010] maximum, and the maximum velocity, 8.42 km/sec, is subparallel to the [100] concentration, as was expected. The maximum ΔV_p is 0.51 km/sec, and maximums ΔV_s and ΔV_{ss} are both 0.14 km/sec. The fabric and body wave velocity anisotropy are therefore somewhat stronger than those for the experimentally deformed specimens.

John Day lherzolite from the Canyon Mountain complex, Oregon. A somewhat different fabric has been determined by H. G. Avé Lallemant (personal communication, 1971) for the John Day lherzolite from the Canyon Moun-

tain complex in Oregon. The specimen is a highly serpentinized lherzolite in contact with dunite having the same olivine fabric (H. G. Avé Lallemant, personal communication, 1971). The olivine fabric (Figure 8*a*) has a strong [100] maximum parallel to the long dimensions of the olivine grains normal to which are well-developed [010] and [001] girdles with maximums; no foliation is evident in this specimen. Such mineral elongation lineations in peridotites and dunites are commonly parallel to [100] concentrations that we believe form parallel to σ_3 [*Avé Lallemant and Carter*, 1970; H. G. Avé Lallemant, personal communication, 1972], This

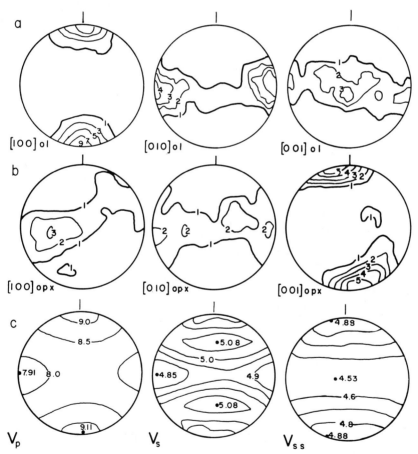

Fig. 8. Pole figures and calculated seismic velocities for the John Day lherzolite from the Canyon Mountain complex in Oregon (data provided by H. G. Avé Lallemant, private communication, 1971). Equal-area projections are shown in the lower hemisphere. (*a*) Pole figures for 100 olivine grains with contours in multiples of the expected number for a random distribution; the counting area is 3.8%. The specimen has a mineral elongation lineation parallel to the [100] concentration. (*b*) Pole figures for 41 enstatite grains with contours in multiples of the expected number for a random distribution; the counting area is 8.9%. (*c*) Seismic velocities calculated from the data in (*a*) and (*b*) and the VRH mode (Table 1).

fabric is, therefore, interpreted to be a variant of the previous one (Figure 7a), σ_3 parallel to [100] being the unique stress axis and σ_1 being approximately equal to σ_2.

The strong enstatite fabric (Figure 8b), although apparently related symmetrically to the olivine fabric, is not parallel to it, in contrast to the previous specimen. This inconsistency has also been observed for other peridotites [e.g., *Raleigh*, 1965b; *Avé Lallemant*, 1967; *Nicolas et al.*, 1971] and is not fully understood to date. The relative ease with which olivine recrystallizes could account for this discrepancy [*Helmstaedt and Anderson*, 1969; *Avé Lallemant and Carter*, 1970; *Carter*, 1971]. The rather common parallelism of $[100]_{ol}$ and

$[001]_{opx}$ could be produced, during recrystallization of the olivine, by slip parallel to [001] in the orthopyroxenes and/or bodily rotation of the orthopyroxene crystals that are inherently elongate parallel to [001].

In Figure 8c we plot the variations of compressional and shear wave velocities calculated for this specimen. Because of the strong fabric the maximum ΔV_p (1.2 km/sec) is very high, and ΔV_s (0.23 km/sec) and ΔV_{ss} (0.35 km/sec) are also fairly high. The actual velocities and anisotropy for this specimen would, of course, be much lower because of the extensive serpentinization.

Garnet peridotite Ar-26 from the Alpe Arami, Switzerland. The diagrams in Figure 9 show

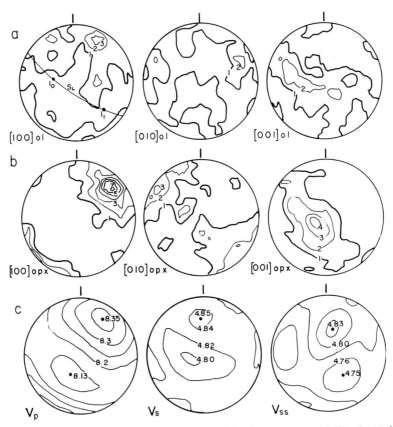

Fig. 9. Pole figures and calculated seismic velocities for garnet peridotite Ar-26 from the Alpe Arami, Switzerland (data provided by J. R. Möckel, private communication, 1971). Equal-area projections are shown in the lower hemisphere. (a) Pole figures for 200 olivine grains with contours in multiples of expected number for a random distribution; the counting area is 2%. The great circle s_L is the layering plane, l_0 is a mineral elongation lineation, and l_1 is a striation lineation. (b) Pole figures for 200 enstatite grains with contours in multiples of the expected number for a random distribution; the counting area is 2%. (c) Seismic velocities calculated from the data in (a) and (b) and the VRH mode (Table 1).

olivine (Figure 9a) and orthopyroxene (Figure 9b) fabrics in a partly altered garnet peridotite specimen, Ar-26, from the Alpe Arami, Switzerland [Möckel, 1969; data provided by J. R. Möckel, private communication, 1971]. The layering plane s_L is defined by pyroxene streaks and mineral elongations, and l_o is a mineral elongation lineation. The striation lineation l_1 may be due to the low-angle intersection of s_L and a later alpine cleavage, but the preferred crystal orientations are related to the earlier layering s_L and lineation l_o [Möckel, 1969]. The olivine fabric is rather weak, but it is nevertheless distinctly different from that found for most lherzolites and for dunites associated with garnet peridotites from Norway [Lappin, 1967]. The main features are a [100] maximum spreading into two partial girdles normal to s_L and a [001] girdle with a maximum near l_o parallel to the layering. As for lherzolite L-62 (Figure 7), the enstatite (Figure 9b) shows a somewhat similar but much stronger fabric. All experimental attempts to produce the [100]-maximum olivine fabric shown in Figures 9a and 10a have been unsuccessful to date [Avé Lallemant and Carter, 1970]. This fabric is predicted by the theory of Kamb [1959a] as it is applied to olivine by Hartman and den Tex [1964] for the case in which interstitial fluids are absent.

Body wave velocities calculated for this specimen are presented in Figure 9c. Because of the weak olivine fabric and the presence of garnet the maximum P-wave velocity anisotropy is only 0.22 km/sec, a maximum velocity of 8.35 km/sec occurring subparallel to [100] concentrations in olivines and orthopyroxenes. Maximum anisotropies for the shear waves are, of course, correspondingly low, i.e., 0.05 km/sec for ΔV_s and 0.08 km/sec for ΔV_{ss}.

Chlorite peridotite Ar-18 from the Alpe Arami, Switzerland. The results for a specimen of chlorite peridotite from the Alpe Arami, Switzerland [Möckel, 1969; data provided by J. R. Möckel, private communication, 1971] are shown in Figure 10. The mode for this specimen is not available, and we have assumed that the mode is similar to that for chlorite peridotite Ar-17 (Table 1), although Ar-17 may be more highly serpentinized than Ar-18. The presence of spinel clouds within amphibole in Ar-18 indicates the earlier presence of clinopyroxene [Möckel, 1969]. The plane s is a schistosity probably parallel to an alpine cleavage and layering plane in which orthopyroxenes and amphiboles are elongated, and s_{01} is a plane containing the long semiaxes of olivine crystals. The mineral elongation lineation l_0 is due to the alignment of amphibole and orthopyroxene, which, according to Möckel, is not parallel to olivine elongations.

The olivine fabric for Ar-18 (Figure 10a) is similar to but stronger than that for Ar-26 and has a [100] maximum normal to s_{01}, which contains [001] and [010] partial girdles. As was observed before, the enstatite fabric is much stronger and clearly related to s and l_0, but it does not appear to be related to the olivine fabric. Thus it appears that the olivine and orthopyroxene fabrics originated during two different deformational episodes, but there is no unambiguous means for determining the relative chronology. Möckel [1969] believes that, because of the preservation of s and l_0, the olivine fabric is the earlier one, but, on the basis of the experimental observations, we prefer the opposite interpretation, i.e., that the orthopyroxene fabric is the earlier one.

Body wave velocity patterns for Ar-18 are dominated by the large percentage and fairly strong fabric of the olivine, the maximum P-wave velocity being 8.49 km/sec, parallel to the [100] maximum. The maximum ΔV_p is 0.48 km/sec, whereas the maximum ΔV_s and ΔV_{ss} are both 0.11 km/sec.

Garnet clinopyroxenites and eclogites

In contrast to work on peridotites and dunites, very little fabric work has been done on crustal and mantle garnet clinopyroxenites and eclogites. To our knowledge, the only published reports of fabrics of clinopyroxenes in such rocks are those for xenoliths in kimberlite-bearing breccia pipes in Arizona and Utah [Kumazawa et al., 1971]. Partial fabrics of four specimens were presented in that study, and, although the strength of the fabric appears to be moderate, the calculated and measured anisotropies are low (discussed more fully later). According to Watson and Morton [1969], the eclogite inclusions in the kimberlites in Arizona may have come from an in situ layer in the deep crust or the upper mantle.

For our studies we have selected one crustal eclogite (GAL-143, supplied by D. E. Vogel)

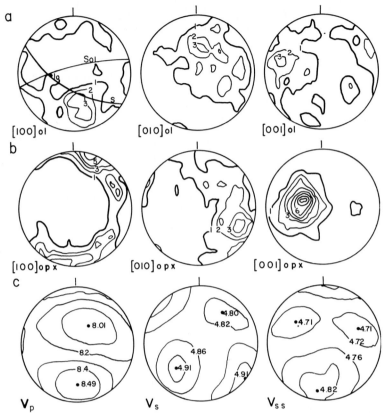

Fig. 10. Pole figures and calculated seismic velocities for chlorite peridotite specimen Ar-18 from the Alpe Arami, Switzerland (data provided by J. R. Möckel, private communication, 1971). Equal-area projections are shown in the lower hemisphere. (a) Pole figures for 200 olivine grains with contours in multiples of the expected number for a random distribution; the counting circle is 2%. The plane marked s_{ol} is an olivine shape foliation, and l_o is a mineral elongation lineation. The plane s is an alpine schistosity. (b) Pole figures for 200 enstatite grains with contours in multiples of the expected number for a random distribution; the counting area is 2%. (c) Seismic velocities calculated from the data in (a) and (b) and the VRH mode (Table 1).

and two mantle garnet clinopyroxenite inclusions (OFS-7 and OFS-8) from the Roberts Victor mine, Orange Free State, South Africa (specimens supplied by E. Maske). In the analyses that follow we have treated the clinopyroxene elastically as augite with the constants at STP of *Aleksandrov et al.* [1964]. The crustal eclogite was discussed in the previous paper [*Baker and Carter*, this voume]; this specimen was analyzed mainly for comparison with the mantle specimens. The clinopyroxene fabric in GAL-143 is very strong, having a [010] maximum normal to a strong foliation defined by layering and grain shape and containing well-developed girdles or partial girdles of the remaining subfabric elements measured [*Baker*

and Carter, this volume, Figure 4]. This fabric is similar to the experimentally recrystallized diopside specimen (Figure 6) and thus consistent with the orientation of σ_1 normal to the foliation and with $\sigma_2 \simeq \sigma_3$. Voigt-Reuss-Hill averages for the pyroxene and amphibole fabrics combined with the elastic constants for garnet, weighted proportionately, lead to a calculated maximum ΔV_p of 0.47 for this specimen, ΔV_s and ΔV_{ss} both being 0.12 km/sec.

Mantle garnet pyroxenites OFS-7 and OFS-8 are similar mineralogically and texturally, although proportions of the minerals are different and the mineral layering observed in OFS-7 is absent in OFS-8. The pyroxene and garnet crystals are both large (averaging about 3 mm), and therefore two parallel sections were cut from each rock for the required number of measurements. The anhedral to subhedral clino-pyroxenes are sodic-aluminous augites (Table 2) with an average $c \wedge \gamma = 37°$ and $2V_\gamma = 59°$. They show profuse lamellar mechanical twinning according to the system {100} [001], which is the high-temperature and low strain rate twinning mechanism in clinopyroxenes [*Raleigh and Talbot*, 1967]. The garnets are dominantly pyrope (Table 2), have kelyphitic rims, and are partially altered internally. Chlorite and phlogopite, which are generally interstitial to the garnet and pyroxene and internal alteration products in both garnet and pyroxene, form a substantial part of both specimens (Table 2) and should be the major factors contributing to errors in the calculated velocities. The over-all texture, petrology, and chemistry of these specimens is similar to that described by *MacGregor and Carter* [1970] as 'type 1 eclogites,' although these rocks have been deformed and metamorphosed.

Clinopyroxene fabrics of OFS-7 and OFS-8 are given in Figures 11 and 12, respectively. The fabric for OFS-7 is rather weak and seems to be dominated by a c-axis maximum inclined at $15°$ to the mineral layering s_L. The remaining subfabric elements form girdles or partial girdles around the c-axis maximum. An analysis of this fabric combined with garnet leads to a maximum ΔV_p of 0.17 km/sec, the maximums ΔV_s and ΔV_{ss} being 0.1 and 0.15 km/sec, respectively.

The clinopyroxene fabric for OFS-8 (Figure 12) is also fairly weak, having a maximum of c axes spreading into an east-west partial girdle. Other subfabric elements are weakly arranged in girdles and partial girdles. The maximum compressional wave velocity anisotropy calculated for this specimen is also 0.17 km/sec, ΔV_s being 0.09 km/sec and ΔV_{ss} being 0.04 km/sec. Calculated P-wave velocity anisotropies of these mantle garnet pyroxenites are, therefore, substantially lower than the anisotropy of the crustal eclogite and all except one of the peridotites analyzed in the last section.

RESULTS

As was pointed out by *Avé Lallemant and Carter* [1970], the olivine fabrics produced by syntectonic recrystallization in our axially sym-

TABLE 2.   Electron Probe Analyses of Pyroxenes and Garnets in Garnet Clinopyroxenites*

	OFS-7		OFS-8	
	Pyroxene	Garnet	Pyroxene	Garnet
SiO_2	56.3	43.1	53.6	40.3
Al_2O_3	2.62	21.9	1.81	22.0
TiO_2	0.14	0.18	0.07	0.12
FeO	3.06	9.2	3.31	11.1
MnO	<.1	0.20	0.05	0.41
MgO	17.9	21.5	16.5	18.3
CaO	19.6	3.8	21.4	4.9
Na_2O	1.48		1.29	
Cr_2O_3	<0.1	0.03		0.21
Total	101.1	99.91	98.03	97.34

Cation Proportions Based on 6 Oxygens

Si	1.99	1.53	1.99	1.52
Al	0.01		0.01	
Al	0.10	0.92	0.069	0.92
Fe	0.09	0.27	0.101	0.33
Mg	0.95	1.14	0.91	1.03
Ca	0.75	0.15	0.85	0.20
Na	0.10		0.093	
Ti	0.003	0.005	0.002	0.004
Cr		0.001		
Mn		0.005	0.003	0.013

* Averages of four analyses by Louis A. Fernandez.

metric tests have common counterparts in naturally deformed peridotites (e.g., Figure 7). For natural deformations, however, two of the principal stresses need not be equal, and, if they are not equal, maximums should develop in the girdles or a three point maximum fabric should result, as is observed for the Twin Sisters dunite [*Baker and Carter*, this volume, Figure 3]. To date, we have definite experimental evidence only that [010] in olivine orients parallel to σ_1, but other observations on the experimentally and naturally deformed specimens [*Avé Lallemant and Carter*, 1970] and preliminary extrusion experiments (H. G. Avé Lallemant, personal communication, 1972) suggest that [100] orients preferably parallel to σ_3. We therefore interpret the John Day lherzolite olivine fabric (Figure 8) as having originated by recrystallization during nearly uniform extension ($\sigma_1 \simeq \sigma_2 > \sigma_3$) parallel to the [100] concentration. *Lappin* [1971] has emphasized the few fabrics (also pointed out by *Avé Lallemant and Carter* [1970]) in which [001] axes are parallel to mineral elongation lineations and fold axes. Lappin equates lineations parallel to both [100]

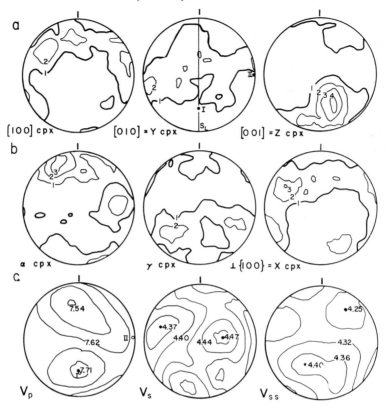

Fig. 11. Pole figures and calculated seismic velocities for garnet clinopyroxenite specimen OFS-7 from the Roberts Victor Mine, South Africa (specimen supplied by E. Maske). Equal-area projections are shown in the lower hemisphere. (*a* and *b*) Pole figures for 86 clinopyroxene grains with contours in multiples of the expected number for a random distribution; the counting area is 4.4%. The north-south plane in the center sphere of (*a*) shows the orientation of mineral layering, and points I and II show the orientations of specimens cored for seismic measurements (Table 3). (*c*) Seismic velocities calculated from the data in (*a*) and (*b*) and the VRH mode (Table 1).

and [001] to fabric *b* axes, which he assumes are parallel to σ_2. His interpretation lacks verification except for possible analogies with some open crustal folds in quartzitic and calcitic sequences [*Dieterich and Carter*, 1969]. As was mentioned above, the origin of the [100]-maximum (normal to foliation) fabrics is not yet understood, but such fabrics are not common and hence should not affect our major conclusions.

Data for fabrics produced experimentally by syntectonic recrystallization of the pyroxenes are fewer, but again similar fabrics have been observed in naturally deformed peridotites and garnet clinopyroxenites. The orthopyroxene fabrics produced are similar to the olivine fabrics (Figures 4 and 5), and this correlation has also

been observed in peridotites L-62 (Figure 7) and Ar-26 (Figure 9) as well as in other peridotites. However, there are also several peridotites for which the olivine and orthopyroxene fabrics do not coincide (e.g., Figures 8 and 10). This discrepancy probably can be accounted for generally by recrystallization of the olivine during a later (lower temperature) deformational episode, as was suggested by the experimental evidence.

The clinopyroxene fabric of crustal eclogite GAL-143 [*Baker and Carter*, this volume, Figure 4] is similar to that produced experimentally (Figure 6), and this fabric probably originated by recrystallization during compression normal to the foliation. The amphibole fabric of GAL-143 is similar and probably also originated by

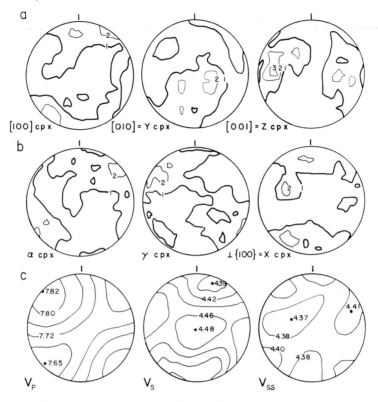

Fig. 12. Pole figures and calculated seismic velocities for garnet clinopyroxenite specimen OFS-8 from Roberts Victor Mine, South Africa (specimen supplied by E. Maske). Equal-area projections are shown in the lower hemisphere. (*a* and *b*) Pole figures for 75 clinopyroxene grains with contours in multiples of the expected number for a random distribution; the counting area is 4.8%. (*c*) Seismic velocities calculated from the data in (*a*) and (*b*) and the VRH mode (Table 1).

this mechanism during the retrograde metamorphism of the rock [*Vogel*, 1967]. In contrast, the weaker clinopyroxene fabrics in garnet clinopyroxenites OFS-7 and OFS-8 (Figures 11 and 12) may be primary fabrics that originated during the accumulation process, in agreement with the interpretation of *MacGregor and Carter* [1970].

In Table 3*a* we give a summary of the *P*-wave velocities and maximum calculated anisotropies in the eight specimens analyzed. We no longer consider the shear wave velocity anisotropies in this paper, since they are low and have not been reported in the seismic refraction studies at sea. In Table 3*b* we compare the calculated velocities and anisotropies with those measured at 25°C and 4 kb by N. I. Christensen. The values are, of course, not strictly comparable, because alteration products in the specimens will

lower measured values, and the values calculated at 225°C and 3 kb should be lower than those measured at 25°C and 4 kb by a few per cent, provided that all cracks are closed. In addition, eclogite and garnet clinopyroxenite inclusions are typically porous, and all pores may not be closed at 4 kb or even at 10 kb (N. I. Christensen, personal communication, 1971), so that the values calculated at STP may be higher than those measured. Nevertheless, the correlation is excellent for lherzolite L-62, which shows very little alteration. The measured values and anisotropy for the Twin Sisters dunite are substantially lower than those calculated on the basis of 100% olivine, as is expected because of the 12% serpentine and the 5% other phases in the specimen [*Christensen and Ramananantoandro*, 1971]. Similarly, the measured values for garnet clinopyroxenite specimen OFS-7 are lower than

TABLE 3a.   Compressional Wave Velocities, km/sec

Specimen	ρ (Calculated)	Maximum	Minimum	Mean	ΔVp (Maximum)
Peridotite Velocities Calculated at 225°C and 3 kb					
Lherzolite, L-62	3.315	8.42	7.91	8.17	0.51
Lherzolite, John Day	3.305	9.11	7.91	8.51	1.20
Garnet peridotite, Ar-26	3.311	8.35	8.13	8.24	0.22
Chlorite peridotite, Ar-18	3.311	8.49	8.01	8.25	0.48
Dunite, Twin Sisters	3.300	9.27	7.88	8.58	1.39
Garnet Clinopyroxenite Velocities Calculated at STP					
Mantle, OFS-7	3.627	7.71	7.54	7.63	0.17
Mantle, OFS-8	3.710	7.82	7.65	7.74	0.17
Crust, GAL-143	3.652	7.86	7.39	7.63	0.47

those calculated, probably because of the presence of micas and garnet and pyroxene alteration products and possible voids, but, surprisingly, the measured anisotropy is appreciably higher. The reason for this result is not yet known, but we note that a similar one was obtained at 4 kb for a specimen of garnet clinopyroxenite by *Kumazawa et al.* [1971, specimen GR-3]. The single measurement from crustal eclogite GAL-143, a very compact rock, is in excellent accord with the calculated value. Therefore, when allowances are made mainly for alteration products and also for other phases and possible voids, the correspondence between the measured and calculated values is rather good.

The data of Table 3 and those from other sources are presented graphically in Figure 13, in which we plot mean P-wave velocity against maximum (or nearly so) velocity anisotropy. We have selected specimens from the literature only for which fabric data are available and for which the measured anisotropies should be near maximum. Two generalizations can be drawn immediately from Figure 13. The first is that the mean velocity for the garnet clinopyroxenites (7.8 km/sec) is, on the average, lower by 0.5 km/sec than that for the peridotites and dunites (8.3 km/sec) even though their average measured density is higher (3.40 versus 3.29 g/cm³), a result anticipated by *Jackson and Wright* [1970]. The average mean velocity of eclogite and garnet clinopyroxenite specimens (GR-3, GR-3A, GR-23, and ME 1-17) measured by *Kumazawa et al.* [1971] is higher by 0.1 km/sec at 10 kb, and the average ΔV_p is lower by 0.07 km/sec. Mean velocities measured in the uppermost mantle in the northeast Pacific in kilometers per second are: (1) Flora, 8.14; Quartet, 7.98 [*Raitt et al.*, 1969]; (2) Hawaiian arch, 8.16 [*Morris et al.*, 1969]; and (3) British Columbia, 8.07 [*Keen and Barrett*, 1971]. These values and their average, 8.09 km/sec, are between the mean velocities observed for the peridotites and the dunites and for the garnet clinopyroxenites and hence do not provide a

TABLE 3b.   Comparison between Calculated and Measured P-Wave Velocities, km /sec

Specimen	ρ (Measured)	Core 1 Calculated	Core 1 Measured	Core 2 Calculated	Core 2 Measured	Core 3 Calculated	Core 3 Measured	ΔVp Calculated	ΔVp Measured
L-62	3.36	7.95	7.97	8.41	8.45			0.46	0.48
Twin Sisters	3.24	8.0	7.96	8.1	7.91	9.2	8.60	1.2	0.69
OFS-7	3.31	7.71	7.54	7.63	7.26			0.08	0.28
GAL-143	3.45	7.85	7.88						

The measurements were made by N. I. Christensen at 25°C and 4 kb. For core orientations see Figures 7 and 11 and *Baker and Carter* [this volume, Figures 3, 4a, and 5c.]

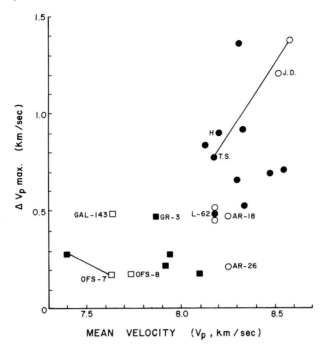

Fig. 13. Mean compressional wave velocity versus maximum P-wave velocity anisotropy for measured (solid symbols) and calculated (open symbols) velocities for peridotites and dunites (circles) and garnet clinopyroxenites and eclogites (squares). The unmarked solid circles represent measurements by *Christensen* [1966, 1971] and *Christensen and Ramananantoandro* [1971] on specimens of Twin Sisters dunite, except for the circle in the lower left, which is from serpentinized Mount Addie dunite. The solid circle H is from measurements at STP by *Kasahara et al.* [1968] on the Hidaka Horoman dunite from Japan; all other measurements were made at 4 kb and 25°C. With the exception of OSF-7 (measurements by N. I. Christensen), the solid squares are measurements on eclogites and garnet clinopyroxenites by *Kumazawa et al.* [1971], who also calculated anisotropies with the method of *Kumazawa* [1964] but did not take the garnet content into account.

means for deciding whether peridotite or garnet clinopyroxenite dominates.

The second generalization is that maximum P-wave velocity anisotropies for the peridotites and the dunites (averaging 0.76 km/sec) are, on the average, higher by about 0.5 km/sec than those for the garnet clinopyroxenites (averaging 0.29 km/sec). If we disregard crustal eclogite GAL-143 and specimen GR-3, which has a fabric [*Kumazawa et al.*, 1971, Figure 7] similar to that of GAL-143 and may also be a crustal rock, the average anisotropy for the garnet clinopyroxenites drops to 0.22 km/sec. Thus even the maximum P-wave velocity anisotropies of the mantle garnet clinopyroxenites are incapable of producing the 0.3- to 0.7-km/sec anisotropies observed in the northeast Pacific. The peridotites

and dunites (with the exception of Ar-26), however, have maximum anisotropies in the range 0.48–1.37 km/sec and are therefore easily capable of producing the anisotropies observed. The maximum calculated anisotropies (and mean velocities) will, of course, be lowered by serpentinization and the presence of other phases. Both measured and calculated maximum values will be reduced by an amount depending on the in situ orientation of the fabric in the upper mantle, a topic discussed next.

Avé Lallemant and Carter [1970] discussed in a preliminary way the orientations of olivine fabrics and anisotropies produced with reference to the two most seriously regarded hypotheses advanced to explain plate motions. These hypotheses are (1) that the displacements result from drag by creep of convecting material below (model 1) and (2) that the displacements are due to rigid translation by the push from injecting hot material in the vicinity of ridges, by the pull from the relatively cold and dense sinking slab near trenches, gravity sliding, or by a combination of these forces (model 2). For model 2, rigid translation is opposed by drag due to creep in the mantle below. In Figure 14a we have reproduced the schematic flow pattern, orientations of planes of maximum shearing stress, and fabrics expected from syntectonic recrystallization for the convection model [Avé Lallemant and Carter, 1970, Figure 7a]. (The idealized fabrics shown in Figure 14a are based on the assumption that the crystals are oriented with respect to the principal stress axes rather than the principal strain axes. Inasmuch as the stress and strain tensors coincide in the experiments to date, it is not possible to distinguish between these two possibilities; extrusion experiments are currently underway to determine whether stress or strain controls the orientations. If the crystals are oriented with respect to the principal strain axes during simple shear, the foliation plane would be inclined to the $\sigma_2 - \sigma_3$ (zero shearing stress) plane, and thus slip could take place parallel to the [100] maximum accompanying recrystallization, as is commonly observed. This possibility does not, however, alter appreciably the analyses in this section or the conclusions that follow.) The shearing sense for the rigid translation model (model 2) is just reversed, so that the orientations of α = [010] = σ_1 and γ = [100] = σ_3 would be opposite those shown in Figure 14a [Avé Lallemant and Carter, 1970, Figure 7b]. We also show schematically on the figure the approximate location of the 500°C isotherm (assuming that the plate is about 50 km thick and the velocity is about 4 cm/yr), as suggested by studies of Langseth et al. [1966] and Oxburgh and Turcotte [1968] for the area near the ridge, Clark and Ringwood [1964] for the ocean basin, and McKenzie [1969b], Griggs [1972], and Toksöz et al. [1971] for the descending slab near the trench.

Syntectonic recrystallization is expected to be the dominant flow mechanism below the 500°C isotherm, as was indicated by the experiments (Figure 2). The fabric shown in the upper right of Figure 14a originates during the ascent of mantle peridotite or dunite and could be 'frozen in' and carried along passively thereafter. The fabric shown in the center would

be expected for steady-state flow above 500°C between the ridge and the trench. The fabric in the upper left could arise from bodily rotation of the frozen-in fabric shown in the center or by continued recrystallization above 500°C in the downgoing slab. Inasmuch as the seismic refraction observations have been made between the ridge and the trench in the uppermost mantle of the Pacific plate at about 225°C and 3 kb, we might expect to sample fabrics having orientations near that shown in the upper right of Figure 14a, although the orientations could range to that shown in the center of the figure.

The maximum P-wave velocity anisotropies, calculated in the horizontal plane, are shown in Figure 14b as a function of tilt about concentrations of [001] axes in the dunite and the two lherzolites regarded as being typical of mantle tectonites. The anisotropies given are the differences in velocities parallel to [001] concentrations (normal to the plane of the diagram in Figure 14a) and the horizontal velocity in the east-west direction. The values, with increasing inclination to the horizontal of γ = [100] concentrations, from the ridge to the trench (right to left), are taken as positive if the velocities normal to the ridge are greater than those parallel to it, as was observed in the refraction studies. For comparison, the inset in Figure 14b shows the ΔV_p between [001] and the east-west direction ($\perp$ ridge) as a function of inclination of [100] for single-crystal forsterite at 225°C and 3 kb. It is evident that the fabrics for both the Twin Sisters dunite and the John Day lherzolite could easily account for the observed 0.3- to 0.7-km/sec anisotropy for the range of orientations considered ([100] $\wedge$ horizontal = 0°–45°). For lherzolite L-62 it appears that the correct sign and magnitude of the anisotropy would be given only if the frozen-in fabric (Figure 14a, right) were sampled.

Inasmuch as the shearing sense and hence the fabrics are reversed for the rigid translation model (model 2), the curves in Figure 14b corresponding to fabric orientations for this model, from the ridge to the trench, should be read from left to right. The observed anisotropies could be produced only near the limiting orientation shown in the center of Figure 14a (but reversed) or under other very special circumstances. Therefore, if syntectonic recrystallization is the dominant mineral orienting process, our results clearly favor the convection model. A similar conclusion, based on geological con-

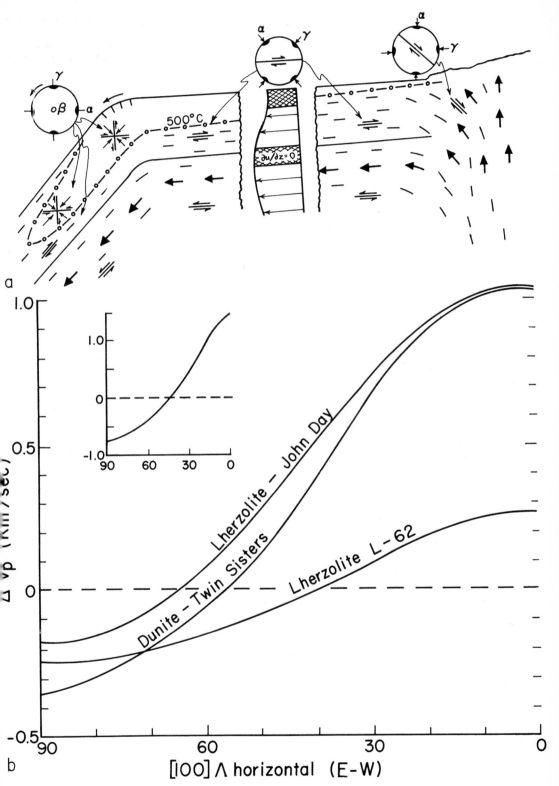

Fig. 14. (a) Schematic flow model showing orientations of planes of maximum shearing stress and fabrics expected for syntectonic recrystallization during thermal convection [*Avé Lallemant and Carter*, 1970] (see the text). (b) Seismic *P*-wave velocity anisotropy as a function of inclination of [100] concentrations to the horizontal for three natural specimens and (inset) olivine single crystal (see the text).

siderations and the results of *Avé Lallemant and Carter* [1970], was reached by *Christensen* [1971] in his study of the Twin Sisters dunite.

Finally, *Francis* [1969] proposed that plastic deformation in olivine according to the system {0kl} [100], during the ascent of convection cells near ridges, might be the dominant mineral orienting process. In this instance, for very large homogeneous strains the [100] axes would tend to align parallel to the least principal stress axis, σ_3. Fabrics produced in this manner and sampled by the refraction studies would also be frozen in, since the system {0kl} [100] is expected to operate only above about 300°C at geological strain rates (Figure 2). Therefore, if large plastic deformation is the dominant mineral orienting process, the fabrics expected would be the same as those for syntectonic recrystallization during convection. (The grain shape fabric should, in general, be quite different and would have to be considered in the seismic velocity calculations.) However, there is no evidence for the strong plastic deformation required to produce the preferred orientations in any of the specimens that we have studied. *Nicolas et al.* [1971] have found such evidence for specimens near the periphery of the Lanzo massif and a few other peridotite bodies in Europe, but the strong plastic deformation may have post-dated steady-state flow in the upper mantle.

On the basis of similarities in fabrics and deformational processes in mantle peridotites and dunites and those observed in the steady-state experiments, we conclude that power law creep governs flow over most of the upper mantle. The dominant deformational process is believed to be syntectonic recrystallization, and the preferred crystal orientations produced by this process are regarded as the primary cause for the seismic P-wave velocity anisotropy observed in the upper mantle of the northeast Pacific. Maximum anisotropies found for mantle garnet clinopyroxenites are insufficient to produce the observed magnitudes (0.3–0.7 km/sec), whereas, in general, those found for peridotites and dunites can easily account for the observations. Hence the upper mantle, at least in the northeast Pacific, is probably composed dominantly of peridotite and/or dunite. If syntectonic recrystallization, accompanied by plastic deformation and recovery, is indeed the dominant creep mechanism in the upper mantle, our results support the hypothesis that thermal convection is the primary driving force for the sea floor spreading process.

Acknowledgments. We are indebted to N. I. Christensen for measurements of compressional wave velocities in cores from some of our specimens. J. R. Möckel kindly supplied us with his data for Ar-18 and Ar-26, and H. G. Avé Lallemant supplied the data for the John Day lherzolite. E. den Tex provided lherzolite L-62, and E. Maske supplied the mantle garnet clinopyroxenites. L. A. Fernandez permitted us to use his unpublished microprobe data on the minerals of the garnet clinopyroxenites. We benefitted greatly by discussions, suggestions, and criticisms from H. G. Avé Lallemant, I. Y. Borg, N. I. Christensen, H. C. Heard, A. Nicolas, and C. B. Raleigh.

This research was supported by National Science Foundation grant GA-15412.

Steady-State Flow in Polycrystalline Halite at Pressure of 2 Kilobars

HUGH C. HEARD

Lawrence Livermore Laboratory, University of California
Livermore, California 94550

The differential stress-strain (σ-ϵ) behavior of annealed isotropic aggregates of halite (NaCl) has been determined at a confining pressure of 2 kb, temperatures of 23°–400°C, and strain rates $\dot{\epsilon}$ of 10^{-1}–10^{-8} sec^{-1}. Jacketed samples were extended up to 12% permanent strain in an externally heated constant $\dot{\epsilon}$ apparatus fitted with an internal force gage. All σ-ϵ curves measured at the lower T and the higher $\dot{\epsilon}$ show pronounced strain hardening. Steady-state flow is much enhanced, and strengths are reduced as either T is increased or $\dot{\epsilon}$ is decreased. Values of σ measured at 10% ϵ range from 470 bars at 23°C and $\dot{\epsilon} = 10^{-4}$ sec^{-1} to 16 bars at 400°C and $\dot{\epsilon} = 10^{-7}$ sec^{-1}. Intracrystalline glide is associated with strain hardening at high σ, whereas polygonization by dislocation glide and climb is responsible for steady-state flow at all lower σ. In the region where steady-state flow prevails, all data are consistent with Weertman's models of creep by dislocation climb and well fit by $\dot{\epsilon} = A \exp{(-E/RT)}\sigma^N$, where $A = 3 \times 10^{-6}$ sec^{-1}, $E = 23.5 \times 10^3$ cal/mole, and $N = 5.5$. In nontectonic regions of the earth's crust, if the average $\dot{\epsilon}$ in massive halite bodies is taken to be between 10^{-11} and 10^{-15} sec^{-1}, the predicted steady-state flow σ ranges between 8 and 48 bars. Burial of the halite between 2 and 3.5 km in a sedimentary sequence would then be required to initiate flow resulting from gravitational instability. Equivalent viscosities for this flowing salt are expected to vary from 10^{18} to 5×10^{21} poises nearly independently of temperature.

In many sedimentary basins throughout the world, polycrystalline halite or salt occurs both as relatively thin widespread layers interbedded with other evaporites and sediments and as more massive pure bodies. Observations made on the salt from the sedimentary sequences that have never been deeply buried indicate little internal disturbance, but, in bodies that at one time had been at depths greater than a few kilometers, large-scale deformations have been well documented for over a century [*O'Brien*, 1968; *Murray*, 1968]. Highly deformed salt bodies occur as anticlines, spines, domes, or other diapiric structures most commonly in juxtaposition with fluid-saturated argillaceous sediments and carbonates. Although in some cases the intrusion may be partly due to high lateral stresses, most evidence suggests that the upward movement of the salt results from the density contrast between the lower-density salt and the overlying sediments, as was first suggested by *Arrhenius* [1912]. It should also be mentioned that the over-all characteristics of these large-scale salt structures are not only governed by the rheological behavior of the salt but are also strongly dependent on the mechanical response of the overlying and enclosing sediments [*Nettleton*, 1934].

The common occurrence of large amounts of hydrocarbons, sulfur, and other minerals adjacent to these diapiric structures provides a strong incentive to understand their complete genesis and develop some predictive capability. More recent proposals to use massive salt bodies for storage cavities for hydrocarbons and radioactive waste products necessitate accurate prediction of the rheological properties of the in situ salt.

The most important operative deformation mechanism responsible for the large strain in these bodies is thought not to be fracture within or between adjacent halite crystals but rather to be intracrystalline slip accompanied by recovery processes, recrystallization, or both [*Clabaugh*, 1962; *Muehlberger and Clabaugh*, 1968; *Schwerdtner*, 1968]. Recent deformation studies [*Carter and Heard*, 1970] of halite single crystals tested at pressures comparable to depths of about 9 km in the crust and over a wide range of temperatures and strain rates show that dodecahedral slip, accompanied by polygonization (subgrain formation by dislocation climb), is to be expected at rates and temperatures consistent with the inferred conditions of deep salt flow. Slip on the cubic and octahedral planes was found to be much less important, and neither annealing nor syntectonic recrystallization was observed.

The work reported here was undertaken to explore the mechanical behavior as well as the mechanisms of deformation of halite aggregates over a wide range of environmental conditions as a logical step between the observed single-crystal properties and the prediction of in situ salt behavior in nature. Both compression and extension tests were conducted on annealed artificial aggregates of pure salt at confining (lithostatic) pressure of 2 kb, temperatures T of $23°-400°C$, strain rates $\dot{\epsilon}$ of $10^{-1}-10^{-8}$ sec^{-1}, differential stresses σ of 470–16 bars, and strains ϵ of $\leq 12\%$. From the observed flow behavior, equations that relate $\dot{\epsilon}$ and T to σ for steady-state flow in the regions of dominant glide and polygonization are derived. These equations are then used to infer the in situ rheological behavior of the salt.

PREVIOUS WORK

The slip systems active in halite single crystals tested at atmospheric pressures are well documented. Depending on the orientation of the principal stress field to the crystallography, and in order of decreasing importance, these systems are $T = \{110\}, t = \langle 1\bar{1}0 \rangle$; $T = \{100\}, t = \langle 110 \rangle$; and $T = \{111\}, t = \langle 1\bar{1}0 \rangle$ [Reusch, 1867; Mügge, 1898; Tammann and Salge, 1927; Buerger, 1930b; Dommerich, 1934; Wolff, 1935; Stepanov and Bobrikov, 1955]. All these slip systems have been observed in halite crystals deformed at pressure of 2 kb, temperatures of $\leq 500°C$, and strain rates of $10^{-1}-10^{-8}$ sec^{-1} [Carter and Heard, 1970]. At temperatures in excess of about 250°C at moderate strain rates, slip became progressively less important as polygonization by dislocation climb (subgrain formation) became increasingly dominant.

Unconfined single-crystal and polycrystalline halite aggregates that have been deformed over a range of temperatures also show this transition from slip at the low temperatures (often complicated by fracture) to polygonization and recrystallization at the higher temperatures [Phil-

lips, 1962; Blum and Ilschner, 1967; Stokes, 1966; Burke, 1968]. There have been relatively few studies investigating the stress-strain-temperature-time dependent flow properties of polycrystalline aggregates in which uncertainties resulting from fracture have been minimized [Handin, 1953; Serata and Gloyna, 1959; LeComte, 1965; Thompson, 1965; Burke, 1968]. Several of these studies were conducted under a superposed hydrostatic pressure, but only a few were able to correlate their mechanical results with an observed flow mechanism. From these studies it is generally agreed that, at temperatures of about 0.5–0.8 T/T_m melting temperature ($\sim 200°-500°C$), steady-state flow (secondary creep) involving polygonization is a thermally activated process having an energy near 30 kcal/mole. At somewhat lower temperatures (and at higher stress levels), slip, with consequent dislocation interaction, is dominant with activation energies of 15–20 kcal/mole. For steady-state flow, $\dot{\epsilon}$ appears to be proportional to σ^5, and for slip, it seems proportional to exp ($b\sigma$). The results presented here encompass a very wide range of experimental conditions and are broadly consistent with most previous investigations.

EXPERIMENTAL PROCEDURE

Test apparatus and measurement. The triaxial deformation apparatus used in this series of experiments is identical to that employed by *Carter and Heard* [1970] for deformation of halite single crystals and similar to that described by *Heard* [1963]. Briefly, the apparatus consists of an externally heated pressure vessel fitted with an internal force transducer to measure axial loads applied to a right circular cylindrical sample. Widely ranging strain rates at the specimen are accomplished by a synchronous motor driving a variable-speed transmission coupled to the loading piston through an efficient ball screw assembly. The CO_2 confining pressures were measured with a manganin coil and maintained at 2.00 kb during all tests, the accuracies being estimated at $\pm 0.5\%$. Temperatures were measured with chromel-alumel thermocouples, and gradients along the 4- to 5-cm-long samples were $\leq 2°C$ at the highest test temperature (400°C); accuracies are estimated at $\sim 1°C$. After calibration the signal from the internal force gage corresponded to an accuracy of 1–2 bars axial stress on the 2- to 2.5-cm-diameter test samples.

The true stress-strain curves were derived by calculating differential stress from recorded values of force and computed cross-sectional areas of the sample after correcting for the force borne by the 0.25-mm Pb or Al jacket and assuming that all deformation was homogeneous, with no volume change. The resulting values were plotted against axial strain calculated from piston displacements (corrected for apparatus distortion) and initial sample length. Each curve is based on some 10–15 such points; Figures 1–5 illustrate typical curves with plotted points. Strain rates are calculated from the loading piston velocity and the sample length. If the sample strains remain small, as in these tests,

average $\dot{\epsilon}$ is constant within a few per cent. Reported values of $\dot{\epsilon}$ (Table 1) are calculated near 10% strain and constant at that point to about 1%. The elapsed times of deformation for 10% strain in the typical tests summarized in Table 1 range from about 0.6 sec for the shortest experiment to about 73 days for the longest one. Comparison of the data in Table 1 shows that, at identical conditions, reproducibility is quite good.

Most of the tests performed on the polycrystalline halite were in extension; i.e., if compressive stress is taken to be positive and the principal stresses are σ_1, σ_2, and σ_3, extension is that stress state at which the confining pressure is $\sigma_1 = \sigma_2 > \sigma_3$. These experiments can then be envisioned as tensile tests conducted

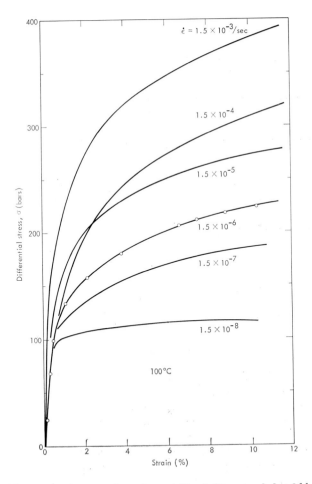

Fig. 1.   Differential stress-strain curves for polycrystalline halite extended at 2 kb, $\dot{\epsilon} = 1.5 \times 10^{-3}$
to 1.5×10^{-8} sec^{-1}, and 100°C.

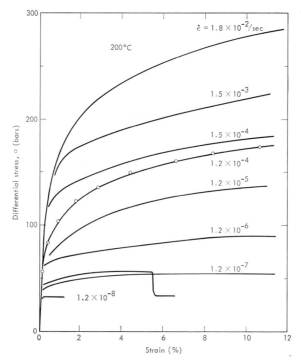

Fig. 2. Differential stress-strain curves for polycrystalline halite extended at 2 kb, $\dot{\epsilon} = 1.8 \times 10^{-2}$ to 1.2×10^{-8} sec^{-1}, and 200°C.

under a superposed high hydrostatic pressure. A few experiments were performed in compression, where $\sigma_1 > \sigma_2 = \sigma_3$ is the confining pressure. No differences in mechanical behavior between these two types of test were noted (Table 1).

Starting material and technique. Natural halite most commonly occurs with elongated crystals having dimensions usually in the range of about 1–10 cm and often possesses a strong fabric. The difficulty of obtaining natural fine-grained, homogeneous, and isotropic aggregates then necessitated the preparation of large blocks for the fabrication of the test cylinders. Many different techniques were tried with little success. Ultimately, excellent-quality billets about 8 cm in diameter by 15 cm long were prepared by hydrostatic compaction of Baker reagent grade NaCl at 1.7 kb and 130°C for 2 hours in rubber bags. Test cylinders were machined (dry) from these blanks and stored in a desiccator over CaCl₂. Porosities ranged from 1 to 1.5%. These samples were then jacketed in thin-wall Al tubes and annealed for 15 hours at 2 kb and 500°C. This treatment resulted in complete

recrystallization of the aggregate to an interlocking mosaic of equant polygonal crystals with an average size of 2–3 mm and a range of 0.5–5 mm. Final densities ranged from about 99.5 to 99.7% of the theoretical value. Overall, the appearance of the sample changed from milky-white and opaque to clear and nearly transparent.

After the annealing process all samples tested at 400°C and about half of those tested at 300°C (Table 1) were cooled under pressure to the desired temperature, and the deformation was begun. In all other cases the annealed cylinders were cooled under pressure to room temperature, rejacketed in thin-wall lead tubes, and reheated at 2-kb pressure to the desired temperature. The additional rejacketing procedure, with the consequent possibility of surface damage to the individual crystals, apparently had no effect on mechanical behavior, at least when compared at 300°C.

Microchemical and spectrometric analysis of the starting material showed that the following elements were present in amounts of <10 ppm after the initial compaction by hot pressing:

Ag, Al, Ba, Ca, Cd, Co, Cr, Cu, Fe, Mg, Mn, Ni, Pb, Sn, V, and Zn. The concentration of Mo, Si, and Ti occurred at the level of 10–20 ppm. Analysis of the NaCl aggregate after the annealing treatment and a test sample after deformation at 300°C (lead jacket) for 3 weeks gave virtually identical results. Analysis of natural halite from the Grand Saline dome, Texas, gave identical results except for 3000 ppm Ca. An analysis of material from the Tatum dome, Mississippi, is also quite similar, except for the presence of 40 ppm Fe, 13 ppm Mg, and 2300 ppm Ca [*Schlocker*, 1963]. The high levels of Ca in each of the latter cases can be reasonably ascribed to traces of anhydrite. The concentration of H_2O in the warm-pressed NaCl was 55 ppm. After annealing at 500°C for 15 hours this value decreased to 45 ppm, and after deformation at 300°C for a duration of 3 weeks it was found to be 20 ppm. For comparison, the two natural halite samples ranged from 40 to 50 ppm H_2O.

Extension experiments were carried out at 200°C and 10^{-4} sec^{-1} on three mutually orthogonal samples of the polycrystalline aggregate after annealing in the usual way to test for directional anisotropy in the starting material. Comparison of the complete stress-strain curves from the three orthogonal tests showed the stress differences to be within about ±3% of the averaged value at all strains up to 10%. Two additional tests on samples prepared in the usual orientation (parallel to the billet axis) also gave a reproducibility of about 3% for the three identical tests. Similar values for the reproducibility are apparent from several groups of samples at other conditions listed in Table 1.

EXPERIMENTAL RESULTS

Typical differential stress-strain curves for the annealed halite aggregates tested in extension are illustrated in Figures 1–5 for a broad range of strain rates at 100°, 200°, 248°, 300°, and 400°C. These data, as well as other similar data not shown, are summarized in Table 1. By comparing Figures 1–5, strain hardening can be observed to be greatly decreased with decreasing strain rate or increased temperature. Steady-state flow is closely approximated during most of the deformation at 10^{-8} sec^{-1} and 100°C (Figure 1). The onset of steady-state flow occurs at somewhat higher rates at the

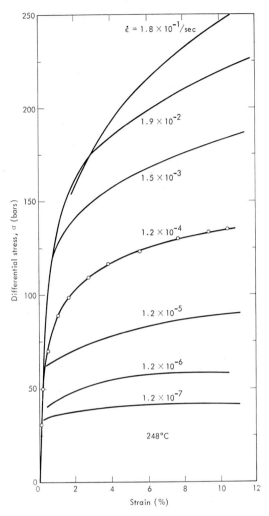

Fig. 3. Differential stress-strain curves for polycrystalline halite extended at 2 kb, $\dot{\epsilon}$ = 1.8 × 10^{-1} to 1.2 × 10^{-7} sec^{-1}, and 248°C.

intermediate temperatures. Finally, at 400°C strain hardening seems dominant only at the highest rates investigated. Near the maximum strains shown in these figures the millionfold decrease in strain rate lowers the measured stress by about a factor of 10 as the temperature is increased from 200° to 400°C. Much less dramatic changes can be noted at the lower temperatures where strain hardening is important (Figure 1 and Table 1).

In Figures 6 and 7, some of the stress-strain results (Figures 1–5) are compared as a function of temperature at two strain rates, 10^{-5} and 10^{-7} sec^{-1}. In either case, strain hardening

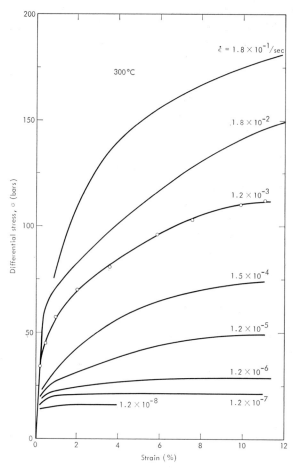

Fig. 4. Differential stress-strain curves for polycrystalline halite extended at 2 kb, $\dot{\epsilon} = 1.8 \times 10^{-1}$ to 1.2×10^{-8} sec^{-1}, and 300°C.

nearly vanishes as the intermediate temperatures are reached. Furthermore, moderate increase of the test temperature markedly lowers the strength. At 10^{-5} sec^{-1} the stress at 10% ϵ at 400°C is only 7% of the 23°C value; at 10^{-7} sec^{-1} the value at 400°C is <5% of that at room temperature. Steady-state flow is closely approximated at temperatures of about 200°C and above at the highest measured strains.

Deformation mechanisms. As was noted above, translation gliding on {110}, {111}, and {001} has been observed in halite single crystals similarly deformed at strain rates of 10^{-1}–10^{-8} sec^{-1} and temperatures of 25°–500°C [*Carter and Heard*, 1970]. In that study, only slip on one or more of these three systems (depending on crystal orientation) was observed at the low to intermediate temperatures and

was closely accompanied by strong to moderate strain hardening. At the intermediate to higher temperatures, depending somewhat on strain rate, steady-state flow was noted. In this region, although glide was still present, polygonization (subgrain formation) by dislocation climb became dominant.

In the present work, representative samples of the polycrystalline halite were selected from both the region of strong strain hardening and the region of steady or nearly steady-state flow in order to characterize the deformation mechanisms. Individual crystals were prepared from the aggregates for examination with conventional optical, etch pit, and X-ray techniques.

N. L. Carter has examined a suite of deformed samples from each behavioral region described above. The procedures used earlier for

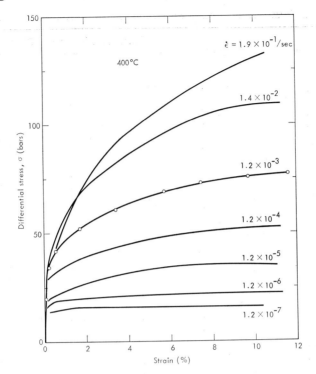

Fig. 5. Differential stress-strain curves for polycrystalline halite extended at 2 kb, $\dot{\epsilon} = 1.9 \times 10^{-1}$ to 1.2×10^{-7} sec^{-1}, and 400°C.

single crystals [*Carter and Heard*, 1970] were duplicated to compare etch pit configurations and densities directly with those reported earlier. The difficulties of preparing suitably cleaved crystals from the fine-grained aggregates and controlling the type of etch permitted only qualitative comparisons, however.

Etch pit patterns typical of crystals from the polycrystalline halite deformed in the strain-hardening region to strains of ~10% are shown in Figure 8a. For reference, the conditions of deformation were 100°C and $\dot{\epsilon} = 10^{-6}$ sec^{-1}; the stress-strain curve is shown in Figure 1. Three distinct linear patterns are evident from Figure 8a, E-W, N-S, and NW-SE. The first two patterns are interpreted as resulting from dislocations aligned along two conjugate planes of the easiest glide system {110} ⟨110⟩ normal to the etched surface {001}. The most prominent NW-SE features are cleavage steps. The fainter NW-SE lineations at the top center in Figure 8a are probably the result of screw dislocations in the two {110} planes dipping at 45° to the viewing surface. Thus the extension direction

would lie in the etched surface and strike NE-SW. Note that no polygonal structures are evident. The halite crystal shown in Figure 8b was deformed at 400°C and $\dot{\epsilon} = 10^{-4}$ sec^{-1}, well within the steady-state flow region at 10% strain (Figure 5). In this case, good polygonal structures are evident, the subgrain boundaries being regions where closely spaced dislocations emerge. The interiors of these subgrains have a dislocation density much lower than those of comparable areas of Figure 8a. The depletion from polygon interiors and development of closely packed walls forming subgrain boundaries is believed to be caused by the climb of dislocations normal to the slip plane by vacancy diffusion [e.g., *Cahn*, 1951; *Amelinckx*, 1954]. These structures can be directly compared with similar results on large single crystals [*Carter and Heard*, 1970, Figures 9 and 10 and Plates 4 and 5].

R. L. Braun has also examined the deformed material and reported results consistent with those of Carter. Figure 9 shows a series of Berg-Barrett topographs taken by Braun with a Weissmann double-crystal diffractometer using

TABLE 1.   Deformation Tests on Annealed Polycrystalline Halite at 2000-Bars Pressure

Experiment	Temperature, °C	Strain Rate, sec⁻¹	Differential Stress σ at Selected Strains, bars			
			4%	6%	8%	10%
846	23	1.53×10^{-5}	390	424	449	470
808	23	1.53×10^{-7}	316	335	345	350
809	23	1.52×10^{-8}	264	288	303	(312)
800	100	1.53×10^{-3}	321	347	368	384
801	100	1.52×10^{-4}	245	273	293	309
802	100	1.52×10^{-5}	231	251	264	274
803	100	1.53×10^{-6}	182	201	214	222
804	100	1.53×10^{-7}	155	170	180	184
847	100	1.54×10^{-8}	112	115	116	116
824	200	1.78×10^{-2}	233	254	269	280
798	200	1.50×10^{-3}	190	201	212	220
823	200	1.53×10^{-4}	169	177	184	188
813	200	1.52×10^{-4}	158	169	176	182
844	200	1.52×10^{-4}	156	170	181	(190)
842*	200	1.53×10^{-4}	150	165	175	183
843*	200	1.54×10^{-4}	152	165	174	181
841†	200	1.52×10^{-4}	152	170	183	193
845†	200	1.51×10^{-4}	152	167	178	188
797	200	1.15×10^{-4}	146	158	167	173
796	200	1.15×10^{-5}	115	125	132	137
789	200	1.16×10^{-6}	79	84	88	90
805	200	1.15×10^{-7}	54	55	55	55
855	200	1.15×10^{-7}	(56)	(56)	(56)	(56)
856	200	1.16×10^{-7}	60	(60)	(60)	(60)
857	200	1.15×10^{-7}	56	(58)	(58)	(58)
857	200	1.15×10^{-8}		34	(34)	(34)
854	200	1.16×10^{-8}	(33)	(33)	(33)	(33)
829	248	1.84×10^{-1}	193	215	232	246
827	248	1.85×10^{-2}	185	198	210	220
826	248	1.54×10^{-3}	157	168	176	182
812	248	1.15×10^{-4}	117	125	131	134
810	248	1.15×10^{-5}	78	83	87	90
859	248	1.15×10^{-6}	55	57	59	59
807	248	1.15×10^{-7}	40	41	41	41
831	300	1.84×10^{-1}	140	156	167	176
828	300	1.82×10^{-2}	102	118	131	142
783	300	1.15×10^{-3}	86	96	104	110
814	300	1.54×10^{-4}	58	66	71	74
858	300	1.15×10^{-4}	70	78	84	88
782	300	1.18×10^{-5}	39	44	48	49
787	300	1.18×10^{-6}	27	28	30	31
806	300	1.15×10^{-7}	21	21	22	22
840	300	1.15×10^{-8}	16	16	(16)	(16)
830	400	1.86×10^{-1}	96	110	121	130
825	400	1.39×10^{-2}	86	98	105	109
784	400	1.15×10^{-3}	63	70	74	75
780	400	1.21×10^{-4}	45	49	51	52
781	400	1.18×10^{-5}	32	35	35	35
788	400	1.16×10^{-6}	21	22	22	22
832	400	1.15×10^{-7}	16	16	16	16

Parentheses indicate extrapolated values.
* Samples oriented at 90° in plane perpendicular to all other test specimens.
† Compression tests.

monochromatized CuKα radiation and the (200) reflection [*Braun et al.*, 1969]. Each topograph was taken as the reflecting crystal was oscillated through its complete range of reflection. The corresponding X-ray rocking curves for each crystal are shown in Figure 10. The annealed undeformed starting material is shown in Figure 9a. This photograph is relatively homogeneous in texture and exposure and, with the corresponding sharply peaked rocking curve shown in Figure 10a, indicates a relatively strain-free crystal with no appreciable domain structure present. The topograph shown in Figure 9b is made from a crystal deformed 12% at 100°C and $\dot{\epsilon} = 10^{-5}$ sec^{-1}. The presence of significant distortions in the crystal is apparent from the highly irregular nature of the light and dark areas. The rocking curve of this crystal shows significant peak broadening, characteristic of strained material (Figure 10b). The rocking curve illustrated in Figure 10c is that of a crystal deformed at the same temperature as that in Figure 10b but at a 10^3 lower rate (100°C and $\dot{\epsilon} = 10^{-8}$ sec^{-1}). The curve of the more slowly strained crystal shows a distinct peak splitting, an indication of polygonization with the misorientation between the two domains of about 13'. Finally, the last topograph and corresponding rocking curve shown in Figures 9c and 10d, respectively, are those of a crystal deformed at 400°C and $\dot{\epsilon} = 10^{-5}$ sec^{-1}. Well-developed polygonal structure is evident both from the distinct domains shown in the dark spots (Figure 9c) and from the multiple sharp intensity maximums (Figure 10d). Note from Figures 8–10 that both temperature and strain rate exert a marked effect on whether the strain occurs by glide with dislocation pile up or by glide with polygonization.

Correlation of experimental results. In the region of moderate to high temperature and at low to moderate strain rates, where the slope of the stress-strain curves (Figures 1–5) is slightly positive or approaches zero, all microscopic and X-ray observations are consistent with the climb of dislocations out of their glide planes to form closely packed walls that separate domains of slight misorientation ($<1°$). This behavior is observed in polycrystalline halite from these experiments at 0.35–0.63 of the melting temperature T/T_m, depending on the rate of deformation. The equivalence of temperature and strain rate in affecting the

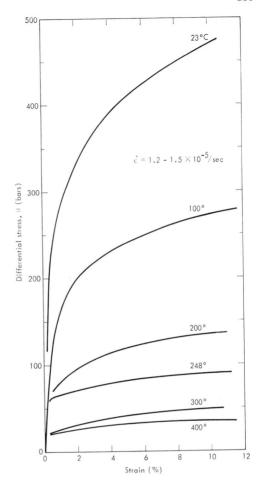

Fig. 6.   Differential stress-strain curves for polycrystalline halite extended at 2 kb and 23°–400°C.

flow stress (Figures 1–5 in comparison with Figures 6 and 7) suggest that this process is likely to be thermally activated. Figures 2–5 indicate that the steady-state flow stress does not appear to be linearly related to either strain rate or log strain rate when compared at constant temperature.

Many materials deformed at $T/T_m > 0.4$ and at intermediate stress levels exhibit virtually identical behavior. Some examples from rocks include marble [*Heard and Raleigh*, 1972], olivine aggregates [*Carter and Avé Lallemant*, 1970; *Post*, 1970; *Raleigh and Kirby*, 1970], ice [*Glen*, 1955], and halite [*Burke*, 1968; *Carter and Heard*, 1970; *Heard*, 1970].

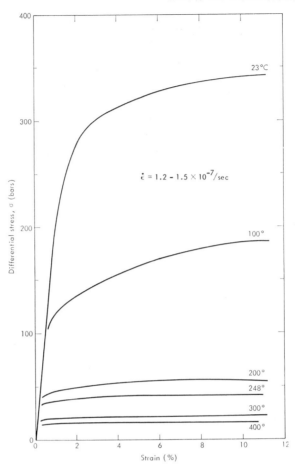

Fig. 7.   Differential stress-strain curves for polycrystalline halite extended at 2 kb and 23°–400°C.

Weertman [1955, 1957, 1968] has proposed that steady-state flow in this temperature region is dependent on the rates of glide and climb of dislocations around barriers in their glide planes. Dislocations of the same sign interact and pile up at the barriers, strain hardening resulting, and the process of dislocation climb normal to the glide plane is regarded as the recovery process. His mechanism requires that vacancies be created at sources and annihilated at sinks with equal ease, and thus a net mass transport would occur. Weertman's models lead to an equation that is usually simplified to the form

$$\dot{\epsilon} = A \, \exp \, (-E/RT)\sigma^N \qquad (1)$$

where $N = 3$, 4.5, or 6, depending on the specific dislocation mechanism assumed to prevail. Nearly all materials exhibiting steady-state flow

at these temperatures and intermediate stresses can be closely fit by (1).

Figure 11 is a plot of log σ versus $-$log $\dot{\epsilon}$, showing the steady-state flow stress or the stress at the highest common strain (10%) from each of the curves summarized in Table 1. All steady or nearly steady-state flow data are grouped in the lower part of Figure 11, and the tests having dominant slip with accompanying strain hardening are shown in the upper part; the indicated boundary between these regions occurs at a stress of about 100 bars at the higher rates and ranges to near 160 bars at the lowest rates. The 25 data points from the polygonization domain were least-squares fit to (1) but treated in logarithmic form: $\ln \dot{\epsilon} = \ln A - E/RT + N \ln \sigma$. The computer program accepts $\dot{\epsilon}$, T, and σ as input and calculates the best values for A, E, and N. In this procedure, deviations in $\dot{\epsilon}$ are

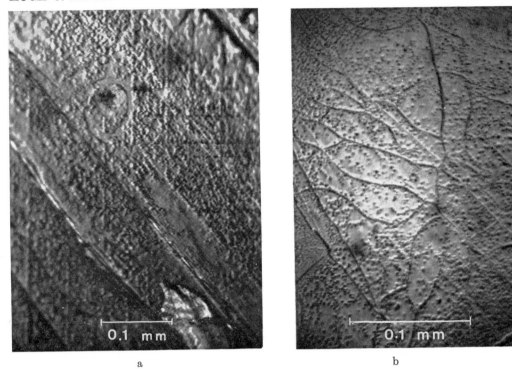

a   b

Fig. 8. Photomicrograph of etched halite grains from aggregates. (a) 11% ϵ, 100°C, and $\dot{\epsilon} = 1.5 \times 10^{-6}$ sec⁻¹. The N-S, E-W, and fainter NW-SE (top center) linear etch pit traces are interpreted as glide bands parallel to {110} ⟨1̄10⟩ system. Note that no polygonal structures are evident. (b) 11% ϵ, 400°C, and $\dot{\epsilon} = 1.2 \times 10^{-4}$ sec⁻¹. Note the well-developed polygonal structure [cf. *Carter and Heard*, 1970, Plates 5 and 6].

minimized. Results from this calculation are listed in Table 2, and the deviations are expressed at the 95% confidence level. The solid-line isotherms shown in Figure 11 at 100°–400°C satisfy this equation at 10% ϵ.

As originally derived by Weertman, the term A in (1) is not constant but slightly dependent on T and shear modulus μ. In addition, μ also appears as $(1/\mu)^N$. Weertman's complete equation is thus

$$\dot{\epsilon} = A\mu/T \exp\left(-E/RT\right)(\sigma/\mu)^N \quad (2)$$

It has been shown in other materials (i.e., marble [*Heard and Raleigh*, 1972]) that the effects of the additional T and μ terms tend to cancel. Analysis of the same steady-state flow results (above) by (2) yields $N = 5.47 \pm 0.43$; and $E = 23.81 \pm 1.86 \times 10^3$ cal/mole when calculated at 10% ϵ (95% confidence level). Comparison of these E, N values with the last column in Table 2 shows excellent

agreement, well within the confidence limits. Thus the results listed in Table 2 by the simpler (1) are preferred.

One might argue that only those results showing steady-state flow should be included in calculating the values for A, E, and N, because Weertman's model assumes only steady-state flow at constant stress. The results are not affected significantly, because these parameters are relatively independent of strain and thus of the degree of strain hardening (Table 2). The numerical values at 4, 6, and 8% ϵ were based on the stresses from the stress-strain curve, and the results were calculated by (1) as outlined earlier.

The empirical equation describing steady-state flow by dislocation climb with consequent polygonization is thus

$$\log \dot{\epsilon} = -5.6(\pm 0.8) - 23.5(\pm 1.9)/2.30RT$$
$$+ 5.5(\pm 0.4) \log \sigma \quad (3)$$

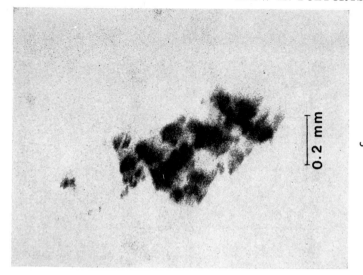

Fig. 9. Berg-Barrett X-ray topographs of halite crystals from polycrystalline aggregates. (*a*) Annealed starting material (undeformed). (*b*) Halite extended 11.5% at 100°C and $\dot\epsilon = 1.5 \times 10^{-5}$ sec^{-1}. Distortions due to local glide (curvilinear traces) rotate irregular regions in the crystals, and thus there is no common reflection position. (*c*) Halite extended 11% at 400°C and $\dot\epsilon = 1.2 \times 10^{-5}$ sec^{-1}. Well-developed polygonal structure within halite crystal is evidenced by discrete equant reflecting domains.

where A is in seconds^{-1}, E is in cal/mole $\cdot$ 10^3, and σ is in bars (10^6 dynes/cm^2).

At the higher stress levels where slip predominates, (3) clearly cannot fit the observed data shown in Figure 11. Examples of such deviation from power law behavior are commonly reported for most materials tested at the higher stress levels, where glide processes that result in strong strain hardening are most important. Common rocks showing this behavior include marble [*Heard and Raleigh*, 1972] and halite [*Burke*, 1968; *Carter and Heard*, 1970]. In this region, $\dot{\epsilon}$ is typically proportional to exp $(B\sigma)$ or sinh $(B\sigma)$, but in some cases a power law with very large values for N cannot be ruled out. *Weertman* [1957], *Ree et al.* [1960], and *Nabarro* [1967] have advanced theories consistent with these observations.

All test data from the slip region in Figure 11 were fit to an equation of the form

$$\dot{\epsilon} = A \exp(-E/RT) \sinh(B\sigma) \qquad (4)$$

by using the procedures outlined earlier. For comparison these data were also fit to (1). The deviations in A, E, and B associated with (4) were found to be only about half of those calculated for A, E, and N from (1), and thus (4) seems clearly favored. The equation best fitting the results of flow from the slip (dislocation pile up) region thus becomes

$$\log \dot{\epsilon} = 4.5(\pm 0.7) - 26.0(\pm 2.3)/2.30RT$$
$$+ \sigma/46.9(\pm 3.0) \qquad (5)$$

The dashed isotherm curves in Figure 11 satisfy (5). Values for A, E, and B were also calculated by using the stresses taken at 4, 6, and 8% ϵ, as was done earlier. Deviations were found to be only slightly greater than those shown in (5) and, when compared with Table

2, these constants appear to be affected by the increased strain hardening at low strains even less than those calculated for steady-state flow.

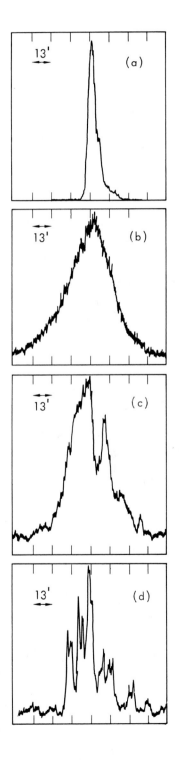

Fig. 10. (Opposite) X-ray diffraction rocking curves of halite crystals; (200) reflections. Intensity is plotted on the ordinate versus the angle on the abscissa. (*a*) Undeformed (annealed) halite starting material. (*b*) Specimen 802 of halite extended 11.5% at 100°C and $\dot{\epsilon} = 1.5 \times 10^{-5}$ sec^{-1}. Note the strong peak broadening as compared with that in (*a*). (*c*) Specimen 847 of halite extended 10% at 100°C and $\dot{\epsilon} = 1.5 \times 10^{-8}$ sec^{-1}. Note the sharper discrete peaks as compared with those in (*b*). (*d*) Specimen 781 of halite extended 11% at 400°C and $\dot{\epsilon} = 1.2 \times 10^{-5}$ sec^{-1}. Multiple diffraction peaks sharper than those in (*c*) and comparable to those in (*a*) are present.

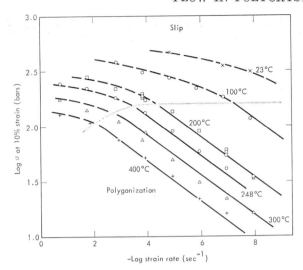

Fig. 11.   Log differential stress σ at 10% ϵ versus $-\log \dot{\epsilon}$. The solid isotherms at the bottom fit (3) in the polygonization (dislocation climb) region. The dashed isotherms at the top fit (5) in the slip (dislocation pile up) region.

DISCUSSION

Burke [1968] has conducted a series of constant stress (creep) tests at atmospheric pressure and 365°–741°C on polycrystalline halite. His results in the secondary creep (steady-state flow) region are closely fit by (1), E being 38 ± 6 kcal/mole (1 standard deviation) at temperatures ranging between 365°C and about 550°C. For temperatures in excess of about 550°C his value for E increases to 48 ± 6 kcal/mole. Burke's measured value for E and the observed creep rates in the high-temperature region correlate well with E and diffusivity D measured for lattice diffusion of the slower-moving Cl ion in single and polycrystalline NaCl, $49-51 \times 10^3$ cal/mole [*Laurent and Bénard,* 1958; *Laurance,* 1960]. At temperatures of $<550°C$ there are no measurements for E or D for either Cl$^-$ or Na$^+$ ion diffusion in polycrystalline NaCl to compare with Burke's E for creep. However, in single crystals the E

characteristic of Na$^+$ diffusion is $18-23 \times 10^3$ cal/mole in this region, whereas E for Cl$^-$ is $39-60 \times 10^3$ cal/mole, depending on the initial dislocation concentration [*Aschner,* 1954; *Barr et al.,* 1960]. The work of *Laurent and Bénard* [1958] on NaCl and other alkali halides, as well as that of *Coble and Guerard* [1963] on Al$_2$O$_3$, indicates that grain boundaries can significantly enhance anion diffusion but do not affect cation diffusion in these materials. Therefore, the Na$^+$ ion may become the slowest diffusing species in the polycrystalline material and thus may be rate controlling.

In the present work at temperatures of 100°–400°C, E for steady-state flow, 23.5×10^3 cal/mole, is much lower than that found by Burke. One might suggest that some of this difference could be a result of the 2-kb difference in confining pressure between the two investigations. However, E for diffusion is generally raised by pressure P through the product PV,

TABLE 2.   Calculated Values of A, E, and N from the Simplified Weertman Flow Equation (1)

Constant	4% ϵ	6% ϵ	8% ϵ	10% ϵ
Log A	-5.42 ± 1.17*	-5.55 ± 1.03	-5.61 ± 0.89	-5.58 ± 0.82
E	25.81 ± 2.84	24.44 ± 2.45	23.80 ± 2.06	23.46 ± 1.92
N	6.08 ± 0.72	5.75 ± 0.60	5.58 ± 0.49	5.47 ± 0.43

* Deviations shown are calculated at the 95% confidence level.

where V is the activation volume associated with the migration of the diffusing species. Data are not available for V of Cl⁻ diffusion in NaCl, but V for Na⁺ vacancies has been measured at 8 cm³/mole [*Pierce*, 1961; *Yoon and Lazurus*, 1972]. Although V for Cl⁻ diffusion could be somewhat larger, in either case it is unlikely that the product at 2 kb would increase E by more than a few thousand calories per mole. This effect then could only increase the discrepancy between the two measurements. The activation volumes associated with transient creep in NaCl are also consistent with the estimate given above [*Davis and Gordon*, 1968].

An additional possibility is that the E for steady-state flow or creep is not constant near the region of temperature overlap of the two sets of data, 365°–400°C, which in turn would imply the presence of two or more contributing mechanisms, the higher-temperature mechanism having the higher E. Such behavior is common in many materials [e.g., *Sherby and Burke*, 1967]. At temperatures of 29°–300°C and confining pressure of <1 kb, *Le Comte* [1965] derived values of E of 12.5–30 kcal/mole in polycrystalline NaCl. Direct comparison with this latter work is not possible, since varying degrees of transient behavior were observed at each temperature; true secondary creep was not observed in any case.

The stress exponent N of 5.5 ± 0.4 from (3) is in excellent agreement with that reported by *Burke* [1968], 5.0 ± 0.6, calculated at 1 standard deviation (5.5 ± 0.1 if the highest temperature value is excluded). Both values of N are close to the 4.5 value predicted on the basis of *Weertman*'s [1957, 1968] dislocation glide and climb models.

The transition from (1) to (4) at the highest stress levels has been observed to take place at $\dot{\epsilon}/D$ of about 10^9 cm⁻², where D is the diffusivity of the rate-controlling diffusing species, in many diverse materials [*Dorn*, 1957; *Williams*, 1964; *Sherby and Burke*, 1967]. *Burke* [1968] finds this transition at 10^9–10^{10} cm⁻² in polycrystalline halite. There are no data for D available for either Cl⁻ or Na⁺ in polycrystalline halite measured in the lower temperature range, but if we assume that the Na⁺ ion is rate controlling and the measured diffusivity for Na⁺ in single-crystal NaCl approximates that characteristic of the polycrystalline material, then $\dot{\epsilon}/D$ here would be

about 10^9–10^{11} cm⁻² at 100°–400°C and thus in reasonable agreement.

Then, on the basis of (3) and (5) we can calculate the flow stress expected during the deformation of large halite bodies in nature. In Figure 12 isotherms that satisfy these equations are shown at 50°–400°C. The dashed curves at high stresses at each temperature fit (5) and are curved on log σ versus $-$log $\dot{\epsilon}$ coordinates in the strain-hardening (slip) region. The solid lines with a slope of 5.5^{-1} at intermediate stresses fit (3) in the steady-state flow (polygonization) region.

In the $\dot{\epsilon}$ region of interest here, i.e., near a 'representative' geologic strain rate of 3×10^{-14} sec⁻¹, it is possible that (3) will no longer hold at the higher temperatures and Nabarro-Herring flow will become dominant. The Nabarro-Herring creep equation predicts a linear dependence of σ to $\dot{\epsilon}$ by

$$\dot{\epsilon} = abD\sigma/l^2RT \qquad (6)$$

where a is a constant of the order of 5, b is the atomic volume, and l is the average grain diameter [*Nabarro*, 1948; *Herring*, 1950]. The transition in flow behavior from (1) to (6) has been observed to occur at $\dot{\epsilon}/D$ values of about 10^2 cm⁻² in several metals, and this transition occurs at 10^2–10^3 cm⁻² at very high temperatures in halite [*Kingery and Montrone*, 1965; *Burke*, 1968]. If we take this transition in $\dot{\epsilon}/D$ to be 10^2 cm⁻² and assume values for D characteristic of Na⁺ as above, it appears that Nabarro-Herring creep can occur only at temperatures in excess of about 170°C at 3×10^{-14} sec⁻¹. The dashed lines of slope 1 at 200°–400°C in Figure 12 illustrate this behavior at the lower values of σ and $\dot{\epsilon}$. The shaded region to the right in Figure 12 represents one estimate of the range of temperature and the average rates of deformation at which massive halite bodies flow in the Gulf coast region [*Carter and Heard*, 1970]. The maximum and minimum $\dot{\epsilon}$ of 10^{-11} and 10^{-15} sec⁻¹ are based on closure rates of existing cavities in salt domes as well as on crude estimates of gross strains and durations for the emplacement of these bodies. Limits to the temperatures (200°–50°C) are set by the regional geothermal gradient along with the observed depth to the buried salt and the depth at which the driving stress, result-

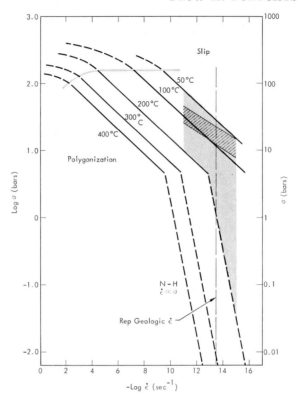

Fig. 12.   Log differential stress σ at 10% ϵ versus $-\log \dot{\epsilon}$; extension of Figure 11 by (3), (5), and (6) to low $\dot{\epsilon}$ characteristic of probable flow in natural halite. The short-dashed curves at low σ show the region of possible Nabarro-Herring (Newtonian) flow, $\dot{\epsilon} \propto \sigma$. The shaded area delineates the range of expected T, $\dot{\epsilon}$, and σ in flowing polycrystalline salt. The hatched region gives a more restricted estimate (see the text).

ing from density contrasts of the halite-over-burden, approaches 0. It should be emphasized that the characteristic $\dot{\epsilon}$ of localized regions in the salt body could be much higher than that assumed in Figure 12, especially at the initiation flow. The minimum $\dot{\epsilon}$ in certain regions could also be much lower when the driving stress decays to very low values.

Thus from Figure 12 it appears that the stress responsible for massive salt flow most probably ranges from about 80 to 0.03 bars. Furthermore, the predominant part of this flow can be described by (3) and is probably of the steady-state type, dislocation glide and climb being the dominant mechanism of deformation at all except possibly at the highest temperatures and lowest rates. Nabarro-Herring flow may occur at the higher temperatures at the lowest stresses, but it should be emphasized that such behavior was never observed in this work. The steady-

state flow stresses expected over the ranges of T and $\dot{\epsilon}$ of interest here are summarized in Table 3.

The dependence of equivalent viscosity on strain rate $\eta = \sigma/3\dot{\epsilon}$ [Griggs, 1939] is shown in Figure 13 for temperatures of 50°–400°C. These η values were based on data summarized in Figure 11 and extrapolated as solid curves to low deformation rates by (3) and (6). (In the Nabarro-Herring region, the viscosity $\eta = \sigma/\dot{\epsilon}$.) The dashed extensions at the higher $\dot{\epsilon}$ extension of each isotherm show the halite response in the strain-hardening region (5). It is apparent from Figure 13 that over a range of $\dot{\epsilon}$ from 1 to 10^{-14} sec^{-1} the equivalent viscosity of halite is only slightly influenced by temperatures up to about 200°C; only at the lowest rates and the highest temperatures in the Nabarro-Herring creep region can η be expected to be strongly tempera-ture dependent. The range of equivalent viscosi-

TABLE 3.   Steady-State Flow Stress Calculated from (3) and the Assumption that Nabarro-Herring Creep Occurs at $\dot{\epsilon}/D < 10^2$ cm^{-2}

Temperature, °C	Steady-State Flow Stress σ, bars		
	$\dot{\epsilon} = 10^{-11}$ sec^{-1}	$\dot{\epsilon} = 3 \times 10^{-14}$ sec^{-1}	$\dot{\epsilon} = 10^{-15}$ sec^{-1}
50	82.0	29.0	15.0
100	33.0	11.3	6.1
200	9.8	9.0×10^{-1}*	3.1×10^{-2}*
300	2.9*	8.7×10^{-3}*	2.9×10^{-4}*
400	1.9×10^{-1}*	5.8×10^{-4}*	1.9×10^{-5}*

* Nabarro-Herring region, $\dot{\epsilon} \propto \sigma$.

ties common to massive halite bodies during natural deformation is shown as the shaded area at the upper right and corresponds to the region of expected stresses emphasized in Figure 12. Expected values for η range from 3×10^{17} to 5×10^{21} poises, depending on the temperature and assumed strain rate (Table 4). This range is in good agreement with estimates of the probable viscosities ($> 10^{17}$–10^{19} poises) based

on limited observations in natural halite bodies [Odé, 1968].

Even closer limits to the steady-state flow stresses, and hence to the equivalent viscosities in massive halite undergoing deformation in nature, can be inferred from Figure 14. Figure 14a shows the observed maximum and minimum thermal profiles for the Gulf coast area and the relationship of σ to T (based on (3)) at the

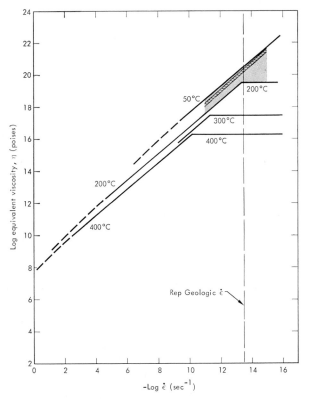

Fig. 13.   Log equivalent viscosity η versus $-\log \dot{\epsilon}$ based on (3), (5), and (6). The shaded region shows the range of probable T, $\dot{\epsilon}$, and η in flowing polycrystalline salt. The hatched area gives a more restricted estimate (see the text).

TABLE 4. Equivalent Viscosities at Various Assumed Geologic Strain Rates

Temperature, °C	Equivalent Viscosities η, poises		
	$\dot{\epsilon} = 10^{-11}$ sec^{-1}	$\dot{\epsilon} = 3 \times 10^{-14}$ sec^{-1}	$\dot{\epsilon} = 10^{-15}$ sec^{-1}
50	2.7×10^{18}	3.2×10^{20}	5.1×10^{21}
100	1.1×10^{18}	1.3×10^{20}	2.0×10^{21}
200	3.3×10^{17}	3.0×10^{19}*	3.0×10^{19}*
300	2.9×10^{17}*	2.9×10^{17}*	2.9×10^{17}*
400	1.9×10^{16}*	1.9×10^{16}*	1.9×10^{16}*

The viscosities are based on (3) and the assumption that Nabarro-Herring creep occurs at $\dot{\epsilon}/D < 10^2$ cm^{-2}; calculated from $\eta = \sigma/3\dot{\epsilon}$.
* Nabarro-Herring region, $\eta = \sigma/\dot{\epsilon}$.

highest, intermediate, and lowest assumed average values for $\dot{\epsilon}$ for massive halite flow. The right-hand ordinate shows depth on a nonlinear scale, whereas the left-hand ordinate shows the stress difference expected from the top to the base of a halite anticline or diapir at that depth due to the negative buoyancy between the salt (2.20 g/cm^3) and the overlying sediments (2.2–2.5 g/cm^3). It is assumed here that the sediment density can be sufficiently well approximated

with *Dickinson*'s [1953] average shale density-depth relationship. At depths less than about 0.9 km the shale has a lower density, but at all greater depths the halite density is less, and thus in this region it is in unstable equilibrium.

It is expected that, as a thick halite layer is progressively buried in a sedimentary sequence, it must move downward between A and B. This halite first becomes unstable at about 0.9 km but will not undergo pervasive deformation

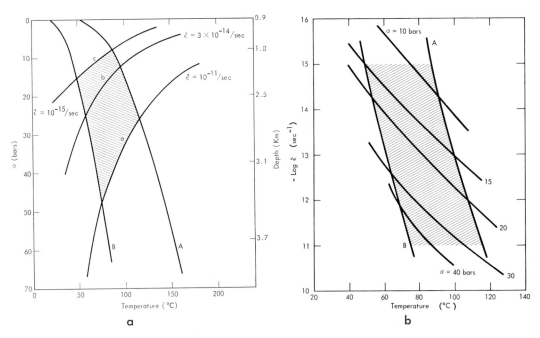

Fig. 14. (*a*) Expected *T*, σ, and depth relationships for Gulf coast province. Note that the left ordinate (σ) is linear, whereas the right ordinate (depth) is nonlinear. The hatched area bounded by A, B, *a*, and *c* limits the possible ranges of *T*, $\dot{\epsilon}$, σ, and thus the η for large-scale flow in natural salt (see the text). Curves A and B represent the observed maximum and minimum thermal profiles, respectively, for the Gulf coast area [*Nichols*, 1947; *Moses*, 1961]. Constant contours *a*, *b*, and *c* are based on (3). (*b*) Projection of thermal profiles from (*a*) on to the log $\dot{\epsilon}$–T plane with constant σ contours calculated from (3). The hatched region truncated at $\dot{\epsilon} = 10^{-11}$ sec^{-1} and $\dot{\epsilon} = 10^{-15}$ sec^{-1} corresponds to the similar region in (*a*).

until curve c is reached. It is expected that most deformations would occur between curves c and a. Burial could continue, however, without flow until the halite is somewhat deeper than that shown by curve a, but any minor perturbation on the surface of the salt could trigger rapid local flow, and the average σ and $\dot{\epsilon}$ would migrate upward to between a and c. If sedimentation ceased or erosion occurred, flow rates would become negligible as the conditions in the body move upward beyond c (Figure 14a). In Figure 14b the hatched area projected on the $\dot{\epsilon}$, T plane corresponds to a similar region from Figure 14a and is bounded by the in situ thermal gradients A and B and the maximum and minimum $\dot{\epsilon}$, a, and c. The constant σ contours shown range from 10 to 40 bars and emphasize the extreme sensitivity of the flow rate to the driving stress.

This argument indicates that flow could begin as the halite becomes buried 1.8–2.3 km (6000–7500 feet), depending on the local geothermal gradient, and continuous flow can be expected when burial approaches 2.7–3.4 km (9000–11,000 feet). These figures are based on the assumption that the intruded sediments behave essentially as a fluid and that the repeat frequency of the triggering mechanism necessary to initiate large-scale flow (e.g., local faulting) is rapid in comparison with the sedimentation rate. Obviously, failure to meet either condition will make it necessary to depress the halite further before flow can begin. This inferred range of depths (and σ) for flow compares favorably with values based on field observations from the German North Sea region. *Romanes* [1931] postulated that the overburden had to be at least 7000 feet thick before growth could begin. *Fulda* [1927] concluded that burial of halite to 11,000 feet would result in diapirism.

A replotting of the limiting σ and T from the region bounded by curves A, B, a, and c in Figure 14 is indicated by the shaded region in Figure 12. The possible range of σ and T for halite flow thus becomes very restricted, and Nabarro-Herring flow would not be expected. At 10^{-11} sec^{-1} the maximum σ is 48 bars (76°C), and at 10^{-15} sec^{-1} the minimum σ is 8 bars (87°C). The narrower range of σ and T from Figure 14, plotted in Figure 12, also yields a restricted range for η, as is illustrated in Figure 13. Maximum equivalent viscosities thus range from 10^{18} to 5×10^{21} poises.

Acknowledgments. N. L. Carter and R. L. Braun deserve much credit for their careful work in examining the experimentally deformed halite. Carter etched and photographed (Figure 8) individual crystals from the aggregates and compared the results with an earlier study [*Carter and Heard,* 1970]. Braun made Berg-Barrett topographs as well as X-ray rocking curves of halite crystals prepared from several of the deformed aggregates (Figures 9 and 10). I gratefully acknowledge permission to publish these results. J. R. Stevens performed the chemical analyses. Discussions of the flow in massive halite with N. L. Price were most helpful.

This work was performed under the auspices of the U.S. Atomic Energy Commission.

Experimental Recrystallization of Ice Under Stress[1]

BARCLAY KAMB

California Institute of Technology, Pasadena, California 91109

Polycrystalline ice in high-temperature creep ($-5°$ to $0°$C) under deforming stresses of a few bars undergoes rapid syntectonic recrystallization and large changes in texture and fabric that have prominent effects on flow. A simple original texture of small equant polyhedral grains recrystallizes at a shear strain γ of about 0.04 to a complex texture of coarse irregular interlocking grains with sutured boundaries and abundant strain shadows. Recrystallized grain size increases as stress decreases and temperature increases. Included air bubbles cause no detectable impediment to recrystallization or grain boundary migration. In simple shear (applied in torsion) the recrystallized c-axis fabric consists of two maximums, one at the pole of the shear plane and the other about 20° away from the shear direction. Fabric perfection mainly depends on total strain rather than time and develops progressively over the strain range $\gamma \approx 0.04$–0.3. At a given total strain the fabric depends only weakly on stress: the maximums become somewhat weaker, and the angle between them increases slightly as the shear stress decreases from 4.1 to 1.9 bars. Under compressive stress superimposed across the shear plane the fabric maximums elongate and join to form a small-circle girdle, whose center reorients toward the compression axis. From the form of the fabric pattern it is possible to infer the stress character (as distinct from stress level and mean stress) over the full range of possible variation for a general state of stress. Also, it is notably possible to distinguish between simple and pure shear and to distinguish the shear plane from its conjugate plane in simple shear. The recrystallization fabric developed under simple shear has monoclinic symmetry and is incompatible in symmetry and orientation with the symmetries of stress, strain rate, and total strain but compatible with the symmetry of the velocity field measured in a coordinate system in which the stress remains fixed. This conflicts with simple thermodynamic theories of crystal reorientation and also a basic assumption in the theoretical formulation of ice flow laws. A combined mechanism of thermodynamically controlled crystal reorientation modified by intracrystalline plastic flow is qualitatively compatible with the fabric. The favored crystal orientations indicated experimentally are those which would have nearly maximum elastic strain energy if the stress were homogeneous in the polycrystalline aggregate, but because of intracrystalline plastic anisotropy it is likely instead that the stress is not homogeneous and that crystals in the preferred orientations have minimum elastic strain energy. Statistical fluctuations in perfection and orientation of fabric elements suggest that the recrystallization-reorientation mechanism is cooperative in nature. The crystal reorientations cause many-fold accelerations in creep rate as the recrystallization fabric develops. The seemingly steady creep observed between initial transient creep and the onset of recrystallization does not represent steady structural conditions, which require recrystallization. Fabrics observed in glacier ice deformed in pure shear to an equivalent strain of $\gamma \approx 0.6$ are compatible with the experimentally produced fabrics, but the commonly observed multiple-maximum fabrics of glacier ice are entirely incompatible. An experimental test shows that the incompatibility is not due to a qualitative difference between crystal reorientation at and below the melting point. An orientation-controlling mechanism that becomes operative only at high strains ($\sim$1) and supersedes the mechanism operative in the experimental strain range appears to be indicated; this implies that truly steady structural conditions are not established at the lower strains. The complexity of the flow-induced and flow-affecting structural changes that occur in ice suggests caution in interpreting experimental creep results for other, more complicated geological materials. Nonsteady creep is probably the rule in nature, except in flow situations involving very large strains, comparable to glacier flow.

---

[1] Contribution 2112 from the Division of Geological and Planetary Sciences, California Institute of Technology, Pasadena, California; contribution #30 from the Swiss Federal Institute for Snow and Avalanche Research, Weissfluhjoch, Davos, Switzerland.

PURPOSE

The flow of ice in nature, as observed in glaciers, leads to substantial changes in structure, texture, and fabric that appear to be controlled by the conditions and the history of deformation and resemble, in a general way, changes that occur in the deformation of rocks and metals (for a summary, see *Kamb* [1964]). The strong *c*-axis fabrics developed in glacier ice show considerable detail and complexity and thus offer the possibility of a means for inferring and interpreting the history of ice deformation. Efforts to establish the relationships between natural fabrics and deformation history by studying field examples [*Rigsby*, 1960; *Allen et al.*, 1960; *Kamb*, 1959b, and later unpublished work] make it evident that a full understanding requires, in addition, an experimental approach in which fabrics are produced under controlled laboratory conditions. In the present study, artificial ice specimens were subjected to flow conditions similar to those in rapidly flowing glaciers, and the changes in texture and fabric were followed. Syntectonic recrystallization occurs readily under these conditions, and strong well-defined fabrics are produced. The study investigates the effects of state of stress, time, and temperature on the recrystallization textures and fabrics. An extensive earlier laboratory investigation by *Steinemann* [1958] demonstrated in considerable detail the textural changes that occur in the recrystallization of ice, and information on this subject has also

Fig. 1. Ice specimen A14 after deformation. The ink line on the surface of the specimen was initially vertical, parallel to the cylinder axis.

been given by *Rigsby* [1960] and *Shumskii* [1958]. In these earlier studies, only relatively limited and partly conflicting data on fabric changes were obtained, and the present work is therefore primarily aimed at providing a suite of experimental recrystallization fabrics extensive enough to allow a study of the consistency and systematics of crystal reorientation by syntectonic recrystallization. The experimentally produced fabric patterns are compared with natural fabrics of glacier ice and also, in a general way, with theoretical predictions.

EXPERIMENTAL METHODS

The experimental work was conducted at the Swiss Federal Institute for Snow and Avalanche Research, Weissfluhjoch, Davos, Switzerland.

Material. Ice specimens, an example of which is shown in Figure 1, were prepared by filling a container with fresh snow, saturating the snow with water, and allowing the water to freeze. The polycrystalline aggregate formed consists of a fine-grained mass of equant grains 0.5–0.9 mm in diameter. Grain boundaries are planar, grain shapes are polyhedral, and in cross section the grains appear as polygons, most of which have four to seven sides. This type of texture, shown in Figure 2, represents an approach to minimum grain boundary surface area for the given average grain size or grain packing density determined by the packing of the nucleating snow crystals. It is here called tessellate texture and resembles the grain textures of soap bubble foams and polycrystalline metals.

The white opacity of the experimental specimens (Figure 1) is due to abundant fine bubbles of air that formed in the ice from air trapped among the original snow crystals and also dissolved in the water. The bubbles are 0.1–0.3 mm in diameter and are present at an estimated concentration of about 14 mm^{-3}. Glacier ice contains air bubbles in comparable concentration, and, since the work was directed primarily toward the interpretation of natural fabrics with the help of experimental ones, air-containing artificial ice was used in preference to air-free ice. In this respect the starting material differs from that of *Steinemann* [1958].

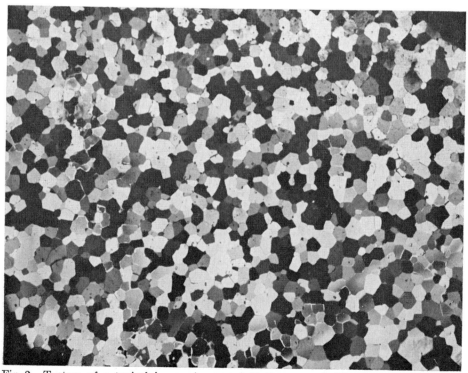

Fig. 2. Texture of a typical ice specimen as initially prepared. The length of the frame corresponds to 2 cm.

Torsion-compression experiments. Most of the deformation experiments were conducted in simple shear or in simple shear with compression superimposed perpendicular to the shear plane. The deformation was produced by subjecting cylindrical specimens to combined torsion and axial compression. Simple shear is of interest primarily because this mode of deformation (or something approximating it) occurs commonly in glacier ice showing strong recrystallization fabrics [*Kamb*, 1959*b*; *Kamb and Shreve*, 1963]. The use of torsion allows arbitrarily large shear deformations to be accomplished without the complicating end effects that limit the usefulness of simple shear apparatus.

The deformation geometry is shown schematically in Figure 3. Ice specimens were machined to hollow cylinders 4–5 cm long with outside and inside diameters of 8.0 and 3.0 cm, respectively, and attached by melting to toothed grips for loading in the torsion machine (Figure 1). The torsion-compression apparatus used was the same as that built and used by *Steinemann* [1958] in his extensive experiments.

Torsional and axial strains of the specimens were measured by dial gage micrometers that followed the motion of the grips. Torsional strain γ and axial longitudinal strain e could be detected at a level of about 3×10^{-5}. Because the grips showed a tendency to come loose and slide past the specimen in the later stages of most torsion experiments, ink lines were ruled parallel to the cylinder axis on the outer surface of each undeformed specimen in A8 and subsequent experiments to provide a record of the total torsional strain experienced by the specimen independently of the motion of the grips, as is shown in Figure 1.

Axial stress σ can be calculated directly from the axial dead-weight loading. Obtaining the torsional shear stress τ from the torsional loading torque requires consideration of the radial distribution of shear stress in the torsion specimen, since the shear strain varies linearly with

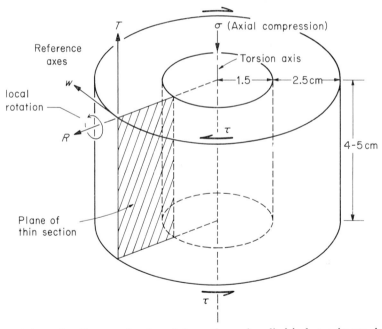

Fig. 3. Schematic diagram showing deformation of cylindrical specimens in torsion apparatus. The orientation of a typical thin section plane is shown, and reference axes, which define the stress orientation in relation to the thin section, are given: T is the axis of torsion and compression, R is the radial direction, and w is the shear direction. Although there is relative rotation of the top and the bottom of the specimen around axis T in the torsion process, the bulk rotation discussed in the text is what accompanies simple shear locally at any point in the specimen and is a rotation around the local axis R at that point, in the sense shown.

TABLE 1. Physical Conditions of the Recrystallization Experiments

| Experiment | Duration t, days | Temperature T, °C | Shear Stress τ, bars | Normal Stress σ, bars | Shear Strain | | Compressive Strain $-e_T$ |
					Rate $d\bar\gamma/dt$, 10^{-7} sec^{-1}	Total $\bar\gamma$	
A10	59.0	−1.2	1.9	0.0	0.2	0.13	0.13
A8	2.8	−1 to −2.5	4.1	0.0	2.8	0.12–0.24	0.00
A7	1.1	−2.5	4.1	1.3	3.1	0.05	0.003
A6	13.5	−4±	4.1	1.3	1.1	0.40	0.01
A5	42.0	−2.5	3.0	2.5	0.7	∼0.27	0.06
A3	5.0	−4	4.9	3.8	1.4	∼0.16	0.05
A14	1.9	−1.2	4.1	6.3	3.4	0.15	0.082
A13	3.0	−1	3.0	11.3	8.9	0.22	0.42
C1	2.0	0	0.0	2.4	0	0	?
C4	2.3	0	0.0	6.4	0	0	0.31

radial distance r from the torsion axis. If we assume, because of the high nonlinearity normally found in the stress versus strain rate relation for ice, that τ is independent of r, then τ in kilograms per square centimeter is 7.9×10^{-3} times the applied torque in kilogram centimeters for the specimen geometry used. Shear stresses are quoted here on this basis. If, on the other hand, the nonlinearity were as low as $n = 2$ in the flow relation $\dot\gamma = b\tau^n$, where b is a constant, the shear stress at $r = 2.75$ cm, midway between the inner and the outer surface of the hollow cylinder, would be consistently larger by a factor 1.2. The range of possible uncertainty here is not large enough to cause concern in the present experiments. Temperature in the cold laboratory was controlled to an accuracy of about ±1°C most of the time, although larger excursions occurred occasionally.

The torsion-compression experiments for which recrystallization data are presented here are summarized in Table 1. The strain rates listed are those observed during the second stage of creep, as defined in the section on flow behavior. The total shear strain was obtained from the external ink lines where available and otherwise was estimated by extrapolating to the full duration of the experiment from the creep observed in the initial creep period, before sliding against the grips began. The shear strain rate $d\bar\gamma/dt$ and the total shear strain $\bar\gamma$ are applicable to ice midway between the inner and the outer wall of the specimens, at $r = 2.75$ cm. The sampling of different total strains was extended somewhat by using the radial dependence of the shear strain

and also, in experiment A8, by taking advantage of inhomogeneous deformation that occurred in the specimen (Figure 4). Results for the top and bottom halves of specimen A8 are labeled A8T and A8B below; the total shear strain in A8B was twice as great as that in A8T.

Compression experiments at 0°C. To compare the results of the torsion experiments, conducted at temperatures of −1° to −5°C, with natural fabrics formed in temperate glacier ice at the melting point, it is desirable to test a basic aspect of the temperature dependence of the recrystallization process, namely, whether

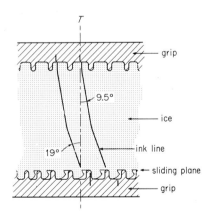

Fig. 4. Sketch of the surface of specimen A8 after deformation, showing distortion of ink lines that were initially vertical. Note the pattern of detachment of the ice specimen from the toothed grip at the bottom, leading to development of a sliding surface. Incipient sliding has begun at the top.

recrystallization at the melting point produces effects fundametally different from those produced at temperatures distinctly below the melting point. Torsion tests cannot be used for this purpose, because the grips fail to hold the specimens at 0°C. A device was therefore built to subject specimens as nearly as possible to pure shear at an ambient temperature of 0°C. The device, shown schematically in Figure 5, consists of a trough of rectangular cross section into which a closely fitting platten slides, delivering a compressive load to a cube-shaped specimen confined in the trough. The specimen is a cube 5 cm on edge. The entire apparatus is submerged in an ice water bath to maintain the melting temperature. Two experiments of this type, for which recrystallization data are given below, are listed in Table 1.

Because of pressure melting on the loaded faces of the specimen, it is not possible to measure the plastic strain by following the motion of the platten. For strain markers four small brass pins were implanted in the horizontal midplane of the specimens, about 1 cm

from the center of each of the four vertical faces of the cube. Measurement of the pin separations before and after the experiment provided rough but definite values for the two longitudinal strain components transverse to the main compression axis; the third strain component, in the direction of the main compression, was obtained from the incompressibility condition. The strain for experiment C1 was not directly measured but was estimated from that for C4 by assuming a nonlinear flow law with $n = 3.5$.

At a compressive stress of 6 bars (experiment C4) the rate of pressure melting on the loaded faces is so large that the specimen is melted down to about half its initial thickness in 10 hours. To continue the recrystallization experiment for a longer time, the specimen was removed from the apparatus after about 7 hours loading and cooled to $-10°C$. Extra slabs of ice were frozen on to the top and the bottom of the specimen to make a cube-shaped sandwich in which the original ice formed the center. The specimen could then be returned to the

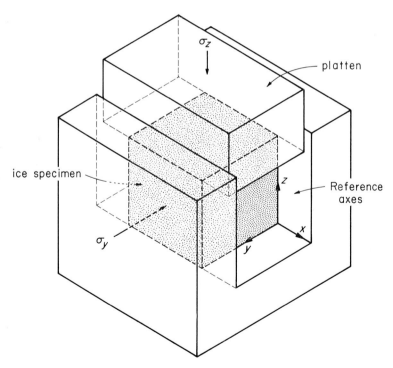

Fig. 5.　Diagram of compression device used for deforming ice specimens at 0°C. Stress axes are indicated and labeled x, y, and z in conformity with those in Figure 17a.

apparatus and loaded for another 7–10 hours at 0°C. This process was repeated 7 times in experiment C4.

Reference stress axes for specimens compressed at 0°C are labeled as follows (Figure 5): axis z is the direction of main compression, perpendicular to the loading platten, axis y is the direction of lateral constraint, and axis x is the direction of free expansion along the length of the rectangular trough. The principal stress σ_z is determined directly by the loading, σ_x is 0, and σ_y is controlled by the nature of the lateral constraint, as is discussed later.

Thin sectioning. After the completion of each experiment the specimen was immediately cooled to −10°C, and thin sections were cut. For torsion specimens the sections were cut along radial planes through the cylinder axis. Such planes have the advantage that reference axes related to the loading have the same orientation in all parts of the section. The reference axes are the direction of torsion and compression T, the radial direction R, and the shear direction w, which is perpendicular to R and T and lies in the plane of simple shear wR (Figure 3). Specimen C1 was sectioned parallel to the yz plane, and C4 parallel to the xz plane. To insure independent sampling of the texture and fabric, separate sections were spaced as far apart as practical.

Each thin section was photographed under crossed polaroids shortly after preparation and stored at −20°C thereafter. Later inspection and comparison with the photographs showed that no textural changes occurred subsequent to thin sectioning. The sections thus preserve the texture and fabric as it was during deformation. Post-tectonic recrystallization was not studied.

Fabric measurement and presentation. Orientations of crystal c axes were measured under the polarizing microscope with a standard four-axis universal stage. For ease of handling, the sections were mounted on the stage without hemispheres. Measured c-axis inclinations were corrected for light refraction by the methods described by *Kamb* [1962]. Thin sections of torsion specimens were measured in traverses parallel to the T axis, so that all the ice in a given traverse lies at approximately the same radial distance from the torsion axis and has therefore experienced approximately the same shear strain. The approximate cross-sectional

area of each crystal measured was recorded, and for many specimens the individual crystals measured were identified by number on the thin-section photographs.

The measured c-axis orientations were plotted in equal-area projection (lower hemisphere), symbols indicating the crystal cross-sectional areas. Since it developed that all crystals, large and small, conformed well to a common overall pattern of preferred orientation, the distinctions among crystals of different areas were disregarded in reducing the plots to contoured fabric diagrams. Counting and contouring of the orientation density was done by following in a general way the method suggested by *Kamb* [1959*b*, p. 1908]. For each diagram a counter area A was chosen so that, in the regions of the diagram where significant concentrations of c axes occurred, the average number of points falling within the counter would be about 10 or somewhat larger. This procedure assures that in these regions statistical fluctuations in counted orientation density should be only about a third or less of the average density. The area A used is stated (as a fraction of the hemisphere area) below each fabric diagram given, together with the number of measured c axes (written as 626 c). In diagrams for undeformed specimens the contour interval is chosen to be twice the standard deviation expected for no preferred orientation [*Kamb*, 1959*b*, p. 1908]. For deformed specimens, solid contours are drawn at orientation densities of 2, 4, 6, $\cdots$% per 1% of hemisphere area; contours at 1 and 3% are shown by dashed lines in some diagrams. Areas of density of $>2\%$ (or $>1\%$ if the 1% contour is shown) are shaded lightly, and those of density of $>4\%$ (or $>3\%$ if the 3% contour is shown) are shaded more heavily.

The expected standard deviation of the contours varies in accordance with the number of c axes measured and the orientation density in the major fabric maximums. In general, it is about a third of the average density in the maximums, and it can be calculated in detail for each diagram from the data given. In drawing the contours, local density irregularities comparable to the expected standard deviations were smoothed somewhat, on the assumption that they are not statistically significant. The smooth appearance of the contours in comparison with the results of conventional contouring procedures reflects the effort to avoid statistically insignificant information in the diagrams while obtaining the maximum resolution of detail consistent with adequate counting statistics.

FLOW BEHAVIOR

The deformation record for experiment A14, given in Figure 6, shows typical features that allow three creep stages to be distinguished. An initial stage of transient decelerating creep is followed by a stage in which the creep rate becomes stabilized and essentially steady. This second stage continues to a total shear strain $\bar{\gamma}$ of about 0.03 (± 0.005 in different specimens). It is followed by the third stage, in which the creep rate accelerates. For some specimens a more or less continual increase in creep rate occurs during the third stage, whereas for others a creep rate acceleration at the onset of the third stage is followed by an interval of steady creep, after which further accelerations occur. Figure 6 illustrates the latter type of behavior: a threefold increase in creep rate near $\bar{\gamma} = 0.025$ is followed by essentially steady creep to $\bar{\gamma} \approx 0.1$, where a twofold acceleration occurs, followed again by approximately steady creep. The three creep stages were designated primary, secondary, and tertiary creep by *Steinemann* [1958, pp. 20, 30]. Steinemann's description does not emphasize the continual or repeated creep rate accelerations that occur in the third stage, although it recog-

nizes that some increase in creep rate does occur during this stage, after the initial acceleration.

The shear strain rates $d\bar{\gamma}/dt$ listed in Table 1 are those measured during secondary creep. The selection of measurements is not extensive enough to define the flow law of the material in detail, but some significant features are evident in the data. A strong correlation between temperature and flow rate is seen when the $d\bar{\gamma}/dt$ values for A6 and A3 (at about $-4°C$) are compared with the value for A8 (at about $-1.5°C$). Comparison of A8 and A10 shows a marked nonlinearity in the dependence of $d\bar{\gamma}/dt$ on τ; interpreted in terms of a flow law of the type $\dot{\gamma} = b\tau^n$, where b is constant at a fixed temperature, it corresponds to $n \approx 3.5$. The increase in $d\bar{\gamma}/dt$ for A14 and especially for A13, in comparison with that for A8, demonstrates the strong effect that superposition of an independent stress component (the normal stress σ) has on the effective viscosity of ice as measured in shear. The effective shear viscosity is much less in A14 (0.3×10^{13} poises) than in A8 (1.5×10^{13} poises), even though the shear stress used is lower in A14 than in A8. This effect is an additional consequence of the nonlinear rheology and agrees quantitatively with the expected behavior of a generalized Newtonian fluid [*Nye*, 1953].

These observations and the numerical data in Table 1 agree well in general with the more extensive flow measurements of *Steinemann* [1958]. However, *Steinemann* [1958, p. 32] concluded that the generalized Newtonian fluid formulation of *Nye* [1953] is not adequate to account for the differences in flow behavior for different states of stress (different ratios of σ/τ).

In experiment A8 the ink line marked on the specimen surface revealed a markedly nonhomogeneous deformation, the lower half of the specimen having undergone a shear strain twice that of the upper half (Figure 4), even though the entire specimen was subjected to the same loading for the same period of time. Observation of this type of behavior shows that caution is needed in interpreting the flow of polycrystalline ice specimens.

Grip sliding. In some of the observed creep curves a large acceleration in the apparent creep rate reflects the onset of a sliding process of the specimen past one or both of the grips. Adhesion between ice and grips tends to be

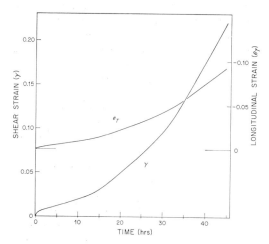

Fig. 6. Deformation history of specimen A14 showing torsional shear and axial shortening as a function of time. Note the creep rate accelerations at 14 and 31 hours.

Fig. 7. A portion of specimen A3 after deformation, showing deformed air bubbles. The length of the frame corresponds to 3.5 cm. The serrations on the upper edge are places at which the teeth of the grips engaged the specimen. The line at the base of the serrations is the sliding surface of the grip against the specimen.

lost at about the time that syntectonic recrystallization starts, and a plane of sliding frequently develops between the ice and the grips, tangent to the crests of the teeth. Such a sliding plane is visible in Figure 7, and a sliding plane at an early stage of development is shown in Figure 4. The ice remaining in the spaces between the teeth executes rotational motions, somewhat like roller bearings. Cohesion across the sliding plane is not entirely lost, and the applied torque is still transmitted to the specimen, but the sliding causes the rate of grip rotation to increase many times over that due to torsional deformation of the specimen alone. Axial compression inhibits the sliding; for example, no sliding occurred in experiment A14 (Figure 1).

This phenomenon offers a possible mechanism for basal sliding of polar glaciers, where the ice is everywhere below the melting point. Unlike other proposed mechanisms [*Weertman*, 1967], sliding of the type observed here does not depend on the melting of the ice. (At the ambient temperatures used, stresses of several hundred bars would be necessary to cause pressure melting.) However, the inhibiting effect of normal pressure appears to limit the applicability of this sliding mechanism to the shallower parts of polar glaciers.

Bubble deformation. In the torsion tests, ice deformation causes a well-defined flattening of air bubbles contained in the ice, as is shown in Figure 7. There is considerable variation in the amount and the direction of elongation of individual bubbles, showing that local inhomogeneities affect the bubble flattening, but there is a consistent over-all pattern. The orientation of the average elongation direction is given in Table 2 for specimens in which it could be observed; the angle given, $w \wedge E$, is the angle between the elongation direction E and the shear direction w. The estimated elongation direction corresponds approximately to the direction of the extensional principal strain axis e_1 ($w \wedge e_1$ in Table 2). The agreement is within 4° except for specimen A8, in which the elongation was not well shown. The degree of bubble flattening (axial ratio of ~3:1 in experiments A5 and A13) is somewhat larger than that expected from the bulk strains of the specimens, but a reliable evaluation of this discrepancy requires a detailed statistical study of bubble shapes.

In the deformation experiments at 0°C the air bubbles respond in a rather different way: instead of becoming flattened, they migrate through the ice. This type of behavior will be discussed in a separate paper.

TABLE 2. Orientation Information for Fabric and Deformation Features in Syntectonically Recrystallized Ice Specimens

		1	2	3	4	5	6	7	8	9
Experiment	Figure	σ/τ	$w \wedge E$, deg	$w \wedge e_1$, deg	$T \wedge \sigma_3$, deg	$T \wedge$ $\frac{1}{2}(M_1+M_2)$, deg	$T \wedge M_1$, deg	$w \wedge M_2$, deg	$M_2 - M_1$, deg	$D \perp wT$, deg
A10	15a	0			45	40	0	11	79	
A8	12	0	~30	40	45	38	-6	19	65	
A7	13a	0.3			40	~35	~0	~20	~70	
A6	14b	0.3	36	37	40	35	-3	24	63	
A5	16a	0.8	39	35	34	30	1	29	62	70
A3	16b	0.8	36	36	34	29	4	30	68	70
A14	16c	1.5			26	31	5	23	72	60
A13	16d	3.8	17	16	14	11	16	52	54	60
C4	17a	∞			0	0	37	53	74	75

1. Ratio of compressive stress σ to shear stress τ in simple shear.
2. Angle between shear direction w and direction of bubble elongation E.
3. Angle between w and extensional principal strain axis e_1.
4. Angle between torsion axis T and compressive principal stress axis σ_3.
5. Angle between T and bisector of acute angle between M_1 and M_2.
6. Angle between pole T of shear plane and fabric maximum M_1.
7. Angle between pole w of conjugate shear plane and fabric maximum M_2.
8. Acute angle between maximums M_1 and M_2.
9. Diameter of small-circle girdle, measured perpendicular to wT plane.

TEXTURAL CHANGES DURING FLOW

The texture of syntectonically recrystallized ice is illustrated in Figures 8–10. In place of the planar grain boundaries of the starting material (Figure 2), grain boundaries in the recrystallized specimens have become highly irregular. The irregularity is not simply a suturing superimposed on a pattern of boundaries inherited from the original tessellate texture but instead represents a thorough reorganization in which the original equidimensional grain shapes have been altered to highly irregular interlocking or interdigitating shapes reminiscent of the pieces of a jigsaw puzzle. Extensive grain boundary migration is indicated by the textural reorganization and also by a general coarsening of grain size from that of the starting material.

The extent of grain coarsening varies inversely with the applied stress over the range studied. Thus the texture of A10 (Figure 8), recrystallized under a shear stress of 1.9 bars, is much coarser than that of A8 (Figure 9), deformed at 4.1 bars.

Neither time nor total strain has a prominent influence on texture over the range studied in these experiments. Thus A7, A8, and A6 show similar grain sizes after deformation times of 1.1, 2.8, and 13.5 days, respectively, under the same shear stress. This finding implies that the recrystallization textures have reached an essentially steady state in the times involved. At a shear strain of 0.05 in A7 (Figure 10) some relict tessellate grains from the original texture remain, but the ice as a whole has already recrystallized to an interlocking texture.

Recrystallized grain shapes appear to show a slight elongation parallel to the shear plane, but the effect is largely obscured by the irregularity of the grains.

A pronounced effect of temperature on the recrystallization texture is seen in a comparison of C4 and A8, which were deformed under comparable effective shear stresses (3.7 and 4.1 bars, respectively) for comparable times; C4 was deformed at 0°C, and A8 at about −1.5°C. The grain size in C4 is somewhat

Fig. 8. Texture of specimen A10 after deformation for 59 days under a shear stress of 1.9 bars. The length of the frame corresponds to 2.5 cm. The torsion axis is oriented parallel to the length of the frame (left and right). The small dark circles and points are air bubbles.

Fig. 9. Texture of specimen A8, after deformation for 2.8 days under a shear stress of 4.1 bars. The torsion axis is oriented parallel to length of frame, which corresponds to 4 cm. Note strain shadows in the large grain near the bottom of the photograph.

larger (by about 50%) than that in A10 (Figure 8), which was deformed at a lower stress for a much longer time, and it is thus substantially larger than that in A8 (Figure 9). A correspondingly great sensitivity of texture to temperature is not evident in a comparison of A6 and A8, although the similarity in the grain sizes of the two samples might be interpreted as the result of the compensating effects of longer deformation time and lower temperature in A8. The marked increase in grain size in C4 suggests that the mobility of grain boundaries is greatly enhanced when the melting point is reached.

Many of the grains show strain shadowing in bands parallel to the trace of the c axis (Figure 9), which represents kinking by translation gliding on the (0001) plane. Strain shadows are abundant in ice deformed both at 0°C (C4) and below the melting point.

Similar textural features were observed in the more detailed study by *Steinemann* [1958]. The relationship between grain size, stress, and temperature was summarized by *Rigsby* [1960].

SYNTECTONIC RECRYSTALLIZATION FABRICS

Simple shear. In the ice specimens as initially prepared, c-axis orientations are nearly random, as is indicated by Figure 11. Recrystallization during deformation produces well-defined c-axis fabrics, which are shown in Figures 12–17. The type of fabric developed under simple shear is illustrated in Figure 12. There are two maximums in the c-axis orientation density, the first maximum, M_1 (Figure 12), oriented approximately normal to the shear plane (or parallel to the torsion axis T) and the second maximum, M_2, inclined about 20° to the shear direction w (the point at the center of Figure 12). Generally, M_1 is somewhat stronger than M_2, and the peak orientation density is higher in M_1 than in M_2 by about 2% per 1% of area. Maximum M_2 is located approximately along the great circle between w and T, and its location in relation to the sense of shear can be described by saying that it is rotated away from w in the direction corresponding to the bulk rotation experienced locally by the deforming ice. This rotation, which occurs about axis R, is clarified in Figure 3. The sense of rotation is shown in Figure 12 by symbols representing the shear stress vector acting across the shear plane wR. The minus sign at the top represents the stress vector acting across the plane wR from above on the material below wR, and the plus sign at the bottom similarly represents the stress vector acting from below on the material above wR. The minus sign represents a vector directed toward w, perpendicularly down into the plane of the diagram, and the plus sign represents a vector in the opposite direction, toward $-w$, up out of the diagram.

Maximum M_2 typically shows a marked elongation perpendicular to the wT plane. In Figure 12, M_2 at the height of the 2% contour is about 28° wide in the wT plane and about 65° wide perpendicular to that plane. The elongation can be explained to some extent by the observation that in individual samples of about 100 c axes, from single thin sections, the position of the M_2 peak varies somewhat from section to section, scattering 5°–15° on either side of the wT plane. Combination of results from several thin sections thus pro-

duces a composite M_2 peak elongated in the direction of scatter. Evidence of a tendency for the M_2 maximums in different experiments to scatter somewhat out of the wT plane can be seen in Figures 12, 14, 15, and 16. These figures suggest a tendency for M_2 to be displaced out of the wT plane preferentially toward $+R$. Such an apparent tendency might be due to a systematic orientation error in the cutting of the thin-section planes.

Maximum M_1 in Figure 12 appears elongated perpendicular to wT, but this effect, although present to some extent, is exaggerated by the peripheral distortions in the equal-area projection. At the height of the 2% contour, M_1 in Figure 12 is about 41° wide in wT and about 54° wide perpendicular to wT.

The relatively few c axes that do not fall within M_1 or M_2 tend to lie in the region spanned by the acute angle between the two maximums. When the ½% contour is drawn in Figure 12, a ring-shaped area of weak orientation density is defined, forming a small circle containing M_1 and M_2 and centered over the bisector of the acute angle between them. The region around R is almost devoid of orientation

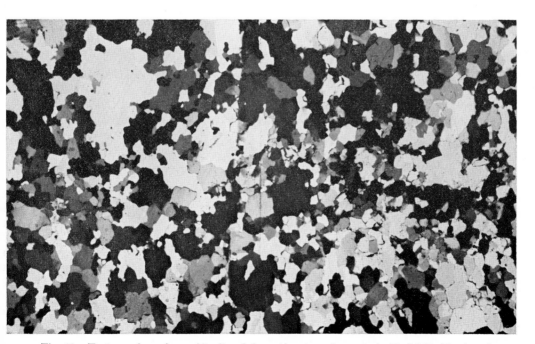

Fig. 10.   Texture of specimen A7 after deformation to a shear strain $\bar{\gamma}$ of 0.05. The length of the frame, which is parallel to the torsion axis, corresponds to 4 cm. Fine crystals in the right-center part of the figure are relicts from the original texture.

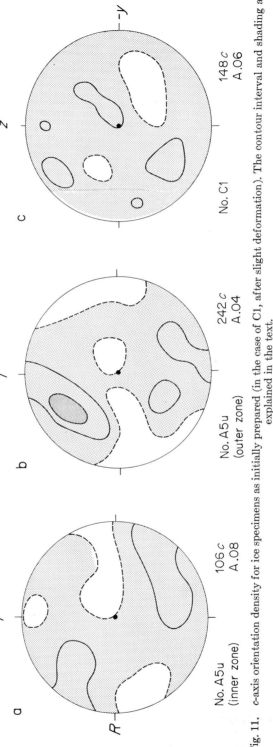

Fig. 11. c-axis orientation density for ice specimens as initially prepared (in the case of C1, after slight deformation). The contour interval and shading are explained in the text.

density, as is the region in the obtuse angle between M_1 and M_2.

The relationships between the type of c-axis fabric shown in Figure 12 and observations of previous workers are as follows. The fabric diagram of a shear experiment of *Steinemann* [1958, Figure 60; reproduced in *Kamb*, 1959*b*, Figure 15] was interpreted by him as showing two maximums symmetrically disposed about T in the TR plane 25° from T and a third maximum in the Tw plane 50° from T. Resolution among the three maximums is not good, and the statistical validity of the individual maximums is not certain. It appears that the first two maximums taken together may correspond to M_1 as observed here but with abnormal elongation in the TR plane. The third maximum may correspond to M_2, but it is poorly resolved and 20° closer to T than is M_2 as found here. Its orientation in relation to the sense of shear is not known. *Rigsby* [1960, p. 603] produced a two-maximum fabric by subjecting an ice specimen to shear deformation with repeated reversals in the sense of shear. The maximums are oriented approximately perpendicular to the two conjugate shear planes of the specimen, and in this sense both play a role analogous to M_1 in the present experiments. (The angle between Rigsby's maximums is, in fact, 82°, so that perpendicularity to conjugate shear planes can only be approximate.) Because of the shear reversal the maximums cannot be related to a sense of shear. From the pattern in Figure 12 we would expect that repeated reversals of the sense of shear would cause M_2 to oscillate back and forth about w and to lie on the average near w, the position found by Rigsby. *Shumskii* [1958, Figure 3] showed a similar but somewhat more diffuse and less clearly portrayed pattern of two broad maximums from an ice specimen deformed for 20 days in shear. The two maximums appear centered at points corresponding to T and w in the present experiments. A relation of the maximums to the sense of shear was not given.

Shumskii [1958, p. 247] stated that c axes forming the maximum near T represent newly formed relatively unstressed crystals, whereas the c axes near w represent relict crystals from the original texture, which are highly deformed. In the present study no such distinction was found between crystals in M_1 and those in M_2:

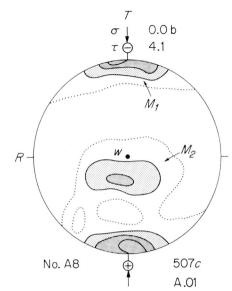

Fig. 12. c-axis fabric of specimen A8 showing typical features of fabrics produced by syntectonic recrystallization under simple shear. The stress axes T, R, and w are explained in the text and Figure 3. The symbols at top and bottom show the sense of the shear stress vector, minus down into the figure (toward w), plus up out of the figure (away from w), as is explained in the text. The number of c axes measured is designated by c; the fractional area of the counter used in preparing the contoured density diagram, as explained in text, is designated by A. The contours are at densities of 2, 4, and 6% per 1% area of hemisphere; the dotted contour is at ½% per 1% area.

the latter grew to sizes comparable to the former, and strain shadows were, in fact, 4 times more abundant in M_1 than in M_2. However, as a specific test of Shumskii's interpretation, a special ice sample (A11) was prepared in which all the c axes of the original crystals were directed approximately radially outward (toward R) in the cylindrical specimen. After 2 days under an average shear stress of 5.5 bars the specimen began to recrystallize, as was indicated by acceleration in the creep rate. The new crystals formed predominantly in the grain boundary regions between remnants of the original crystals, producing a brick and mortar texture. Numerous crystals in orientations corresponding to both M_1 and M_2 were found in the newly recrystallized material.

Dependence of fabric on strain and time. Progressive development of recrystallization fabrics under simple shear is shown in Figures

13–15, obtained from experiments carried out to different total strains. Diagram A8B (Figure 14a) was obtained from the more highly sheared bottom half of specimen A8 (Figure 6), and diagrams A8T (Figure 13b) and A8T′ (Figure 15b) were obtained from the top half in traverses at radial distances of 2.6 and 3.3 cm, respectively, corresponding to $\gamma = 0.11$ and to $\gamma = 0.13$.

Experiments A7 and A8T (Figure 13), at $\bar{\gamma} = 0.05$ and at $\bar{\gamma} = 0.11$, show early stages in the development of the recrystallization fabric. Although larger c-axis samples would be needed to define these fabrics fully, the beginning of a preferred concentration of axes around T and in the area between $+w$ and $-T$ is evident. The fabric in A7 is weak, but the concentration of axes near T and the deficiency of axes near R are statistically significant. In A8T the disappearance of axes in the region between $+T$ and $+w$ has become significant, and the concentration near T has increased. By $\bar{\gamma} = 0.24$ (A8B, Figure 14a) the fabric has become much stronger, and maximum M_2 has appeared clearly. The contrast between A8T (Figure 13b) and A8B (Figure 14a) is noteworthy, since these fabrics come from the same specimen but from regions of different total strain (Figure 4). Further strengthening and sharpening of the pattern is evident at $\bar{\gamma} = 0.40$ (A6, Figure 14b), the largest shear strain reached in these experiments.

The transition from the relatively weak fabric in A8T to the well-developed fabric in A8B is bracketed more sharply by comparing A8T (Figure 13b) and A8T′ (Figure 15b), which are c-axis samples from traverses at two different radial distances in the top part of specimen A8 and therefore differ only slightly in total strain. The fabric in A8T′, at $\bar{\gamma} = 0.13$, is fully as well developed as that in A8B; this finding suggests that a major step in fabric reorganization occurs at a shear strain of about 0.12. Again, the fabric in A5 (Figure 16a), for which $\bar{\gamma}$ is estimated as 0.27 (Table 1), is almost as strong as that in A6, at $\bar{\gamma} = 0.40$, and it is significantly stronger than that in A8B, at $\bar{\gamma} = 0.24$. These results suggest that the fabric development proceeds in a stepwise fashion, distinct reorganizations occurring at shear strains near 0.04, 0.1, and 0.3. It is, however, alternatively possible that the fabric develops continuously but that individual regions containing several hundred crystals are subject

to large statistical fluctuations in the extent of crystal reorientation at a given total strain. Thus A8T′ and A5 may have sampled regions in which the fabrics became perfected more rapidly than they do on the average, and A8T and A8B may represent regions in which fabric development was retarded in relation to the average. More extensive experiments are necessary to choose between these two alternative interpretations of the data in Figures 13–15.

The similarities in strength and perfection of fabrics A5 (Figure 16a) at $\bar{\gamma} \sim 0.27$ and A6 (Figure 14b) at $\bar{\gamma} = 0.40$ suggest that a steady state is reached in the development of the fabric pattern at a shear strain of about 0.3. (In A5 the equivalent shear strain, if the contribution of the longitudinal strain is taken into account, is 0.30.) Likewise, in comparing A13 with A14 there is only a modest increase of fabric strength, in spite of the much greater strain (especially e_T) for A13; this evidence suggests that a steady state is being reached.

The position of M_1 in simple shear ($\sigma = 0$) remains steady as strain increases. Maximum M_2 remains roughly steady in position from A8T′ to A8B, but the angle $w \wedge M_2$ increases about 8° from A8B to A6. Part of this increase (about 5°) should be due to stress reorientation in A6 ($\sigma = 1.3$ bars causes $T \wedge \sigma_3$ in Table 2 to change by 5°). In the present experiments there is thus no definite indication that $w \wedge M_2$ increases progressively with increasing strain, as would be required to explain the angle $w \wedge M_2 \sim 40°$ in *Steinemann's* [1958, Figure 60] diagram, for which a shear strain $\bar{\gamma}$ of 0.75 was given. A progressive increase in $w \wedge M_2$ with strain would imply that a structural steady state is not reached for shear strains γ of ~ 0.5. Further experimental study of this important point is needed.

Since shear strain and time increase together, the dependence of fabric on strain in Figures 13–15 can also be viewed as a dependence on time. In experiment A8, however, the different strains and fabrics developed under a given stress and in the same time; this experiment indicates that time alone is not the determining variable.

Dependence of fabric on shear stress. Figure 15 compares the fabrics of two samples deformed to the same total shear strain but at different shear stresses. The fabrics are similar

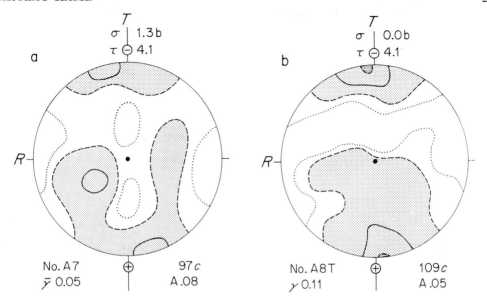

Fig. 13. *c*-axis fabrics of ice deformed in simple shear to shear strains $\bar{\gamma}$ of 0.05 and 0.11. The contours are at 0.5% (dotted), 1% (dashed), and 2 and 3% (dashed) per 1% area of hemisphere. The conventions are as in Figure 12 and explained in the text.

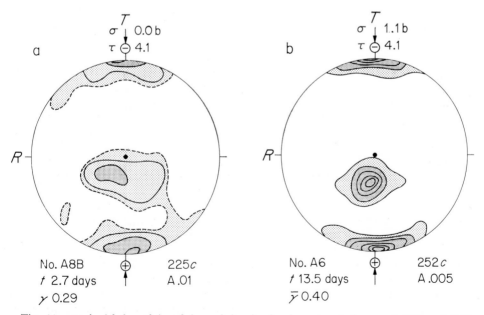

Fig. 14. *c*-axis fabrics of ice deformed in simple shear to strains γ of 0.29 and 0.40. The duration of the deformation for each experiment is designated by *t*. The conventions are as stated in the legends of Figures 12–13.

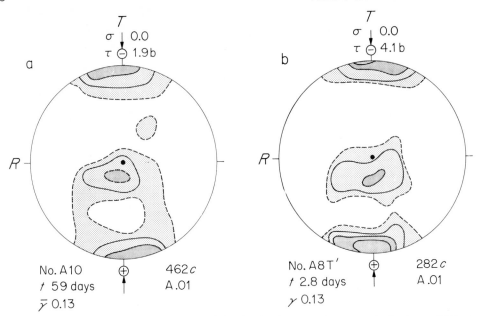

Fig. 15. *c*-axis fabrics of ice deformed in simple shear to the same strain ($\gamma = 0.13$) at shear stresses τ of 1.9 and 4.1 bars. The conventions are as stated in the legends of Figures 12–14 and explained in the text.

in a general way, although the one obtained at the higher stress (A8T′) is somewhat stronger (contrary to the conclusion of Steinemann [1958, p. 46]). The M_2 peak is about 7° closer to w in the experiment at lower stress (A10). Also, the weak small-circle girdle between M_1 and M_2 is more strongly developed at the lower stress.

Since the time duration was some twentyfold longer in experiment A10 than in experiment A8, the comparison in Figure 15 might alternatively be interpreted as a test of the effect of time on the recrystallization process for fixed total strain. If stress and strain rate are related functionally by a flow law, so that time and stress are not independent variables at fixed total strain, the distinction between alternative interpretations of Figure 15 emphasizing time or stress is, in fact, academic. In any case the relatively slight differences between A10 and A8T′ in Figure 15, in spite of the very different times used to produce the two fabrics, show that any effects of time alone, as distinct from effects tied to total shear strain, must be rather small.

Although strain shadows were observed microscopically with about equal over-all frequency

(10%) in the grains of specimens A8 and A10, the fraction observed in grains of M_2 was much lower in A10 (5%) than in A8 (28%). This difference may be related to the detected difference in fabric.

Effect of stress state. Figure 16 shows the effect of progressively increasing the normal stress σ superimposed on a shear stress τ in the torsion experiment. Two changes occur in the resulting *c*-axis fabrics. (1) The small-circle girdle connecting M_1 and M_2 becomes enhanced at the expense of the maximums, which weaken and elongate perpendicular to the wT plane, following the girdle. (2) The pattern as a whole rotates around R so as to bring the center of the girdle toward coincidence with T. The first effect is faintly recognizable when $\sigma/\tau = 0.8$ (A3) and becomes definite at $\sigma/\tau = 1.5$ (A14). A clear indication of the rotation effect, maximum M_1 definitely moving away from T, is seen for $\sigma/\tau = 3.8$ (A13). At $\sigma/\tau = 1.5$ (A14), maximums M_1 and M_2 have moved about 5° away from their average positions for $\sigma = 0$, but the amount of rotation appears somewhat less than that for A3, in spite of the larger ratio of σ/τ.

Quantitative information on the orientations

of the maximums in relation to axes T and w, as a function of σ/τ, is given in Table 2. The angles listed as $T \wedge M_1$ and $w \wedge M_2$ are the directly measured angles if the maximums lie in the wT plane; otherwise they represent the component obtained by projecting the positions of the maximums into the wT plane, along great circles through R. The bisector of the angle between M_1 and M_2, called $\frac{1}{2}(M_1 + M_2)$ in Table 2, shows a general correlation with the

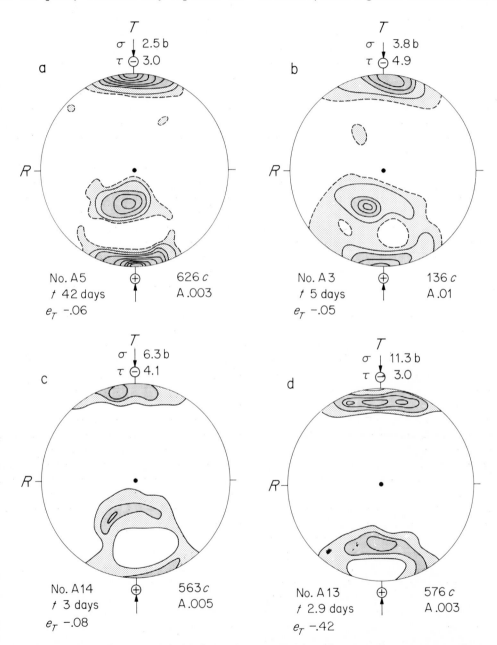

Fig. 16. The effect on c-axis fabrics of the superposition of compressive stress σ on shear stress τ. Stresses are indicated at the top of each diagram. The contours and conventions are as stated in the legend of Figure 13 and explained in the text. The longitudinal strain parallel to the torsion axis is indicated by e_T.

maximum compressive principal stress axis σ_3, whose orientation is given in terms of the angle $T \wedge \sigma_3$ between it and T. However, at $\sigma = 0$ there is a definite discrepancy between the two orientations, which reflects the fact that M_2 is not located at w while M_1 is located at T. The angular separation between M_1 and M_2 (designated $M_2 - M_1$ in Table 2) and the diameter of the small-circle girdle measured perpendicular to the wT plane (designated $D \perp wT$ in Table 2) do not show any clear dependence on stress state, but a small dependence could be masked by the rather large and seemingly random fluctuations in these angles.

The limiting case $\sigma/\tau \to \infty$, toward which the sequence in Figure 16 tends, represents simple uniaxial compression, which was studied by *Steinemann* [1958, Figures 62–65], who gives syntectonic recrystallization fabric data for five stress levels from $\sigma = 3.4$–16.0 bars. Although the fabrics are rather irregular, particularly for the higher stresses, the general pattern is that of a small-circle girdle centered about the pole of compression, in agreement with what is expected from the trend shown in Figure 16. At a stress of $\sigma = 11.0$ bars the diameter of the small-circle girdle in the fabric of *Steinemann* [1958, Figure 65] is about 80°, rather larger than that indicated here in experiment A13 (Table 2). The considerable irregularity in Steinemann's fabrics may reflect poor counting statistics, and the tendency for increased apparent orientation densities toward the center of the diagrams may reflect the fact that the orientations were plotted in stereographic projection. Difficulties with the counting procedure are indicated by the fact that the diagrams show large departures from the required centrosymmetry at the margins.

Compression at 0°C. The c-axis orientation density produced in experiment C1 shows no significant preferred orientation and is given in Figure 11. The specimen showed no visible change from the tessellate texture of the starting material. Specimen C4, deformed slightly longer at a stress about 3 times greater than that for C1, gives a fabric showing substantial crystal reorientation (Figure 17a). There is an incomplete small-circle girdle about z with broad maximums located around the two points where it crosses the xz plane. (The separate closures of the 3% contour in the upper-right and lower-

left quadrants of Figure 17a are not statistically significant enough to imply resolution into two maximums there.)

To interpret this fabric in relation to those of the torsion experiments, it is necessary to estimate the state of stress in C4. If the lateral constraint along the y axis (Figure 5) were such as to produce plane strain, the stress state would be pure shear, with $\sigma_y = \frac{1}{2}\sigma_z$. However, pressure melting on the y faces of the specimen lowers the compressive stress σ_y substantially. From the roughly measured strains, $e_{xx} = +0.18$ and $e_{yy} = +0.13$, the stress σ_y is estimated as $0.1\sigma_z$ on the assumption that the ice deforms as a generalized Newtonian fluid. If the fabric pattern does not depend on the mean stress, the 0°C compression experiment with $\sigma_y = 0.1\sigma_z$ and $\sigma_x = 0$ corresponds to a torsion experiment having $\sigma/\tau = 2.7$, a stress state intermediate between that in experiments A13 and A14. The fabric of C4 is, in fact, compatible with one intermediate between A13 and A14, as can be seen by comparing Figures 16 and 17a, with due regard for the different orientations of the principal stress axes in the different figures (Table 2). The angle between the two maximums in the fabric of C4 is nearly the same as that in the fabric of A14, whereas that in the fabric of A13 is distinctly smaller (Table 2). The strength of the pattern for C4 is distinctly weaker than that for A13 and A14.

It thus appears that syntectonic recrystallization at the melting point produces a fabric pattern that is qualitatively, if not quantitatively, similar to that produced at temperatures a few degrees below the melting point. A full evaluation of the suggested quantitative differences between fabrics produced at the melting point and below requires more extensive experimental material to cope with statistical fluctuations in the fabric data.

The longitudinal strain $-e_z$ required for recrystallization and fabric reorientation is bracketed between the limits 0.009 and 0.31 by experiments C1 and C4. It is thus compatible with the shear strains $\bar{\gamma}$ of 0.04–0.1 required for comparable effects in experiments below the melting point.

EFFECTS OF RECRYSTALLIZATION ON FLOW

The extensive changes in texture and fabric accompanying syntectonic recrystallization of

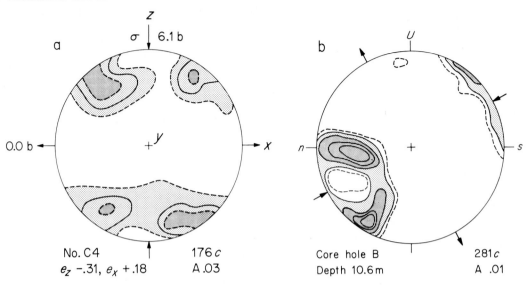

Fig. 17. (a) Fabric of ice syntectonically recrystallized at 0°C. Contours of c-axis orientation density are at 1% (dashed), 2% (solid), and 3% (dashed) per 1% area of hemisphere. Stress axes are labeled to correspond with Figure 5. The arrows represent the principal strains designated e_x and e_z. (b) Fabric of glacier ice from the base of an icefall in Blue glacier, Washington, core-sampled at a depth of 10.6 meters. Contours of c-axis density are at 1% (dashed), 2, 4, 6, and 8% (solid) per 1% area of hemisphere. The core axis is U. Compression and extension axes (inward-directed and outward-directed arrows, respectively) represent pure shear, compression being perpendicular to macroscopic foliation and sub-parallel to ice flow direction.

ice contribute in an important way to the flow process and are responsible for some of the complexities in the flow curves discussed earlier (Figure 6). Since the textural and fabric changes are only partly developed at a shear strain $\bar{\gamma}$ of 0.05 (experiment A7), there is a general correlation between the onset of recrystallization as recognized texturally and the acceleration in creep rate that occurs at the end of the second creep stage ($\bar{\gamma} \sim 0.03$). This acceleration can be attributed to the crystal reorientations produced by recrystallization: the newly formed crystals are in orientations that are relatively favorable for intracrystalline plastic flow, and the orientations preferentially eliminated are those for which intracrystalline plastic flow is most difficult. Single crystals of ice deform readily by translation gliding parallel to the basal plane (0001), whereas other types of intracrystalline deformation, if they occur at all, are very difficult [Griggs and Coles, 1954; Glen and Perutz, 1954; Nakaya, 1958]. Under simple shear, crystals having c axes centered in maximum M_1 (at T) are oriented with their glide plane parallel to the

shear plane wR of the specimen and are thus in the orientation for 'easy glide' [Butkovich and Landauer, 1958]; intracrystalline plastic flow in these crystals contributes directly to the over-all shear deformation of the specimen. The same statement would be true for crystals with c axes at w. Because of the approximately 20° displacement of M_2 away from w, somewhat more resistance to the shear deformation of the specimen is offered by crystals in this maximum than by crystals with c axes at T or w: for a given resolved shear stress on the crystal glide plane the required shear stress across the specimen shear plane is greater by a factor of $1/\cos 40° = 1.3$. Nevertheless, crystals in M_2 are favorably oriented in comparison with the orientations eliminated in recrystallization, those with c axes near R and midway between w and $\pm T$. The orientation at R is the orientation for 'hard glide' [Butkovich and Landauer, 1958], and the orientation midway between $+w$ and $-T$ with the c axis aligned with the maximum compressive principal stress axis is one for which single crystals of ice show no measurable plastic flow and remarkably high strength, ~1 kb (W. F. Brace and B. Kamb, unpublished manuscript, 1968). Crystals with c axis midway between $+w$ and $+T$ could fail in tension by basal cleavage, but, in fact, there was no microscopic evidence of such failure in the specimens listed in Table 1.

As a compressive stress σ is superimposed on the shear stress τ, the changes in the fabric pattern (Figure 16) are such as to promote intracrystalline plastic flow under the altered stress state. The crystal orientations would be maximally effective if, as the stress state approaches uniaxial compression, the diameter of the small-circle girdle in the fabric pattern were 90°. The actual angle is somewhat less, but, again, the orientations eliminated (c axes nearly parallel and perpendicular to the compression axis) are those offering the greatest resistance to intracrystalline plastic flow.

A threefold acceleration in shear-strain rate is achieved with an extent of crystal reorientation indicated in the relatively weak fabric of experiment A7 (Figure 13a). A large scatter remains in crystal orientations away from the easy glide orientations at T and w. The further accelerations in creep rate that occur later during the third creep stage are attributable to the additional steps in sharpening and strengthening the fabric pattern that occur at higher strains (Figure 14). The abrupt transitions to higher creep rates seen in some flow curves (such as those in Figure 6 at a strain $\bar{\gamma} = 0.10$) have an explanation in the seemingly stepwise fabric reorganization implied by the measured fabric diagrams.

Textural and structural changes other than fabric reorientation doubtless also contribute to the complexities in the flow curve of polycrystalline ice, but their effects are less easily assessed. It is not known in detail how the change from tessellate to interlocking grain shapes should affect the flow of the aggregate, although it seems clear that irregularities in the grain boundaries should interfere with grain boundary sliding and thus inhibit one possible deformation mechanism in the flow of the polycrystalline aggregate. A general coarsening of grain size during recrystallization should tend to increase the creep rates. An inverse relationship between grain size and creep resistance has been suggested from experimental indications by *Steinemann* [1958, pp. 24, 34] and is also recognizable in field observations [*Kamb and Shreve*,

1963]. This effect could appear in the present experiments as a correlation between increase of grain size on recrystallization and increase in creep rate. Comparison between experiments A14 and A10 should reveal such a correlation if it exists, because the recrystallized grain size in A10 (Figure 8) is several times that in A14 (similar to A8, Figure 9). However, the creep rate acceleration in A14 (sixfold at a shear strain of about 0.12) is comparable to the creep rate acceleration in A10 (fivefold). It appears that any effect of smaller grain size in A14 is compensated for by the somewhat stronger and sharper fabric pattern that developed at the higher stress level. *Steinemann* [1958, Figure 14] reported a somewhat increased creep rate acceleration at larger stresses and hence smaller recrystallized grain sizes.

The strain-softening property of ice single crystals [*Griggs and Coles*, 1954; *Steinemann*, 1958, p. 13] should contribute to creep rate acceleration in the flow of polycrystalline ice. Since strain softening is due to the accumulation of dislocations in the process of intracrystalline plastic flow, its contribution to the flow of the polycrystal tends to be eliminated by recrystallization.

The development of strain shadows in the individual crystals is direct evidence that the creep process in polycrystalline ice involves intracrystalline plastic flow and results in the accumulation of crystalline defects. The attainment of a true steady state in the creep process therefore requires the operation of a recrystallization mechanism, which generates undistorted crystalline material at a rate that keeps pace with the generation of defects by intracrystalline flow. The process that leads to the fabrics in Figure 14 is thus essential to the attainment of a steady state. The abundance of strain shadows in the specimens shows erratic fluctuations but no systematic increase as total strain increases (A10, shadows observed in 11% of the grains; A8, in 10%; A14, in 23%; A6, in 4%); these data indicate that continual recrystallization keeps pace on the average with the generation of bending distortions by plastic flow. The fabrics developed under simple shear appear to become steady near $\gamma = 0.3$, but some uncertainty about the final steady state remains, as was explained earlier. The available flow data do not extend to high enough strains to show whether a final steady state is reached in the creep rate near $\gamma = 0.3$.

The appearance of a temporarily steady creep rate during the second stage of creep is in a sense deceptive, because conditions cannot in reality be steady: there is progressive accumulation of defects that will ultimately initiate recrystallization. The steady creep rate in this stage may reflect merely a temporary balance between processes tending to harden the aggregate (evidenced by the initial stage of decelerating creep) and processes tending to soften it (in particular, intracrystalline strain softening).

PHYSICAL FACTORS GOVERNING
RECRYSTALLIZATION

Comparative behavior of texture and fabric changes. The experimental results show that, although in a general way the changes in texture and fabric produced by syntectonic recrystallization occur synchronously, in detail they follow a distinctly different course and respond rather differently to the controlling physical variables. The transformation from tessellate to interlocking texture is completed early in the recrystallization process at a shear strain γ near 0.05 well before a strong recrystallization fabric has developed. Beyond $\gamma \sim 0.05$, relatively little further change in the textural pattern and grain size takes place, whereas the fabric pattern continues to develop and strengthen up to a strain of about 0.3. Grain size is strongly dependent on shear stress, whereas fabric shows only a slight dependence when observed at fixed total strain. Grain shapes show only slight anisotropy, if any, in response to anisotropic stress and are not affected appreciably by changes in stress state attainable by variation in the ratio of σ/τ, whereas the fabrics become highly anisotropic as recrystallization proceeds and respond sensitively in orientation and pattern to the stress state. Texture is sensitive to temperature, whereas fabric is not: recrystallization gives a distinctly coarser texture at the melting point than at temperatures only a few degrees below, whereas the fabrics developed under the two conditions are nearly the same.

Control of fabric strength. The perfection of the experimentally produced syntectonic recrystallization fabrics depends primarily on the total specimen strain and only secondarily on temperature, time, or stress level. The experiments indicate that the fabrics become somewhat weaker at lower stress levels and at higher temperatures very near or at the melting point, but more extensive experimental material is necessary to establish these effects with certainty. Peak orientation density depends to some extent on the state of stress simply because the favored orientations, confined into discrete maximums under simple shear, become necessarily spread out over girdles under uniaxial compression, owing to the rotational symmetry.

Statistical nature of the process. In comparing the fabric data for experiments of the same or similar strains and stress states, or even from within individual specimens, marked differences in strength, perfection, or orientation of fabric maximums suggest that the course of recrystallization is subject to statistical fluctuations that are substantially larger than those that could be caused by sampling statistics alone. The fluctuations presumably occur in response to random local variations in the texture or fabric of the initial material, and they suggest that the recrystallization process is cooperative in nature, so that original fluctuations become amplified. Evidence for a phenomenon of this kind is found in the inhomogeneous deformation that occurred in specimen A8 (Figure 4). Stepwise development of fabric strength as a function of strain can explain some of the anomalous features in the measured fabric diagrams without recourse to statistical fluctuations, but, even if this explanation is valid, it appears probable from experiment A8 that the implied abrupt fabric reorganizations occur on a statistical basis rather than at rigidly fixed total strains or times.

Effect of air bubbles. Since the changes in texture and fabric that took place here in specimens containing abundant air bubbles were comparable to the changes observed by *Steinemann* [1958] in air-free samples, it follows that the air bubbles do not substantially inhibit the processes of recrystallization and grain growth. *Rigsby* [1960, p. 601] reported a seemingly contrary field observation indicating that air bubbles strongly inhibit grain growth at the melting point. In the present experiments at the melting point, extensive grain growth occurred in 2.3 days in a sample initially containing abundant air bubbles (experiment C4).

If air bubbles acted to inhibit recrystallization

and grain growth by pinning the grain boundaries locally and interfering with their lateral migration, one would expect to find that grain boundaries in the experimental specimens had moved into positions such that a large fraction of the bubbles are located along them. This is not the case, as is shown in Figure 8: most of the bubbles are intracrystalline. A pinning mechanism of this type could not be effective in preventing grain boundary migration in the experiments at 0°C because of the high mobility shown by the bubbles in these experiments.

It seems likely that the stress level in the glacier ice observed by Rigsby was much lower than that in the present experiments, and it therefore remains possible that an inhibitory effect of air bubbles on recrystallization could operate at low stresses. However, *Rigsby* [1960, Figure 16] showed an example of recrystallization and grain growth occurring in bubbly ice in an unstressed state.

Fabric symmetry and orientation. In the fabrics developed under simple shear (Figures 12–15), maximums M_1 and M_2 are of unequal strength and do not lie at 90°. With due allowance for statistically expectable fluctuations of orientation density in the measured diagrams, the indicated fabric pattern has monoclinic symmetry; a single twofold symmetry axis lies to a good approximation along the radial direction R, and a single mirror plane lies along wT. The pattern does not conform to the orthorhombic symmetry required if the crystal reorientations occurred in response to either the stress tensor, the strain tensor, or the strain rate tensor. Even if M_1 and M_2 were of equal strength, their relative orientations are not compatible with the symmetry axes (principal axes) of these tensors. The pattern does conform to the 'movement picture' (velocity field) as this concept is used in structural petrology, but some clarification is required to make this concept meaningful because of the possibility that arbitrary rotations can be added to the velocity field without affecting phenomena based on specimen deformation. Rational choice of a reference coordinate system in which the velocity field should be described can be made when conditions allow a steady-state fabric to develop, namely, when the deforming stresses remain constant; this coordinate system is the one in which the fabric, and hence also the

stresses and velocity field, remains constant. For this choice of coordinate system under simple shear the velocity field has monoclinic symmetry, conforming with the observed fabrics. The agreement here signifies that crystal reorientation in syntectonic recrystallization is controlled by the time progression of deformations experienced by the material rather than by the stress or corresponding momentary deformation increment alone or by the fully integrated deformation (total strain).

In considering theoretically the flow law of ice *Glen* [1958, p. 180] assumed that fabric anisotropy in deforming ice has the same symmetry as the applied stress. The present experiments show that this assumption is not rigorously valid, but they do not indicate the extent to which the observed departures from Glen's assumption should have measurable consequences for the flow law.

The fabric symmetry and geometry make it possible in principle to distinguish a deformation history of pure shear from simple shear, and, in the latter case, to identify the 'shear plane' and sense of shear across it. Simple shear is recognized when nonequivalent maximums M_1 and M_2 are distinguished: M_2 is weaker and shows a greater tendency to elongate perpendicular to the M_1M_2 plane. Identification of the flow pattern within the continuous range of transitional possibilities between simple shear and pure shear requires a quantitative measure of the relative strengths of M_1 and M_2, which is dependent on adequate sampling statistics. When simple shear is recognized, the shear plane is perpendicular to M_1, and the sense of shear across it is the same as the sense of rotation of M_2 away from the shear direction w. In pure shear the maximum principal compressive stress bisects the acute angle between the two equivalent fabric maximums, and the minimum principal compressive stress bisects the obtuse angle (Figure 18).

The possibility of using the fabric pattern to distinguish between the shear plane wR and its conjugate plane TR is noteworthy, because other methods of inferring deformation history, such as fault plane solutions, do not have this capability. The physical distinction between the two planes, which makes it possible in principle to distinguish between them, is the fact that, under simple shear, material particles located in a plane parallel to the shear plane wR remain in this plane as deformation proceeds, whereas particles originally in a plane parallel to the conjugate plane TR rotate out of this plane as deformation proceeds.

Fabric geometry in relation to stress state. As the stress state is modified from simple shear toward uniaxial compression, the c-axis fabric evolves progressively from the two-maximum pattern toward a small-circle girdle via intermediate patterns consisting of two maximums linked by a small-circle girdle. From the geometry of this pattern, and specifically from the ratio of orientation densities at minimum and maximum around the girdle, it is possible in principle to infer the stress character in terms of the ratio of σ/τ. Figure 16 provides data for this purpose but makes it evident that more experimental material would be needed to develop the method on a statistically sound quantitative basis.

At the high value of $\sigma/\tau = 3.8$ (A13, Figure 16d) clear maximums still appear in the small-circle girdle where it crosses the wT plane. The corresponding feature is also visible in the fabric of C4 (Figure 17). These observations show that the fabric pattern is sensitive to small departures from rotational symmetry in the applied stress.

In considering the general dependence of fabric pattern on stress state it is useful to describe a general state of stress in terms of three parameters: (1) mean stress, (2) maximum principal stress difference $\sigma_3 - \sigma_1$, which we can call 'stress level,' and (3) relative value of the intermediate principal stress, expressed as $(\sigma_2 - \sigma_1)/(\sigma_3 - \sigma_1)$, which we can call 'stress character.' There is only a weak dependence of fabric on stress level, as was discussed previously. The present experiments are not suited to testing for a dependence on mean stress, but a strong dependence is ruled out by the similarity of the C4 fabric (Figure 17a) to a pattern intermediate between A14 and A13 (Figure 16c, d): the C4 experiment has a mean stress higher than that of the corresponding torsion-compression experiment (with $\sigma/\tau = 2.7$) by 0.6 bars. If we assume that any dependence of fabric pattern on stress level and mean stress can be neglected, the fabric geometry can be described as a function of a single stress parameter, the stress character, which can be expressed either in terms of the principal stresses or, in the torsion experiment, in terms of σ/τ. The experiments explore σ/τ for stresses σ that are compressive or 0. Experimental fabrics for tensile stresses σ (corresponding to $\sigma/\tau < 0$ with the convention used in Tables 1 and 2) are not available, but from the consistent behavior seen in the available experiments we can predict the fabric patterns expected. They are given in Figure 18 (upper-right) together with diagrams schematically summarizing the experimental results for $\sigma/\tau \geq 0$. Also given in Figure 18 (lower half) is a series of diagrams that predict fabrics expected for triaxial tests in which normal stresses σ_1, σ_2, and σ_3 are applied to the surfaces of a rectangular specimen. The stress character is in this case best expressed in terms of $(\sigma_2 - \sigma_1)/(\sigma_3 - \sigma_1)$, which equals $\frac{1}{2}$ for the central diagram, representing pure shear. The different orientations of adjacent fabric patterns in the upper and lower halves of Figure 18 reflect the different orientations of the principal stress axes, but the patterns are otherwise closely similar. The remaining differences are due to the effects discussed in the previous section, which cause the fabric symmetry to depart from the symmetry of the stress tensor.

CAUSES OF CRYSTAL REORIENTATION

The disparity between fabric and stress symmetry shows that theoretical models of crystal reorientation based on crystal elastic strain energy and related thermodynamic variables are not rigorously applicable, since the strain energies depend only on the state of stress.

It may, however, be possible to explain the reorientation process as one in which the orientations selected in recrystallization are determined originally by thermodynamic conditions, and then become modified by mechanical rotation in the course of subsequent plastic flow of the polycrystalline aggregate. This explanation accounts qualitatively for the discrepancy between fabric and stress symmetries in simple shear: crystals in maximum M_1 have no tendency to rotate mechanically, because their slip plane is parallel to the shear plane wR, whereas crystals originating with orientation at w, which is the point equivalent to M_1 by the stress symmetry, will rotate from w toward M_2 during plastic flow in simple shear. Rotation away from w removes the crystals in M_2 from the

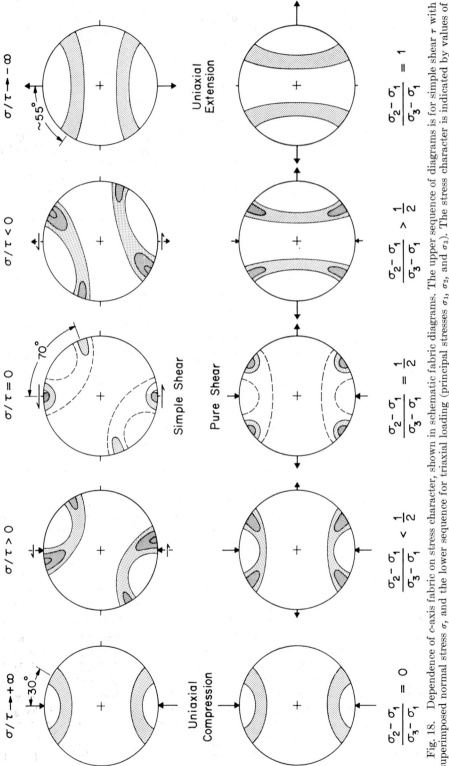

Fig. 18. Dependence of c-axis fabric on stress character, shown in schematic fabric diagrams. The upper sequence of diagrams is for simple shear τ with superimposed normal stress σ, and the lower sequence for triaxial loading (principal stresses σ_1, σ_2, and σ_3). The stress character is indicated by values of σ/τ or $(\sigma_2 - \sigma_1)/(\sigma_3 - \sigma_1)$ and by arrows in the diagrams. The upper middle diagram represents simple shear, and the lower middle, pure shear. Diagrams to the left of the center represent stresses of essentially compressive character, and those to the right of the center, stresses of essentially extensional character. The diagrams for $\sigma/\tau \geq 0$ and for $(\sigma_2 - \sigma_1)/(\sigma_3 - \sigma_1) < \frac{1}{2}$ summarize experimental data schematically. The remaining diagrams are proposed by generalization, and the angle values given are hypothetical.

TABLE 3. Comparison between Rotation Angle θ_R and Departure Angle of M_2 from w

Fabric	Figure	γ	θ_R, deg	$w \wedge M_2$	
				Peak, deg	Average, deg
A8T′	15b	0.13	7.5	18	15
A10	15a	0.13	7.5	12	12
A8B	14a	0.24	13.5	14	18
A6	14b	0.40	22	24	25

supposed orientation of maximum stability equivalent to M_1, and hence the maximum should be weakened in relation to M_1. (The nature of this rotation is explained in Figure 3.)

To test this explanation quantitatively, note that the rotation angle for each crystal is the same as that of its slip plane, so that, if crystals in M_2 undergo a shear strain equal to that of the bulk specimen, the crystal rotation angle θ_R is given by $\theta_R = \tan^{-1}\gamma$, where γ is the bulk shear strain. Table 3 compares the angle between w and M_2 with rotation angles θ_R on this basis for four fabrics. Angles $w \wedge M_2$ are listed both for the position of the summit of the orientation-density peak and for a visually estimated orientation average over the peak to provide a measure of the uncertainty in estimating the peak position; as in Table 2, the angles are measured as projected from R into the wT plane. At the higher strains the rotation θ_R is adequate to reorient crystals from w to M_2, but at lower strains it is inadequate, notably for the fabric in A8T′. The inadequacy is increased when we consider that most of the crystals forming maximum M_2 probably did not originate or become selected by recrystallization at the very beginning of specimen deformation but instead formed later, either during the initial recrystallization marking the transition from the second to the third creep stage or even later. If they originated at a shear strain $\gamma = 0.05$, when the recrystallization fabric was still poorly developed (Figure 13), the available rotation angles are 3° less than the θ_R values in Table 3. The increase in $w \wedge M_2$ from A8B to A6 is consistent with mechanical rotation of M_2, but the lack of increase from A8T′ to A8B is not. Part (5°) of the increase from A8B to A6 should be due to stress reorientation ($T \wedge \sigma_3$ values in Table 2). If crystals originated at w and rotated toward M_2 in a

subsequent history of growth to maximum crystal size followed by disappearance, there should be a correlation between size and orientation along the arc from w to M_2 and beyond, but no such correlation is found. The statistical nature of the fabric reorientation process interferes with an unambiguous decision here, but, in balance, the data in Table 3, and especially the fabric of A8T′, do not support the idea that maximum M_2 represents crystals that originated at w and were rotated into their observed orientation by plastic flow. Some rotation of this type is, however, required in any case, the amount depending on the lifetime of the crystals in M_2, under the recrystallization process.

If the departure of M_2 from w is ignored, the preferred orientations developed experimentally by recrystallization in pure shear are the ones that would make the shear stress across the basal plane of the ice crystal a maximum, if the stress were homogeneous in the polycrystalline specimen. By the methods of Brace [1960] it is readily shown that for pure shear these orientations maximize the elastic strain energy of the ice crystal. Under uniaxial compression the elastic energy is maximized for c inclined 50° to the compression axis [Brace, 1960]. The small-circle girdle produced experimentally has a radius of about 30° (Table 2), but, as with the departure of M_2 from w, a collapse of the radius from 50° to 30° might be explained as the result of mechanical rotation by plastic flow subsequent to crystallization in the energetically favored orientation. Although there is thus an approximate correlation between the observed orientations and those which would maximize the elastic strain energy under homogeneous stress, it cannot be taken as unqualified support for theories of crystal reorientation based on maximum elastic strain energy.

The plastic anisotropy of the ice crystal is much greater than the elastic anisotropy and has the same general form—greatest weakness in shear across the basal plane (the slip plane). Through basal gliding, stresses applied externally to crystals with c axes at T and w in the shear experiments will tend to be relaxed rapidly, whereas crystals with c axes at R and midway between w and $\pm T$ cannot relax the over-all applied stresses by plastic flow and will thus bear a higher proportion of the applied stress than crystals at T and w. If the more highly stressed crystals are less stable, as general thermodynamic considerations indicate, the favored orientations are the ones at T and w. These orientations have minimum stress and minimum elastic strain energy in the nonhomogeneous stress field that results from stress relaxation by plastic flow.

By this reasoning, theories of crystal reorientation that assume homogeneous stress and omit the effects of plastic flow should not be applicable to the recrystallization phenomena in the present experiments. Orientations predicted from Gibbsian thermodynamics [Kamb, 1959a] by assuming intergranular material transfer under homogeneous stress, without intracrystalline plastic flow, differ greatly from the observed orientations; crystals in the predicted orientations would have elastic strain energies near minimum under the assumed homogeneous stress. Although there are theories that predict orientations giving maximum elastic strain energy under the assumption of homogeneous stress, the approximate agreement between these predictions and the present experiments is of doubtful significance if the actual stress state is highly inhomogeneous and especially if the fundamental basis of the theories is unsound [Kamb, 1961a]. The validity of maximum strain energy as the criterion for preferred crystal orientation in twinning transformations of Dauphiné type under fixed stress, although a useful conclusion in its own right, does not provide a basis for its application to recrystallization, in which the special coherence features of the twinning transformations do not operate.

COMPARISON WITH GLACIER ICE

The interlocking grain texture produced by recrystallization in these experiments is qualitatively similar to the texture of typical glacier ice, although on a finer scale, and the starting material texturally resembles firn, although lacking the permeability of firn. The experiments thus reproduce in miniature the textural metamorphism of firn to glacier ice that occurs in nature. Whether the geometry of the experimental recrystallization texture is fully comparable quantitatively to that of glacier ice, particularly its extreme degree of interdigitation [Bader, 1951], is not established here.

The typical c-axis fabric of glacier ice that has undergone substantial strain is well developed in the marginal zones of valley glaciers, where the ice deformation consists primarily of simple shear across nearly vertical planes parallel to the margin. The fabric consists typically of four sharp maximums clustered around the pole of the shear plane [Rigsby, 1960; Kamb, 1959b]. Most observed examples of this fabric type have come from surface samples in which, either because of the approach of the free surface or because of crevassing, the original flow stresses have been relaxed or modified before sampling, so that the fabrics may have been affected by post-tectonic recrystallization. However, fabrics from core and tunnel samples obtained near the bottom of a flowing glacier [Kamb and Shreve, 1963; Kamb and LaChapelle, 1968], where there was no previous stress relaxation and the ongoing strain rates were high (~ 1 year^{-1}) show the same type of fabric and establish that it is the pattern characteristic of glacier ice undergoing rapid deformation. A typical example, from core sampling [Kamb and Shreve, 1963], is shown in Figure 19a. According to measurements of borehole tilting, the ice deformation occurring in the part of the glacier sampled in Figure 19a is approximately simple shear, the pole of the shear plane being at T and the direction of shear across it being at w in Figure 19a. Except for a rotation of 13° around R, the orientation and state of stress, including the sense of shear, are the same in Figure 19a and in the experimental diagrams in Figures 12–15, and corresponding axes are therefore labeled with the same letters. The pattern in Figure 19a has fairly good mirror symmetry about the plane Um indicated on the diagram. The 20° discrepancy between this plane and wT, about which mirror symmetry is expected on the basis of the measured ice deformation, is a fairly typical complication, but for simplicity it will be ignored here, since it

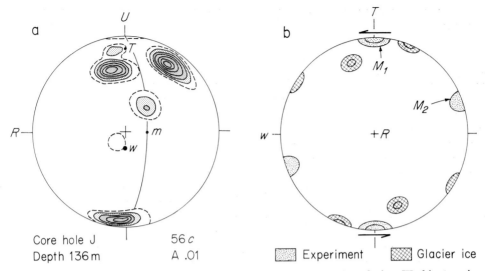

Fig. 19. (a) The c-axis fabric of ice from near the base of Blue glacier, Washington, in core-hole *J*, at a depth of 136 meters. The core axis is *U*, the pole of shear plane parallel to the glacier bed is at *T*, and the direction of shear is at *w*, approximately parallel to the ice flow direction. Plane *Um* is a plane of approximate mirror symmetry in the fabric. Contours are at 1½, 3, 6, 9, ⋯ % per 1% area of hemisphere. (b) The relationship between positions of c-axis maximums in experimental and natural recrystallization fabrics under simple shear, shown in equal-area projection. Maximums in experimental fabrics are dotted, and maximums in natural fabrics are crosshatched. The pole of the shear plane is *T*, the shear direction is *w*, and the sense of shear is indicated by arrows.

could result from a 20° error in orientation angle around the core axis *U*.

The complete contrast between the multiple-maximum type of natural fabric in Figure 19a and the experimental type of fabric in Figures 12–15 is emphasized by plotting the positions of the two sets of maximums on a single diagram, Figure 19b. The only feature common to the two fabrics is monoclinic symmetry. The cause of this great discrepancy is not due to a difference in stress character, because in the full range of experimental fabric types in Figure 18 there is no near equivalent of the natural fabric. Nor is it due to a difference between recrystallization at and below the melting point; this finding is established by the results of experiment C4 and by the observation of multiple-maximum fabrics in the subfreezing ice of polar-type glaciers [*Gow*, 1963; *Vallon*, 1963]. Shear stresses in the torsion experiments are higher than the in situ shear stress of 1.4 bars estimated for the ice of Figure 19a, which is near the maximum of about 1.5 bars generally reached in glacier ice; but the relatively slight change in fabric over the experimental stress range

from $\tau = 4.1$ bars (A8) to 1.9 bars (A10) reasonably precludes a major change in fabric type between 1.9 and 1.4 bars. The grain size of glacier ice showing multiple-maximum fabrics is usually much coarser than that achieved by recrystallization in the experimental specimens, and admittedly the fabric of fine-grained glacier ice is very different [*Kamb*, 1959b, p. 1896], but it is difficult to imagine that macroscopic grain size alone can itself be the cause of these major differences in crystal fabric.

The largest difference in physical conditions between the experiments and the natural situation is in the length of time available for the recrystallization process and correspondingly in the total strain accumulated. The 59-day length of experiment A10 is to be compared with many tens or hundreds of years available in typical rapidly flowing glaciers. To explain the difference in fabrics on this account, we must surmise that a second mechanism exists that is capable of controlling crystal orientations but that requires long-continued recrystallization for its realization. The nature of such a second mechanism will be considered more closely in a separate paper dealing fully with the natural fabrics. Here it is important to note that the existence of a process whereby the second mechanism takes over from the first as deformation continues, and as the proper textural and fabric architecture becomes established, implies that a true steady state has not been reached in the recrystallization process even at the largest strains attained in the present experiments. It also implies that at an early stage in their deformation history the natural fabrics would resemble the experimental ones and that at a later stage the second mechanism is able to generate, select, and amplify crystals in orientations not included in the fabric generated by the first mechanism.

Support for these ideas comes from the type of natural fabric shown in Figure 17b, which has been found repeatedly in core sampling of ice subjected to deformation in approximately pure shear at the base of an icefall in Blue glacier, Washington [Kamb and Shreve, 1963; B. Kamb, E. R. LaChapelle, and C. F. Raymond, unpublished manuscript, 1971]. A somewhat similar fabric is reported by Gow [1963, Figure 7] from a depth of 140 meters in the Ross ice shelf. The example shown in Figure 17b comes from coarse bubbly ice in which there is textural evidence that recrystallization from a finer-grained starting material has recently occurred, and for which a total logarithmic strain in pure shear of about 0.3 (corresponding to $\gamma \approx 0.6$) accumulated in a period of about 1 year. The compression and extension axes judged from the orientation of macroscopic foliation are marked in Figure 17b. As in Figure 19a, there is some incompatibility between the fabric orientation and the independently inferred strain symmetry; it corresponds to a rotation error of 25° around the core axis U and will again be passed over here. The fabric consists of two moderately broad maximums linked by a weak small-circle girdle and is thus similar to the experimental fabrics of specimens A10 (Figure 15a), A14 (Figure 16c), and C4 (Figure 17a). The angle between the two maximums in Figure 17b (57°) is smaller than that in the experimental fabrics cited (79°, 72°, 74°) and is instead comparable to A13 (54°). The maximums in Figure 17b are subequal in strength and stronger than those in A10, A14, or C4, although comparable to those in A13.

Although there are thus some detailed differences between the natural and experimental fabrics, Figure 17b is approximately what is expected experimentally for a stress character deviating somewhat from pure shear in the direction of uniaxial compression. Extensive sampling of situations in other glaciers (Austerdal in Norway and Trift and Great Aletsch in Switzerland) where the deformation pattern is similar but the total strain, time of deformation, and extent of recrystallization are greater, consistently reveals fabrics of four-maximum type (B. Kamb, unpublished manuscript, 1961). In Austerdal glacier, multiple-maximum fabrics are found in ice only 8 years old, deformed to a logarithmic strain of 2.2 in approximately pure shear. It thus appears that the development from the two-maximum fabric of Figure 17b to a multiple-maximum fabric like Figure 19a takes place over a time scale of 1–8 years in natural ice deformation at high strain rates (~ 1 year^{-1}).

Another important class of glacier ice fabrics consists of single-maximum types: in temperate glaciers the broad single maximum of fine ice [Kamb, 1959b, Figures 5–7] and rare coarse ice samples [Allen et al., 1960, Figure 10] and in polar glaciers the sharp single maximum reported at depth in the Antarctic ice sheet by Gow et al. [1968, p. 1013] and in a sample from Greenland by Rigsby [1960, Figure 8]. The present experiments and discussion do not provide direct evidence for the origin of these fabrics, although they raise the possibility that progressive rotation of the M_2 maximum in in simple shear to large strains could produce a single maximum by merging M_2 and M_1 in the experimental fabric.

IMPLICATIONS FOR TECTONOPHYSICS

The panoply of structural and textural phenomena that unfolds in the high-temperature creep of such a compositionally and structurally simple material as ice and that has major effects on the flow behavior of the material should probably serve as a warning against overly simplistic interpretations of creep experiments on other, more complicated geological materials. If other materials behave with a complexity similar to that of ice, results of creep experiments should not be extrapolated much beyond the range of strains explored with regard to either the flow behavior (strain versus time and stress) or the textural and fabric effects of flow. In particular, apparently steady creep behavior observed at early stages in the deformation (strains of a few per cent or less) should not be assumed to continue to strains larger than those observed. This restriction is seriously at odds with the need to carry out the experiments at tectonically appropriate low strain rates: it is obviously not possible in the laboratory to reproduce or approach simultaneously both the strain rates and the total strains that occur in flow processes on geological time scales. If test materials having a structural and textural state preserved by quenching from conditions of deformation in nature are available, a deformation increment beyond the total strain already achieved in nature can be experimentally explored, but, to apply this increment reliably, information on the previous history of deformation and proof of the quenching are needed. The difficulty would be ameliorated if it were empirically true that, in spite of the possible effects of time-dependent processes such as recovery and recrystallization, the structural architecture of materials deforming in creep is determined mainly by the total strain, without regard to the time scale. It would be possible to satisfy both the strain and the strain rate restriction by rapidly prestraining the material to the needed total strain and then reducing the strain rates toward tectonically pertinent values for the actual experimental tests. To some extent the experiments on ice give support to this approach, because the fabric, which greatly affects the flow properties, shows at most a modest dependence on stress or strain rate for given total strain. As far as existing evidence goes, this stress-independent behavior appears to hold even in comparing the experimentally produced fabrics with the very different natural fabrics typical of glacier ice. However, the well-defined inverse relation between stress level and steady-state grain size in syntectonically recrystallizing ice negates the idea of time scale independence for the texture. It would be dangerous to ignore this finding in extrapolating experimental results to geological time scales, because grain size generally affects both the relative contributions of different creep mechanisms (such as diffusion versus dislocation creep) and the creep behavior under a given mechanism. In fact, the recognized effects of grain size on creep are rather modest for ice, but a similar situation cannot be assumed to hold for other geological materials, without experimental confirmation.

Experience with ice suggests that significantly nonsteady creep effects extend to high strains, of the order of 1. If other geological materials behave similarly, only at strains comparable to those in flowing glaciers can we expect a reasonably steady state to be reached in the creep process.

Acknowledgments. I am grateful to Dr. Marcel de Quervain, Director of the Swiss Federal Institute for Snow and Avalanche Research, and to other members of the Institute staff for help in many ways. Field work was made possible by cooperation of the National Park Service, Olympic National Park.

The field work and the preparation of the paper were supported by grants from the U.S. National Science Foundation. The experimental part of the work was made possible by a John Simon Guggenheim Memorial fellowship and by the hospitality and support provided by the Swiss Federal Institute for Snow and Avalanche Research, Weissfluhjoch, Davos, Switzerland.

Deformation of Non-Newtonian Materials in Simple Shear

B. E. HOBBS[1]

Department of Geophysics and Geochemistry, Australian National University
Canberra, A.C.T., Australia

This paper discusses the deformation of fluids in which there is a nonlinear relationship between stress and deformation rate. These fluids comprise four classes, the general class containing the fluids in which the stress at a particular time is determined by the history of all deformed states before that time. These are called simple fluids. The next class contains the fluids in which the stress is a nonlinear function of the rate of deformation and of the higher time derivatives of this tensor. These are called Rivlin-Ericksen fluids. The third class includes the fluids in which the stress is a nonlinear function of the instantaneous rate of deformation. These are called Reiner-Rivlin fluids, and examples, when properly formulated, are the 'power laws' observed for steady-state creep of polycrystalline aggregates. The fourth class includes the fluids in which the stress is a linear function of the rate of deformation. These are called Newtonian fluids. Analytical solutions exist for the stress in terms of the motion only for certain deformation geometries, and examples are presented for inhomogeneous simple shear to illustrate the distributions of velocity, strain, stress, and mean pressure that are dynamically possible in each of these classes of fluids. For deformation by inhomogeneous simple shear all simple fluids behave as Rivlin-Ericksen fluids. For slow deformations simple fluids can behave as elastic fluids; again, for faster deformations they behave as Rivlin-Ericksen fluids. An important experimental point is that in a steady-state shortening experiment all simple fluids behave as Reiner-Rivlin fluids, so that it is impossible to distinguish between these two classes by this type of experiment. Such experiments cannot provide enough information to predict the behavior of simple fluids in deformations with other geometries, such as in simple shear, folding, or convective motion.

There is considerable interest at present in developing constitutive equations that can adequately describe the behavior of geological materials during large permanent deformations of the crust and the mantle of the earth. For the most part recent discussions center on the use of the linear 'Newtonian' relationship between stress and deformation rate or on the use of various empirical power laws observed for the steady-state creep of polycrystalline aggregates.

This paper discusses part of the literature on the deformation of materials with nonlinear relationships between stress and deformation rate. This discussion is undertaken to place the empirical power laws for creep within the general context of nonlinear constitutive equations and to indicate the types of experiments that

are necessary before the behavior of geological materials can be predicted in general deformations or in deformations with a geometry that is more complicated than that of axially symmetric compression.

Experimental investigations [*Heard*, 1963; *Griggs*, 1967; *Carter and Heard*, 1970; *Carter and Avé Lallemant*, 1970; *Goetze*, 1971; *Raleigh and Kirby*, 1971] of minerals and rocks and arguments based on a variety of other observations [e.g., *Crittenden*, 1967; *Stacey*, 1969] indicate that the shear stresses that many geological materials can support during deformation are quite small (of the order of bars or tens of bars) at the slow strain rates (10^{-12}–10^{-14} sec^{-1}) expected in natural deformations. Coupled with this expectation is a general tendency to regard rocks as behaving as fluids under natural conditions of deformation. (A discussion of the concept of a fluid is given by

---

[1] Now at the Department of Earth Sciences, Monash University, Clayton, Victoria, Australia.

Truesdell and Noll [1965, pp. 79–82]. They propose that the concept includes the idea 'that a fluid should not alter its material response after an arbitrary deformation that leaves the density unchanged.') Thus in a variety of recent papers [e.g., *McKenzie*, 1967; *Crittenden*, 1967; *Dieterich and Carter*, 1969; *Torrance and Turcotte*, 1971] the properties of rocks have been likened to those of Newtonian fluids, in which the stress is related linearly to the rate of deformation:

$$T_{ij} = -p\delta_{ij} + 2\eta D_{ij} \qquad (1)$$

Here T_{ij} are the components of the stress tensor, p is the hydrostatic pressure, δ_{ij} is the Kronecker δ with components $\delta_{ij} = 1$ for $i = j$ and $\delta_{ij} = 0$ for $i \neq j$, and D_{ij} are the components of the rate of deformation tensor (see the second section). The term η is the viscosity. If (1) is true, the material is called Newtonian. Otherwise it is called non-Newtonian.

The popular mechanism of producing such a linear relationship for polycrystalline aggregates is Herring-Nabarro diffusional creep [*Herring*, 1950]. However, recent discussions [*Weertman*, 1970; *Green*, 1970] have indicated that this mechanism only leads to a linear relationship between stress and deformation rate if the mean grain diameter remains constant during deformation. If there is a dependence of grain shape on stress, Newtonian behavior does not result.

Recent experimental work [*Heard*, 1963; *Carter and Avé Lallemant*, 1970; *Goetze*, 1971; *Raleigh and Kirby*, 1971] suggests a nonlinear relationship between stress and deformation rate, and the mechanisms leading to such relationships have been discussed by *Weertman* [1968, 1970]. Moreover, it is to be expected that rocks will display phenomena such as stress relaxation (see *Griggs and Blacic* [1965], *Griggs* [1967], and *Hobbs* [1968] for experimental examples), where the stress slowly decays to an equilibrium value after deformation ceases. It is also to be expected that the stress in a deforming rock will depend in some manner on the history of deformation, so that identical rocks taken along different deformation paths to the same strain states are likely to display different stress states.

Hence, any discussion of the stress fields that are likely to exist in deformed rocks should employ nonlinear constitutive equations in which phenomena such as stress relaxation may appear and in which some dependence on the history of deformation is considered.

Although there has been considerable progress in the past 20 years in the development of constitutive equations that embody such requirements (see *Coleman et al.* [1966, pp. 84–92] for a history of such development), there has been very little progress in obtaining analytical solutions for general deformations of non-Newtonian materials (or, for that matter, of Newtonian materials). In general, iterative procedures such as the finite element method employed by *Dieterich and Carter* [1969] have to be used to arrive at the stress state corresponding to a particular deformation. There are, however, a group of simple deformations that do admit of straightforward analytical solutions, and some of these deformations are of geological interest.

This paper examines one of these deformations, namely, inhomogeneous simple shear, to discuss (1) the states of both stress and strain that may exist during such deformation and (2) the ways in which these states and other features of the deformation differ from Newtonian to non-Newtonian materials.

It is also sometimes proposed that, since geological strain rates are small, the constitutive equation of a complicated material may approximate very closely that of a Newtonian material. A secondary aim of this paper is to examine this proposition and to discuss the conditions under which (1) becomes a good approximation for more complicated materials.

In the next section of this paper the ideas governing the formulation of constitutive equations are considered to state the form of these equations and to discuss the principal ways in which various materials differ from Newtonian materials and from each other. In the third section the theory of deformation by inhomogeneous simple shear is outlined, and in the fourth section a number of geologically relevant examples are given. For the most part, only results are given, without elaboration, in the second and third sections; for a rigorous and detailed treatment, see *Rivlin* [1948], *Truesdell* [1952], *Noll* [1955], *Eringen* [1962], *Truesdell and Noll* [1965], and *Coleman et al.* [1966]. The discussion in the second and third sections follows to some extent that of *Markovitz* [1967].

Cartesian coordinates are employed throughout. All deformations are considered to be isothermal, and the material is assumed to be incompressible. The theory in which none of these restrictions is imposed is discussed by *Truesdell and Noll* [1965]. The tensor summation convention is used throughout.

CONSTITUTIVE EQUATIONS

There are a number of fundamental laws of mechanics that provide differential equations that are true for all materials and that, in turn, enable relations between the density, the velocity, the stress, the mass, the heat flux, and other physical quantities to be established [see *Truesdell and Noll*, 1965, pp. 1–2]. These equations are known as field equations. The field equations that are used explicitly in this paper are those expressing the balance of linear momentum:

$$\partial T_{ij}/\partial x_j + \rho b_i = \rho \ddot{x}_i \qquad (2)$$

where x_i are the coordinates of a material point, ρ is the density, $\ddot{x}_i$ are the components of the acceleration, and b_i are the components of a body force such as that due to gravity.

The field equations alone are not sufficient to establish unique relations, and, in addition, some description of the material itself is necessary, defining the way in which the stress is related to the motion (and also, in the general case, to the heat flux vector). Such a relation, true for a particular material under particular conditions, is expressed by the constitutive equations.

A number of rules have evolved governing the establishment of constitutive equations [*Truesdell and Toupin*, 1960, pp. 700–704]. These rules involve various invariance and dimensional requirements, such as a statement that the form of the constitutive equation should be independent of the coordinate system chosen to describe the material. An important rule is that of material indifference, which states that the form of the constitutive equation should not be altered by arbitrary rigid rotations of both the coordinate system and the material. This rule precludes any dependence of the stress on the velocity in the constitutive equations discussed here.

Materials with constitutive equations that depend only on the history of the deformation together with various material constants comprise an important class. If only constant-volume (isochoric) deformations are considered, it is possible to deal only with incompressible materials. For these materials the stress is determined to within a hydrostatic pressure by the history of the deformation. Fluids that are incompressible and have stress state at time t that depends only on the deformation that has taken place before t are called simple fluids, which include Newtonian and Reiner-Rivlin fluids as special examples.

Deformation and motion. As a preliminary to discussing the constitutive equations mentioned above in a little more detail, the following discussion presents certain useful measures and functions of deformation.

It is convenient to label a material point in a body in some specified reference state by the coordinates x_k and to discuss the way in which the coordinates in some other, strained state depend on x_k and time. Thus, if x_k are the coordinates of a material point in the reference state at some time t and if ξ_k are the coordinates of that same material point in another, deformed state at time τ, the deformation gradient at τ is

$$B_{ij}(\tau) = \partial \xi_i/\partial x_j \qquad (3)$$

As a measure of deformation the symmetric tensor J_{ij} is used where

$$J_{ij} = C_{ij} - \delta_{ij} = B_{ki}B_{kj} - \delta_{ij}$$

$$= \frac{\partial \xi_k}{\partial x_i}\frac{\partial \xi_k}{\partial x_j} - \delta_{ij} \qquad (4)$$

where C_{ij} is the Green deformation tensor, which is discussed with examples by *Hobbs* [1972].

For small deformations

$$J_{ij} = 2[E_{ij}(\tau) - E_{ij}(t)] \qquad (5)$$

where E_{ij} is the classical infinitesimal strain tensor.

In all subsequent deformations the reference state is considered to be the unstrained condition of the material, so that in (5), for instance, $E_{ij}(t) = 0$. Also $J_{ij}(t) = 0$.

The tensor J_{ij} is important, since it figures prominently in properly formulated constitutive equations. As is indicated in (4), this tensor is related to C_{ij}, which gives the strain relative to the undeformed state. The principal axes of C_{ij} are the principal axes of strain in the undeformed state, and its proper numbers are equal to the principal quadratic elongations [*Jaeger*, 1969, p. 22]. The tensor that gives the strain relative to the deformed state is of interest to geological applications. The convenient one to use [*Hobbs*, 1972] is c_{km}^{-1}, the inverse of the Cauchy deformation tensor. The principal axes of c_{km}^{-1} are the principal axes of strain in the deformed state, and the proper numbers are again the principal quadratic elongations. If the coordinate system in both the deformed and the undeformed states is Cartesian, the components of this tensor are given by

$$c_{km}^{-1} = \partial \xi_k / \partial x_i \; \partial \xi_m / \partial x_i \qquad (6)$$

If x_k is the velocity of the material point under consideration, the rate of deformation tensor D_{ij} is defined as the symmetric part of the velocity gradient:

$$D_{ij} = \tfrac{1}{2}(\partial \dot{x}_i / \partial x_j + \partial \dot{x}_j / \partial x_i) \qquad (7)$$

Some of the more complicated constitutive equations employ tensors other than D_{ij} to describe the motion, and of these other tensors the most useful ones here are the Rivlin-Ericksen tensors $A_{ij}^{(1)}$, $A_{ij}^{(2)}$, $\cdots$, $A_{ij}^{(N)}$, where $A_{ij}^{(N)}$ is called the Nth Rivlin-Ericksen tensor. These tensors are defined by

$$A_{ij}^{(1)} = 2D_{ij} \qquad (8)$$

$$A_{ij}^{(N+1)} = \dot{A}_{ij}^{(N)}$$

$$+ \left(\frac{\partial \dot{x}_k}{\partial x_j} A_{ik}^{(N)} + \frac{\partial \dot{x}_k}{\partial x_i} A_{jk}^{(N)} \right) \qquad (9)$$

the overdot denoting differentiation with respect to time. Successive Rivlin-Ericksen tensors therefore involve higher time derivatives of the deformation rate tensor, so that their incorporation into constitutive equations enables complicated time-dependent effects to be represented.

It should be noted that the strain rate tensor $\dot{e}_{ij}$ normally introduced in the geological literature is related to D_{ij} by

$$\dot{e}_{ij} = D_{ij} - \left(e_{mi} \frac{\partial \dot{x}_m}{\partial x_j} + e_{mj} \frac{\partial \dot{x}_m}{\partial x_i} \right) \qquad (10)$$

[see *Eringen*, 1962, p. 80]. It is only for small

deformations, in which the asymmetric part of the velocity gradient is of no greater magnitude than D_{ij}, that

$$\dot{e}_{ij} = D_{ij} \qquad (11)$$

Examples of Constitutive Equations

The general form of the following constitutive equations is prescribed by the various rules mentioned above [*Truesdell and Toupin*, 1960, pp. 700–704]. In particular, the principle of material indifference often forces the stated dependence on the tensors D_{ij}, $A_{ij}^{(N)}$, and J_{ij}.

Simple fluids (also called memory fluids). The constitutive equations for simple fluids are

$$T_{ij} = -p\delta_{ij} + \overset{\infty}{\underset{s=0}{F_{ij}}} [J_{ij}(t-s)] \qquad (12)$$

where s is a dummy parameter that measures the time backward from the reference state at time t to the past deformed state at time τ; i.e., $s = t - \tau$. The term F_{ij} is a functional with a value established once all values of J_{ij} are known over the range $-\infty < \tau < t$. A particular example of a functional is a definite integral having a value established once all values of the integrand are known over the range covered by the lower and the upper limits of integration. Thus it can be seen that F_{ij} contains all the information on the history of the deformation of the material. Functionals have a counterpart in the hereditary integrals of classical linear viscoelasticity [*Eringen*, 1967, p. 339], although these integrals generally violate the principle of material indifference. Discussions of functionals are given by *Eringen* [1967, pp. 147–148].

For a compressible fluid the pressure p is determined once the density of the fluid is known. This situation does not exist for an incompressible fluid, and p is then not specified. This indeterminancy is removed by defining

$$p = -\tfrac{1}{3} T_{ii} \qquad (13)$$

where p is what is often referred to as the mean or mechanical pressure in the material. A simple fluid also possesses the maximum possible material symmetry.

Rivlin-Ericksen fluids (also called fluids of differential type). If the stress depends on the velocity gradient and the time derivatives of this gradient, the constitutive equations become

$$T_{ij} = -p\,\delta_{ij}$$
$$+ f_{ij}(A_{lk}^{(1)}, A_{mn}^{(2)}, \cdots, A_{pq}^{(N)}) \qquad (14)$$

where f_{ij} is not necessarily linear and is an isotropic function; i.e., the components f_{ij} are the same in all Cartesian systems. These materials are capable of exhibiting a wide range of time-dependent effects, but they cannot exhibit stress relaxation.

Reiner-Rivlin fluids. If the stress depends only on the velocity gradient, the constitutive equations become

$$T_{ij} = -p\,\delta_{ij} + f_{ij}(D_{kl}) \qquad (15)$$

where f_{ij} has the same significance as in (14). This type of constitutive equation is employed to describe steady-state creep [*Weertman*, 1968, 1970], f_{ij} being a simple power law, generally expressed as a function of the strain rate $\dot{e}_{kl}$ rather than of D_{kl}. As was noted earlier, a dependence on strain rate is present only for certain small strains. It is clear that these power laws are intended to represent only steady-state creep and that (15) cannot represent the complete mechanical behavior of polycrystalline aggregates in general deformations. In particular since (15) expresses a dependence only on the instantaneous value of D_{kl}, the Reiner-Rivlin fluid cannot display a wide range of time-dependent effects. Reiner-Rivlin fluids do, however, possess an intrinsic material constant with the dimensions of time [*Truesdell*, 1952, p. 235], and to this extent these fluids exhibit some time-dependent properties. If this constant is dimensionless, the material is called a Stokesian fluid.

For deformation by steady shortening all simple fluids behave as Reiner-Rivlin fluids [*Truesdell and Noll*, 1965, pp. 472–473]. Thus, if polycrystalline aggregates behave as fluids, the classical type of deformation experiment performed by geologists and metallurgists must lead to constitutive equations of the form of (15), even though the materials may have a far more complicated nature. More complicated types of experiments are necessary to determine if geological materials actually have constitutive equations of the form of (15) or if they behave as more complicated simple fluids in general deformations.

Thus the power laws quoted for geological and metallurgical materials and obtained from steady-state creep experiments are to some extent the result of the experimental technique and not of the material being deformed. The fact that stress relaxation experiments may also be performed on these materials means that, in general, the constitutive equations for general deformations must be more complicated than those for Reiner-Rivlin fluids.

Newtonian fluids (also called Navier-Stokes fluids). If the function f_{ij} in (15) is linear, the constitutive equations reduce to (1). Again, the often quoted dependence of the stress on the strain rate $\dot{e}_{ij}$ for Newtonian fluids is only true for certain small strains. In general, the stress is a linear function of the rate of deformation D_{ij}.

DEFORMATION BY SHEAR

The constitutive equations of simple fluids (12) are so general that they are useless in any real problem. However, some progress can be made by examining particular deformation geometries in which the history of deformation for individual material particles has never been complicated. Most progress has been made by considering deformations in which the history of deformation does not change along the path of material particles. A class of these deformations is called viscometric. For these deformations all simple fluids, no matter how complicated their memory, behave as Rivlin-Ericksen fluids. That is, for viscometric flows the functional in (12) becomes a function, since all complicated memory capabilities the fluid may possess are not required.

A particular example of a viscometric flow is inhomogeneous simple shear, which is treated in detail below for the various materials discussed in the second section. The relevance of this type of deformation to naturally deformed rocks is discussed by *Ramsay and Graham* [1970].

Consider two blocks of material A and B (Figure 1) that move in relation to each other, so that the material between them is deformed by an inhomogeneous simple shear. No faulting at the boundaries of the blocks takes place. We wish to ask: (1) What restrictions are placed on the type of deformation occurring within the shear zone? (2) What is the distribution of strain within the shear zone for these deformations? and (3) What is the distribution

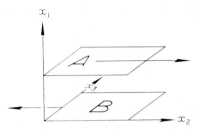

Fig. 1. Cartesian coordinate system for deformation by inhomogeneous simple shear. The two boundaries A and B move in the manner indicated by the arrows. Coordinate x_1 is normal to the shear zone, and x_2 is parallel to the direction of movement.

of stress within the shear zone for these deformations?

If Cartesian coordinates x_k are selected, as in Figure 1, and the motion is steady, the components of the velocity field are

$$\dot{x}_1 = 0$$
$$\dot{x}_2 = v(x_1) \qquad (16)$$
$$\dot{x}_3 = 0$$

where v is a nonlinear function of x_1.

The position ξ_k of a material point at any time τ can be found by solving

$$\xi_1 = 0$$
$$\xi_2 = v(\xi_1) \qquad (17)$$
$$\xi_3 = 0$$

so that, if x_k are the coordinates of the material point at time t,

$$\xi_1 = x_1$$
$$\xi_2 = x_2 - s \cdot v(x_1) \qquad (18)$$
$$\xi_3 = x_3$$

at time τ. As in (12), $s = t - \tau$.

The components of the deformation gradient at time s before t are

$$B_{ij} = \begin{bmatrix} 1 & 0 & 0 \\ -sk & 1 & 0 \\ 0 & 0 & 1 \end{bmatrix} \qquad (19)$$

where $k = \partial v / \partial x_1$.

The deformation rate tensor is

$$D_{ij} = \frac{k}{2} \begin{bmatrix} 0 & 1 & 0 \\ 1 & 0 & 0 \\ 0 & 0 & 0 \end{bmatrix} \qquad (20)$$

and the tensor J_{ij} is given by

$$J_{ij} = \begin{bmatrix} (sk)^2 & -sk & 0 \\ -sk & 0 & 0 \\ 0 & 0 & 0 \end{bmatrix} \qquad (21)$$

The strain in relation to the deformed state can be obtained from

$$c_{ij}^{-1} = \begin{bmatrix} 1 & -sk & 0 \\ -sk & (s^2 k^2 + 1) & 0 \\ 0 & 0 & 1 \end{bmatrix} \qquad (22)$$

The first two Rivlin-Ericksen tensors are, from (8) and (9),

$$A_{ij}^{(1)} = k \begin{bmatrix} 0 & 1 & 0 \\ 1 & 0 & 0 \\ 0 & 0 & 0 \end{bmatrix}$$

$$A_{ij}^{(2)} = 2k^2 \begin{bmatrix} 1 & 0 & 0 \\ 0 & 0 & 0 \\ 0 & 0 & 0 \end{bmatrix} \qquad (23)$$

All other $A_{ij}^{(N)}$ have components equal to 0. Hence the strain at time $(t - \tau)$ is given by

$$J_{ij} = -s A_{ij}^{(1)} + \tfrac{1}{2} s^2 A_{ij}^{(2)} \qquad (24)$$

Expression 24 states that the strain at time $(t - \tau)$ is dependent on only the first two Rivlin-Ericksen tensors and the difference in time s between the time corresponding to the reference state and that corresponding to the deformed state. The strain is independent of the value of t selected as the reference state. In other words, the state chosen as the reference state is not important. The functional F_{ij} in (12) then reduces to a function [Coleman and Noll, 1960], so that (12) becomes

$$T_{ij} = -p \delta_{ij} + f_{ij}(A_{kl}^{(1)}, A_{mn}^{(2)}) \qquad (25)$$

f_{ij} being an isotropic function.

Thus any simple fluid, no matter how complicated its memory response, behaves as a Rivlin-

Ericksen fluid for deformation by inhomogeneous simple shear. Clearly, there are an infinite number of simple fluids that exhibit similar behavior for these deformations, the strain history being such that the variety of complicated memory effects do not have an opportunity to manifest themselves. Since $A_{ij}^{(K)} = 0$ for $k \geq 3$, all Rivlin-Ericksen fluids (14) also behave as in (25) during inhomogeneous simple shear.

Coleman and Noll [1959] show that, for a simple fluid undergoing the flow (16), the matrix of the stress tensor has the form

$$T_{ij} = \begin{bmatrix} T_{11} & T_{12} & 0 \\ T_{12} & T_{22} & 0 \\ 0 & 0 & T_{33} \end{bmatrix} \qquad (26)$$

where it follows from (23) and (25) that the T_{ij} are functions only of k. Since from (13) all the T_{ij} are not independent, these functions can be written:

$$T_{12} = \mu(k)$$

$$T_{11} - T_{33} = \sigma_1(k) \qquad (27)$$

$$T_{22} - T_{33} = \sigma_2(k)$$

where the functions μ, σ_1, and σ_2 are called the viscometric functions. It is the aim of experiment to establish these functions.

The function μ is called the shear stress function, and the functions σ_1 and σ_2 are called the normal stress functions. The function

$$\tilde{\mu} = \mu(k)/k \qquad (28)$$

is the viscosity function. For Newtonian and some other fluids it is a constant.

Expressions 26 and 27 state one of the important ways in which non-Newtonian materials differ from Newtonian fluids: the stress tensor in an inhomogeneous simple shear flow contains nonzero and unequal normal stresses as well as the shear stress T_{12}. For a Newtonian material $\sigma_1(k) = \sigma_2(k) = 0$, so that only the shear stress T_{12} is developed.

In the following paragraphs the dynamically possible forms of (16) are established with the details of the stress field for these flows. When the equations of balance of linear momentum (2) and the observation that the T_{ij} are func-

tions only of x_1 are used [*Truesdell*, 1966, p. 81],

$$T_{12} = -ax_1 + c \qquad (29)$$

follows, where a and c are constants. Thus the shear stress is a linear function of x_1 independent of the function $v(x)_1$ in (16) and of the memory properties of the fluid. The normal stresses are given by [*Truesdell*, 1966, p. 82]

$$T_{11} = ax_2 + b$$

$$T_{22} = \sigma_2(k) - \sigma_1(k) + T_{11} \qquad (30)$$

$$T_{33} = -\sigma_1(k) + T_{11}$$

where the motion is assumed to be steady and the components of the body force are 0.

The dynamically possible velocity profiles are given by the solutions of

$$T_{12} = \mu(k) = \mu(\partial v/\partial x_1) = -ax_1 + c \qquad (31)$$

From the first part of (30) the constant a is seen to be the gradient of the normal stress T_{11} in the direction of flow:

$$a = \partial T_{11}/\partial x_2 \qquad (32)$$

Hence, if a is 0, from (31) the velocity profile is linear, and then for all simple fluids the only dynamically possible flow is a homogeneous simple shear.

Slow flows. Most real materials respond more readily to deformation that has occurred in the recent past than to that which has occurred in the distant past. If this observation is taken into account, *Coleman and Noll* [1960] have shown that, for very slow general flows, simple fluids deform as though their constitutive equations were those of elastic fluids. The next approximation as the flow is speeded up becomes that of the Newtonian fluid (1). For faster flows the best approximation is a Rivlin-Ericksen fluid having the constitutive equation

$$T_{ij} = -p\delta_{ij} + \eta_0 A_{ij}^{(1)}$$
$$+ \alpha_1 A_{ik}^{(1)} A_{kj}^{(1)} + \alpha_2 A_{ij}^{(2)} \qquad (33)$$

in which η_0, α_1, and α_2 are material constants. This fluid, called the second-order fluid, represents the first departure from classical Newtonian behavior, and, if geological materials are to be represented by simple fluids (12), then (33) is potentially a good approximation for slow geological deformations. For viscometric flows *Truesdell and Noll* [1965, p. 489] quote the measure

$$k \ll 1/S_0 \qquad (34)$$

as the condition for which the Newtonian fluid becomes a good approximation for any simple fluid. In (34) the term S_0 is a material constant known as the natural time lapse [*Truesdell,* 1964; *Truesdell and Noll,* 1965, p. 437]. From (34) the constant S_0 would have to be 10^3–10^4 years or less for the Newtonian fluid to be a good approximation in geological deformations. Clearly, there is insufficient knowledge of geological materials to decide if (1), (33), or even higher-order fluids are good approximations of geological materials, but examination of the second-order fluid is of interest, since it is capable of illustrating many of the features so characteristic of non-Newtonian materials. It is emphasized that the criterion for Newtonian behavior (34) holds only for viscometric flows. No such criterion is known for general flows in simple fluids.

For viscometric flows in second-order fluids the viscometric functions (27) are given by

$$\mu(k) = \mu_0 k$$

$$\sigma_1(k) = (2\alpha_1 + \alpha_2)k^2 \qquad (35)$$

$$\sigma_2(k) = \alpha_2 k^2$$

where $\mu_0 > 0$ and $\alpha_1 < 0$ [*Truesdell,* 1965]. The viscosity function is a constant, independent of k, as in the Newtonian fluid, but the normal stress functions are not 0. Thus the second-order fluid differs from the Newtonian fluid in that it is capable of exhibiting normal stress effects as well as some time-dependent effects.

The stress distribution in a second-order fluid undergoing inhomogeneous simple shear is obtained by substituting (35) into (30). Some examples of unstable behavior of second-order fluids are discussed by *Truesdell and Noll* [1965, pp. 506–513].

EXAMPLES OF STRESS AND STRAIN STATES DEVELOPED DURING INHOMOGENEOUS SIMPLE SHEAR

This section provides examples of the velocity profiles, the strains, and the stresses developed during the inhomogeneous simple shear of the various materials discussed in the preceding sections.

The values of the various coefficients used in these examples have been selected to facilitate computation, since experimental values for such coefficients are generally absent. The values attached to the stresses in these examples may not be geologically realistic; however, the character of the stress distribution, the velocity profiles, and the strain distributions is independent of the values chosen (subject to various thermodynamic requirements), so that these examples serve to illustrate the differences and the similarities among the various types of fluids considered above.

Rivlin-Ericksen fluids. As an example of a Rivlin-Ericksen fluid, the second-order fluid defined by (33) is considered. From (29), (30), and (35) the stress distribution is given by

$$T_{12} = -ax_1 + c$$

$$T_{11} = ax_2 + b$$

$$T_{22} = -2\alpha_1 k^2 + T_{11} \qquad (36)$$

$$T_{33} = -(2\alpha_1 + \alpha_2)k^2 + T_{11}$$

The dynamically permissible velocity profiles are, from (31) and (35),

$$\mu_0 v(x_1) = -\tfrac{1}{2}ax_1^2 + cx_1 + d \qquad (37)$$

where d is a constant. Hence, for a nonzero value of a the only flows possible are such that the velocity profiles are quadratics in x_1. The strain at time s from the unstrained reference configuration is then specified by (22). The mean pressure at a given point (x_1, x_2) given by (13) remains constant throughout the motion; lines of equal pressure in the x_1x_2 plane are parabolas:

$$p = -\left[ax_2 + b - \frac{4\alpha_1 + \alpha_2}{3}(-ax_1 + c)^2\right] \qquad (38)$$

The pressure distribution is therefore symmetrical about the axial plane of the fold formed during the deformation. The normal stress on the wall of the shear zone T_{11} differs from the mean pressure by $2(ax_2 + b) - [c^2(4\alpha_1 + \alpha_2)/3]$.

These aspects are illustrated in Figures 2 and 3, where the distributions of strain and

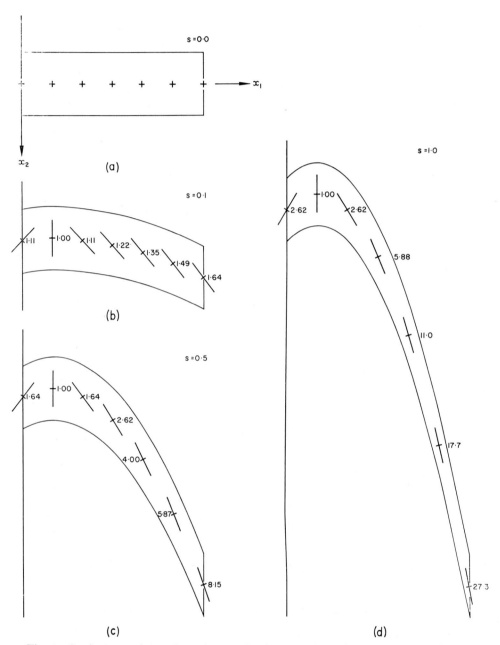

Fig. 2. Strain states for various times s developed during inhomogeneous simple shear of a second-order (Rivlin-Ericksen) fluid (33). In (36), $a = b = c = 1$, $\alpha_1 = 1$, and $\alpha_2 = -10$; in (37), $d = 1$ and $\mu_0 = 1$. The orientation of a principal axis of strain c^{-1} is indicated by the long line. The number at each point is the ratio of the principal axes of strain: c_1^{-1}/c_2^{-1}. The shape and orientation of the undeformed layer are given by diagram a.

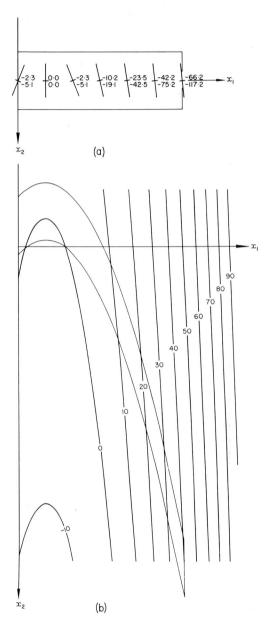

(a)

(b)

x_2

Fig. 3. Stress states for the deformation shown in Figure 2 in a second-order fluid (33). The constants in (36) and (37) have the same values as they do in Figure 2. (a) Principal stresses $(T_1 + p)$ and $(T_2 + p)$ at all times in the deforming body. The orientation of a principal axis of stress is shown by the long line at each point. (b) Distribution of mean pressure p at all times in the deforming body. The deformed layer of Figure 2d is shown superposed on this pressure distribution.

stress are shown for various times s. No restrictions have been placed on the way in which T_{33} varies.

If $a = 0$, the only deformation possible is a homogeneous simple shear, and all the normal stresses and the pressure maintain constant values throughout the deforming body.

Reiner-Rivlin fluids. The best-known example of this constitutive equation is that in which the function f_{ij} in (15) is of the form observed for steady creep of polycrystalline aggregates:

$$T_{ij} = K(D_{ij})^{1/n} \qquad (39)$$

where n is >1 [see *Weertman*, 1970; *Berg*, 1969]. This equation has the property that the dimensions of K are a function of n, so that it is impossible to compare materials with different values of n. Hence (39) cannot be physically realistic.

The simplest form of (15) consistent with the requirements of dimensional invariance [*Truesdell and Toupin*, 1960, pp. 720–722] is

$$T_{ij} = -p\,\delta_{ij} + p_0 f_{ij}(t_0\,D_{kl}) \qquad (40)$$

in which p_0 is a function of temperature having the dimensions of stress and t_0 is a constant with the dimensions of time, known as the natural time. The function f_{ij} is then a dimensionless isotropic function.

As an example, (40) is written in the form

$$T_{ij} = -p\,\delta_{ij} + p_0(t_0\,D_{ij})^{1/n} \qquad (41)$$

so that the dimensions of p_0 are independent of n. It then follows from a theorem concerning isotropic tensor functions [*Truesdell and Noll*, 1965, p. 32] that (41) can be rewritten

$$T_{ij} = (-p + \alpha_0)\,\delta_{ij} + \alpha_1 D_{ij} + \alpha_2 D_{im} D_{mj} \qquad (42)$$

in which α_i are functions of I_D, II_D, and III_D, the three principal invariants of D_{ij}. Thus

$$\alpha_i = \alpha_i(I_D, II_D, III_D) \qquad (43)$$

where

$$I_D = D_{ii}$$

$$II_D = \tfrac{1}{2}(D_{ij}D_{ji} - D_{ii}D_{jj}) \qquad (44)$$

$$III_D = \det D_{ij}$$

where det represents the determinant of the quantity concerned.

In general, the coefficients α_i in (42) are determined by experiment, as is the viscosity in (1). However, it is possible to approximate these coefficients by various polynomials in I_D, II_D, and III_D. Since D_{ij} and $D_{im}D_{mj}$ in (42) are of the order of 1 and 2, respectively, in D_{ij}, it follows that, if f_{ij} in (40) is of the order of N, then α_0, α_1, and α_2 must be of the order of N, $N - 1$, and $N - 2$, respectively, in D_{ij}.

As an approximation, we then write

$$\alpha_0 = (\beta_1 I_D^{1/n} + \beta_2 II_D^{1/2n} + \beta_3 III_D^{1/3n})t_0^{1/n}$$

$$\alpha_1 = (\gamma_1 I_D^{(1-n)/n} + \gamma_2 II_D^{(1-n)/2n}$$
$$+ \gamma_3 III_D^{(1-n)/3n})t_0^{1/n} \quad (45)$$

$$\alpha_2 = (\lambda_1 I_D^{(1-2n)/n} + \lambda_2 II_D^{(1-2n)/2n}$$
$$+ \lambda_3 III_D^{(1-2n)/3n})t_0^{1/n}$$

Since $I_D = III_D = 0$ and $II_D = k^2/4$ for inhomogeneous simple shear, (45) reduces to

$$\alpha_0 = \beta_2(t_0 k/2)^{1/n}$$
$$\alpha_1 = \gamma_2(k/2)^{(1-n)/n}t_0^{1/n} \quad (46)$$
$$\alpha_2 = \lambda_2(k/2)^{(1-2n)/n}t_0^{1/n}$$

The constants β_2, γ_2, and λ_2 have the dimensions of stress.

The viscometric functions for a Reiner-Rivlin fluid are given by

$$\mu(k) = \tfrac{1}{2}k\alpha_1$$
$$\sigma_1(k) = \sigma_2(k) = k^2\alpha_2/4 \quad (47)$$

[*Truesdell and Noll*, 1965, p. 479], and hence the stress distribution in a Reiner-Rivlin fluid undergoing inhomogeneous simple shear is found by substituting (47) into (30) by using (46) and (29):

$$T_{12} = (-ax_1 + c) = \mu(k)$$
$$T_{11} = ax_2 + b$$
$$T_{22} = T_{11} \quad (48)$$
$$T_{33} = T_{11} - k^2\alpha_2/4$$

The viscosity function (28) is

$$\tilde{\mu} = \tfrac{1}{2}\alpha_1 = \tfrac{1}{2}\gamma_2\left(\frac{k}{2}\right)^{(1-n)/n}t_0^{1/n} \quad (49)$$

It is no longer constant, as it is for Newtonian fluids (1) and second-order fluids (33), but it depends on k. However, since $\tilde{\mu}$ is a material constant for a given fluid, it cannot reverse its sign when k reverses sign. In fact, $\tilde{\mu}$ must always be positive so that energy is dissipated and not produced during the deformation [*Truesdell and Noll*, 1965, p. 435]. Hence, n can only be odd. The fact that n is odd also ensures that T_{12} is always real and has the same sign as k [see *Berg*, 1969, p. 271].

The dynamically permissible velocity profiles from (31) are

$$v(x_1) = A(-ax_1 + c)^{n+1} + d \quad (50)$$

where A and d are constants. A problem similar to the one considered here has been treated by *Berg* [1969]; some of his assumptions force the velocity profiles possible in his study to be linear.

Lines of equal mean pressure (13) in the x_1x_2 plane are given by

$$x_2 = -[(p + b)/a] + B(-ax_1 + c) \quad (51)$$

where $B = \lambda_2/3a\gamma_2$. These lines are therefore straight and independent of the value of n.

Carter and Avé Lallemant [1970] have reported a value for n of about 5 for peridotite deformed dry, and the distributions of stress, strain, and mean pressure are shown for $n = 5$ in Figures 4 and 5.

Newtonian fluids. The stress distributions and dynamically permissible velocity profiles in a Newtonian fluid can be obtained by substituting $\alpha_1 = \alpha_2 = 0$ into (33), the constitutive equation for a second-order fluid. From (36) then the stress distribution in a Newtonian fluid undergoing inhomogeneous simple shear is

$$T_{12} = -ax_1 + c$$
$$T_{11} = ax_2 + b \quad (52)$$
$$T_{22} = T_{33} = T_{11}$$

The dynamically permissible velocity profiles are identical to those of the second-order fluid (37). If a is nonzero, a fold develops during the deformation. The mean pressure (13) is not constant, and lines of equal pressure on the x_1x_2 plane are straight lines given by

$$x_2 = -(p + b)/a \quad (53)$$

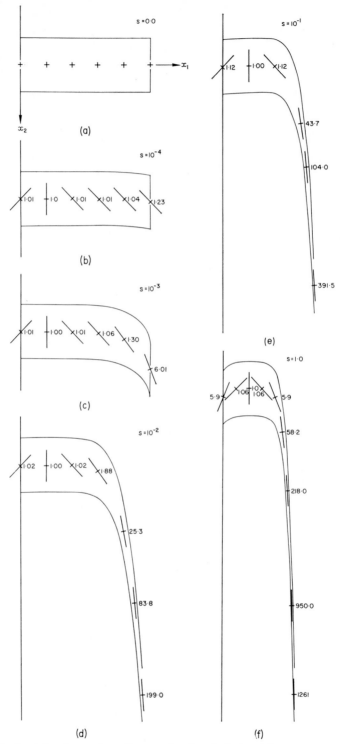

Fig. 4. Strain states for various times s developed during inhomogeneous simple shear of a Reiner-Rivlin fluid (41). In (46), $\beta_2 = \gamma_2 = \lambda_2 = 1$, $t_0 = 1$, and $n = 5$. In (48), $a = b = c = 1$. The orientation of a principal axis of strain (c^{-1}) is indicated by the long line. The number at each point is the ratio of the principal axes of strain: c_1^{-1}/c_2^{-1}. The shape and orientation of the undeformed layer are represented in (a).

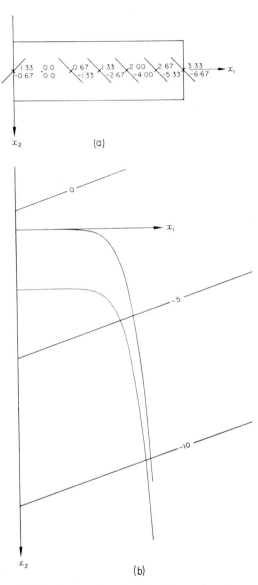

If $a = 0$, the only permissible deformation is a homogeneous simple shear, and the stresses and mean pressure maintain constant values throughout the deforming body.

The distributions of stress and pressure in a Newtonian fluid in which the deformation is the same as that in Figure 2 are shown in Figure 6.

SUMMARY AND DISCUSSION

Four classes of fluids have been considered in this paper. The general class is that of the simple fluids (12), where the stress in a particular deformed state is determined by the history of all the deformed states that the fluid has experienced. These fluids are capable of exhibiting a wide range of time-dependent effects, including stress relaxation. The next class considered comprise the fluids in which the stress is a function of the rate of deformation and of higher time derivatives of this tensor. This class is the Rivlin-Ericksen fluids (14). Again, since the stress depends on time derivatives of the rate of deformation, a large variety of time-dependent effects may be represented, but stress relaxation is not represented. If the stress is a (nonlinear) function of the instantaneous rate of deformation, the fluid is a Reiner-Rivlin fluid (15). These constitutive equations include the power laws often used to represent the behavior of polycrystalline aggregates undergoing steady-state creep. Clearly, only a limited variety of time-dependent effects can be displayed by these materials. If the stress is a linear function of the rate of deformation, the fluid is Newtonian (1). No time-dependent effects can be displayed by this fluid except the effects associated with an intrinsic time constant [*Truesdell*, 1952, p. 231].

For various special deformation geometries much of the generality in these fluids is not needed. Thus for inhomogeneous simple shear all simple fluids behave as Rivlin-Ericksen fluids (25); for deformation by steady shortening all simple fluids behave as Reiner-Rivlin fluids. For such special deformation geometries the complicated memory capabilities of the simple fluids are not required, so that in steady shortening, for instance, it is impossible to distinguish between a Reiner-Rivlin fluid and a far more complicated one.

Fig. 5. Stress states for the deformation shown in Figure 4 in a Reiner-Rivlin fluid (41). Constants in (46) and (48) have same value as they do in Figure 4. (a) Principal stresses ($T_1 + p$) and ($T_2 + p$) at all times in the deforming body. The orientation of a principal axis of stress is shown by the long line at each point. (b) Distribution of the mean pressure p at all times in the deforming body. The deformed layer of Figure $4d$ is shown superposed on this pressure distribution.

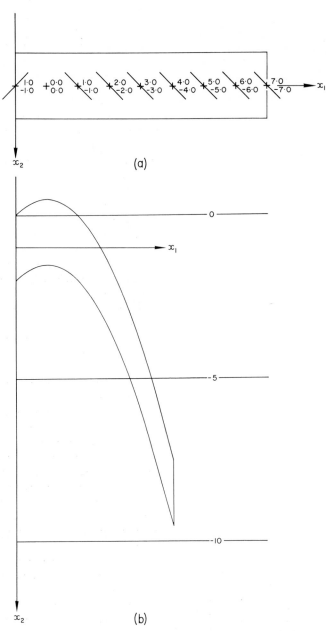

Fig. 6. Stress states for the deformation shown in Figure 2 in a Newtonian fluid (1), where $a = b = c = 1$ and $\mu_0 = 1$. (a) Principal stresses $(T_1 + p)$ and $(T_2 + p)$ at all times in the deforming body. The orientation of a principal axis of stress is shown by the long line at each point. (b) Distribution of the mean pressure p at all times in the deforming body. The deformed layer of Figure 2d is shown superposed on this pressure distribution.

If the rate of deformation is slow enough, all simple fluids behave as Rivlin-Ericksen fluids (33). For even slower flows the behavior is Newtonian (1). Also for slower flows the material behaves as an elastic fluid.

The principal differences between Newtonian fluids and the other fluids in deformations such as inhomogeneous simple shear is that (1) the viscosity function (28) is a constant for Newtonian fluids, but generally it is not constant for the other fluids and (2) the normal stresses are all equal in a Newtonian fluid, whereas generally they are not equal in other fluids. Deformations that are dynamically possible in one class of fluid are generally impossible in other classes. Thus homogeneous simple shear is possible in all classes, but nonlinear velocity profiles in simple shear are generally impossible, except for certain ones prescribed by (31).

For the most part, analytical solutions for velocity and stress distributions do not exist, except for special deformation geometries, examples of which have been presented here for deformation by inhomogeneous simple shear. A number of observations can be made concerning this type of deformation.

1. Folds are not formed unless there is a gradient of the normal stress T_{11} (Figure 1) along the zone in the direction of flow. This gradient in normal stress implies a gradient of the mean pressure in the direction of flow. If this gradient is 0, the deformation is by homogeneous simple shear, and no folds develop.

2. The folds that form are always similar in style, the strain distribution forming a divergent principal plane fan [Hobbs, 1972]. For all but Reiner-Rivlin fluids the velocity profile must be either quadratic or linear. Folds developing from such profiles are certainly widespread in natural shear zones [Ramsay and Graham, 1970], but many natural examples show more complicated fold shapes [Hobbs, 1966a, Figure 8]. These complicated fold profiles are not predicted by any of the fluids considered here for such simple deformations. The characteristic feature of all the velocity profiles predicted is the development of a fold closure opposite to the direction of movement of the wall of the zone (e.g., Figures 2, 4, and 6). This feature is always present unless c in (29) is 0 or negative (a in (29) positive); closures of this type are

characteristic of shear zone boundaries [Ramsay and Graham, 1970].

3. The distribution of mean pressure (13) has been drawn for the various fluids in Figures 3, 5, and 6. This distribution varies from that shown for a Newtonian fluid in Figure 6, where the curves of equal pressure are straight lines normal to the direction of flow, to that shown for a Rivlin-Ericksen fluid in Figure 3, where these curves are parabolas symmetrical about the axial plane of the fold. The pressure gradients in this latter example can be very large. Pressure distributions of this type may be capable of explaining the metamorphic differentiation observed in some deformed rocks where one phase (often quartz) is predominant in the axial plane regions of folds and another phase (often mica) is predominant in the limb regions. In general, however, the mean pressure defined by (13) is different from the thermodynamic pressure, although for Newtonian fluids the two are equal. This situation is discussed by Eringen [1962, pp. 166–171] for Stokesian fluids, but a discussion of the variation in thermodynamic pressure in a simple fluid undergoing deformation awaits further progress in thermodynamics [Truesdell, 1969, p. 188].

All the materials discussed here remain isotropic during deformation and have no specific configuration that can be labeled the undeformed state. If stressed, they will ultimately flow, although a considerable amount of time may elapse before flow begins, depending on the nature of the memory functional in (12). In view of the isotropy requirement it is difficult to see the relevance of the stress distributions shown in Figures 3, 5, and 6 to such questions as the development of a preferred orientation of mica where an anisotropy slowly develops during the deformation. For such questions the anisotropic materials of Ericksen [see Truesdell and Noll, 1965, pp. 520–537] may be relevant. The character of the strain distributions shown in Figures 2 and 4 is independent of a particular constitutive equation, and Ramsay and Graham [1970] have commented on the resemblance between such strain distributions and the patterns of preferred orientation of mica observed in shear zones.

If the constitutive equations of geological materials are those of simple fluids, the conventional type of steady shortening experiment (axially symmetric creep experiment) is not sufficient to establish all the mechanical properties of the material. Hence, the mechanical properties measured in such experiments cannot be used to predict the flow behavior in general deformations such as folding. They do not even suffice to predict the flow behavior during simple shear. In general, other types of experiment, such as nonsteady shortening, torsion, and stress relaxation are required, in addition to the steady creep experiment commonly performed, to establish the mechanical behavior of geological materials in general deformations. That geological materials are manifestly not Reiner-Rivlin fluids (15) is demonstrated by the observations that stress relaxation occurs [*Griggs and Blacic*, 1965].

Model for Aftershock Occurrence

LEON KNOPOFF

Institute of Geophysics and Planetary Physics
University of California at Los Angeles, Los Angeles, California 90024

A model is proposed in which aftershocks are the consequences of the superposition of two physical effects. First, large shear stresses are to be found near and beyond the edge of a rupture surface. Second, shocks can occur delayed in time from the primary shock by the process of tertiary creep and the subsequent lowering of fracture strength under the influence of a supercritical shear stress. On this model most of the aftershocks would tend to cluster around the edge of the fracture surface for a primary shock.

Burridge and Knopoff [1967] described a mechanical analog of earthquake sequences on a fault. Their model was reasonably successful in describing the interaction among events in what they called the 'charging cycle' of earthquakes, i.e., the events that remain in an earthquake catalog when aftershocks are removed from it. The notion of a charging cycle was introduced because the smaller shocks were imagined to serve as necessary events in anticipation of catastrophes and provided a mechanism for introducing deformational energy into a fault region at a quasi-steady state.

What was absent in the model of Burridge and Knopoff was a physical reason for the occurrence of aftershocks. Aftershocks were, in fact, obtained on their model but only through the expedient of introducing a 'viscous coupling' to provide the time delays in the system. In this paper a model of interaction that is more physically reasonable than simple viscosity is discussed to provide a model for aftershock occurrence. It is appropriate to review the Burridge-Knopoff model and note the features that are pertinent to the later discussion.

NEAREST-NEIGHBOR MODEL

Imagine two large massive blocks in relative motion. The contact between the two blocks is governed by friction. The details of the contact are that a linear chain of masses interconnected by springs in a nearest-neighbor configuration rests on one block, which we call 'stationary,' with frictional contacts under the masses (Figure 1). The masses are connected to the moving block through weak springs whose main function is to build up 'tectonic' stress and hence to increase the force of deformation on the masses. The total force on any mass is the sum of the effects of the deformational forces in the three springs connecting to it. When this force exceeds the static friction on one of the masses, that mass begins to move. Evidently, the condition for the occurrence of a shock depends not only on the 'tectonic driving stress,' which acts as a trigger for the event, but also on the prestress along the fault, i.e., the residual stress left in the wake of earlier shocks.

As a mass starts to move, it relatively compresses and stretches the springs connecting it to its neighbors and also reduces the stress in the spring connecting it to the moving block. The compression and stretching of the springs may trigger motion in the neighboring masses. The velocity of rupture depends on the prestress distribution along the fault. Rupture, i.e., the motion of successive masses in the chain, stops when the force of compression or extension in the spring at the end of the ruptured segment is insufficient to overcome the static friction under the next mass. Thus a stress concentration is found at the two ends of the ruptured segment of the fault. This stress concentration contributes to an inhomogeneous stress distribution along the fault; the precise nature of this inhomogeneous distribution de-

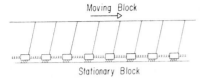

Fig. 1. Schematic model of contact between two fault blocks useful for computation [after *Burridge and Knopoff*, 1967].

pends on the past history of shocks on the fault. The stress concentration left in the wake of any given shock contributes to the conditions governing the time when the next shock will occur in its neighborhood.

Alternative descriptions of this process, which permit us to bypass the discrete mass-spring configuration and to replace it by a continuum, are possible from the point of view of dislocation theory. The static frictions can be considered to be critical stresses governing the migration of dislocations. When the total deformational shear stress, which is the sum of the shear stress derived from the tectonic drive and the local prestress due to the past history of earthquakes, exceeds the local critical shear stress, a dislocation starts to propagate into a region of lower local net stress. (Net static stress is the difference between the critical stress and the total deformational shear stress.) Dislocations can already be found along the fault due to the prestress; i.e., they are the result of earlier shocks. The rupture propagates if the increment in stress due to the shock, plus the total deformational shear stress, exceeds the local critical stress, the latter identified with the static friction. The process stops when the local net stress is too high, i.e., when a barrier exists. After a shock is over, dislocation concentrations or stress concentrations are found beyond both ends of the ruptured segment. The configuration of these stress concentrations is in direct relationship to the stress drop within the ruptured segment and especially near the termini.

As was noted above, no aftershocks can arise in this model except through the artifice introduced by Burridge and Knopoff, i.e., by setting some of the static frictions under the masses to 0 and replacing them with viscous elements. If all the masses in the model were underlain by viscous elements without friction, no earthquakes would occur. To introduce aftershocks based on a physically near-realistic model, the nature of the stress concentration at the ends of the segment and the effects of such stress concentrations on the fault itself are investigated.

STRESS CONCENTRATIONS

The static distribution in the plane of rupture of an earthquake fault is well known. If the end of a hypothetical fault break has a zero radius of curvature, the stress beyond the end of the fault falls off inversely as the square root of the distance from the edge for short distances from the edge. Since nature abhors an infinity, the edge of a real fault probably does not have a zero radius of curvature. In the case of the two-dimensional problem of plane strain, *Starr* [1928] has shown that the shear stress beyond a fault edge of zero radius of curvature, and in its plane, is

$$\sigma_R + [(\sigma_0 - \sigma_R) |x|/(x^2 - a^2)^{1/2}] \qquad |x| > a$$

for a fault of width $2a$, where x is measured from the middle of the fault [also see *Knopoff*, 1958]. The shear stress at infinity is σ_0, i.e., the stress that existed before rupture, and σ_R is the residual stress on the fault break after the shock. Thus a shock with a larger width of fault break surface, $2a$, will have a larger range of penetration of the stress concentration (Figure 2) into the unfaulted medium, σ_0 and σ_R being held constant. The stress drop is $(\sigma_0 - \sigma_R)$, and, if it increases, the range of penetration increases.

This point, then, is the first one at which I propose to modify the model of Figure 1. The

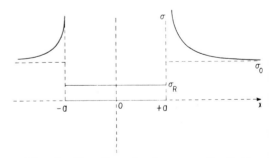

Fig. 2. Two-dimensional stress distribution after a rupture for a fault under a uniform shear stress σ_0. The stress drop on the fault is $\sigma_0 - \sigma_R$ uniformly.

stress concentrations at the ends of a ruptured segment have a considerable range and cannot be concentrated in the single springs immediately adjacent to the ends of the tear. The range is large since the term $(x - a)^{-1/2}$ falls off quite slowly. If the edge of the crack has a small radius of curvature, the infinity in the stress at the edge is removed [*Starr*, 1928]. The long-range stresses are not essential to the model of aftershocks unless a strong stress barrier exists. The presence of a strong barrier is essential to the process of termination of rupture. The long range of the stresses provides a means for tunneling through the barrier into a possibly weaker region.

The mathematical elements of this model are based on the assumption that the stress drop $(\sigma_0 - \sigma_R)$ and the deformational stress σ_0 are uniform and hence that the displacement is nonuniform on the fault. The displacement for such a model varies as $(x^2 - a^2)^{1/2}$. Although we expect the displacement on a real fault to differ from this expression and the stress drop to be nonuniform, the important feature here is that the displacement on the fault, going to 0 at the edge, corresponds to a strong stress concentration beyond the edge.

PRESHOCK CREEP AND FRACTURE

Griggs [1940] showed that the application of a supercritical stress to a rock at room conditions caused it to fracture, but fracture did not occur instantly. The time delay was a monotonically decreasing function of the stress (Figure 3). The fracture was preceded by a period of tertiary creep. Tertiary creep is an accelerated creep, occurring intermediate to the time of fracture and the episode of secondary creep, which is an interval of deformation probably associated with a process of dislocation climb. The processes of tertiary creep and subsequent brittle fracture were known for metals, but Griggs was the first to show that they also occurred in rocks.

Tertiary creep is thought to be due to a combination of the effects of necking and the smoothing of irregularities in grain boundaries. The smoothing is probably a tearing of the irregularities associated with stress concentrations in the grain interfaces, arising from these same irregularities in shape. Fracture is assumed to be the final stage in the response to

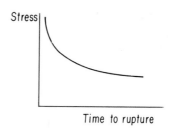

Fig. 3. Schematic curve of shear stress versus time to rupture for an earthquake fault [after *Griggs*, 1940].

the load stress after a significant episode of smoothing of the irregularities in the grain boundaries. Tertiary creep in nonfractured samples is described microscopically by a process in which dislocations move along grain boundaries away from triple junctions among grains.

I assume that tertiary creep followed by brittle fracture is a real phenomenon not only in the laboratory but also on real earthquake faults. Let us suppose that tertiary creep on earthquake faults is an accelerating process associated with the smoothing of irregularities in the fault surface. Large-scale rupture occurs when the fault has been smoothed down sufficiently. A great effort is being made today toward the prediction of the time of occurrence of earthquakes by instrumenting earthquake faults in the hope of observing prerupture creep. I assume, therefore, that this hope has some real basis. Macroscopically, then, on the model of tertiary creep followed by fracture, tertiary creep can be thought of as a gradual lowering of the average friction or critical stress by smoothing of the interface, until it is ultimately reduced to the level of the dynamic friction.

MODEL FOR AFTERSHOCKS

The two main features of aftershock occurrence that the model will be called on to generate will be, first, the observation that the magnitudes of the aftershocks decrease more or less regularly with time after the main shock and, second, that the time interval between aftershocks increases. The first observation is a little more difficult to document than the second one. The first condition states that the large-magnitude aftershocks are likely to occur early in the aftershock sequence. A quantitative statistical observation of this condition has been made for the Kern County (1952) and San Fernando (1971) aftershock sequences (L. Knopoff and J. K. Gardner, unpublished manuscript, 1971). Quantitative statements of the second observation are considerably older. *Omori* [1894] proposed the law of frequency of aftershocks,

$$\dot{n} \sim (t + a)^{-b}$$

Jeffreys [1938] used

$$\dot{n} \sim t^{-1}$$

Omori's relation has had considerable substantiation in one form or another for different values of the constants a and b. Omori actually used the case $b = 1$, and *Hirano* [1924] introduced $b \neq 1$ [*Aki*, 1956].

The model is perhaps best illustrated with the aid of the diagram of Figure 4. What might be a map of a linear earthquake fault is sketched at the top, but there are some deviations from linearity as well as some inhomogeneities in its physical properties along its length. The solid curves in Figure 4a, b, and c illustrate the variations in critical stress along the fault that are due to both effects. When a deviation from linearity is abrupt, the critical stress looks like a sharp pulse, as at point H. Variations in critical stress, not directly shown on the map as deviations from the linearity of the fault, can be presumed to be due to the nonuniformity of the shear strength of the rocks. The dotted curves represent the deformational stress due in part to dislocations on the fault at the times represented in the figures. The net stress, i.e., the difference between the critical stress and the deformational stress, can be considered to be the size of the barrier for the propagation of dislocations, i.e., for tearing.

In Figure 4a the sum of the prestress and the tectonic stress at point A is close to the critical value, and we can assume that a large earthquake is imminent. When the large shock takes place, a large dislocation tears loose from A and propagates to the right, finally stopping at B. Between A and B the stress on the fault drops to a value determined by both the static and the dynamic friction. The dislocation can be considered to be a pulse traveling along the fault as the rupture progresses to the right. Rupture stops when the barrier at B is larger than the stress due to the sum of all the dis-

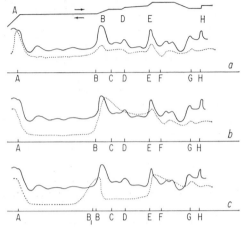

Fig. 4. Map of a hypothetical fault under shear (top line). Geographical displacements in the fault from a straight line appear as variations in critical shear strength (solid curve), as shown at A, B, D, E, and G (below). (a) At A an earthquake is about to occur. (b) After the shock, the deformational stress (dotted curve) is reduced between A and B and raised at the ends of the fractured segment to the left of A and to the right of B. The deformational stress is greater than the strength between B and E. (c) After some delay, rupture takes place between B and E. Further aftershocks may occur at B_1 and F in the wake of the first aftershock.

locations accumulated during the event plus the pre-existing stress at B. The stress is lowered to the right of A, but it is raised to the left of A when compared with the state before the shock. The condition immediately after the shock is shown in Figure 4b.

After the shock a large stress concentration is found at B, which has a large penetration to the right. The stress to the right of the barrier at B may be greater than the critical stress in some region. In this case an aftershock will occur, but there will be a time delay determined by a curve similar to that in Figure 3. Suppose that the greatest overstress, and presumably the shortest time delay, occurs at point C. The friction, i.e., the critical stress, is lowered in the region where the total stress (dotted curve) is greater than the critical stress (solid curve). The aftershock will start at C, which is the place where the rate of lowering of the friction is greatest, and the rupture will propagate in both directions. It may tear through the barrier at D if the dislocation stress added to the prestress exceeds the friction. It finally stops at points B and E, where new stress concentrations are introduced. These stress concentrations have long tails to the left at B and to the right at E. The actual total stress after the first aftershock is shown in the dotted curve of Figure 4c.

New aftershocks might be initiated at points B and F, where the new distribution of stress exceeds the local fracture strength. The event at F might tear as far as G.

The process described here does generate a series of aftershocks with generally diminishing magnitudes, since the total stress beyond a rupture barrier is proportional to the stress drop on the ruptured segment. Since the magnitude of a shock decreases with the width of the faulted segment, among other variables, and since the stress beyond the terminus of a narrower fracture has a shorter range of penetration, according to the formula $|x|/(x^2 - a^2)^{1/2}$, the magnitudes will generally decrease due to the reduction in the total stress. The process is degenerative in the sense that a small shock probably causes its successor to be even smaller. Furthermore, the reverse is also possible in principle, but practically the primary shock will probably have swept away a large contiguous, or almost contiguous, weak region, and thus there are only geometrically smaller weak regions available for the reduction of the stress in aftershocks. The process depends on the statistical distribution of strengths, and hence the decrease in magnitudes will not be monotonic.

The time intervals increase as the magnitudes decrease, according to the creep fracture delay process. Aftershocks of larger aftershocks, i.e., second-generation aftershocks, should cluster in time.

Tertiary creep serves to reduce the stress as well as the fracture strength, so that we cannot imagine that the stress pattern (dotted curve) of Figure 4b, for example, remains constant over the time interval between the time of the end of the rupture of the preceding shock and the time of the start of the next aftershock. If the stress drops too much, owing to post-shock creep, the aftershock sequence will end. Thus the survival of an aftershock sequence depends on the likelihood that the stress drop due to creep is not excessive. This reason need not be the only one for terminating an aftershock sequence. Inhomogeneity serves just as well.

The static stress or strength distribution of Figure 4b also may be a function of time if the healing of fractures is a time-dependent process. Furthermore, there is no reason to assume that the details of the strength distribution between A and B (Figure 4b) are the same after the major shock as they were before this shock (Figure 4a), since there may be a redistribution of gouge or other matter controlling the friction.

Finally, this process must be applied to two-dimensional fault surfaces. Thus we can imagine that the aftershocks are agents in increasing the area of rupture beyond that torn in a large shock. Aftershocks are not prevented from occurring in regions that have already ruptured earlier in the aftershock sequence; an example is shown at point B of Figure 4. Numerical examples of the process will be presented elsewhere.

Acknowledgment. I am deeply indebted to David D. Jackson for his enlightened comments on and criticism of this manuscript.

Pore Pressure in Geophysics

W. F. BRACE

Department of Earth and Planetary Sciences, Massachusetts Institute of Technology
Cambridge, Massachusetts   02139

Many mechanical, electrical, and transport properties of rocks are strongly affected by cavities in the form of cracks. There is some evidence that these cavities are interconnected and persist to a considerable depth. The shape of the cracks, and therefore the properties themselves, depends strongly on the pressure of fluid in the cracks. For rocks containing fluids under pressure these properties are not determined by the depth of burial alone. Many rock properties commonly considered shallow or associated with low pressure may occur at considerable depth in regions of high pore pressure. Such properties include low seismic velocity, high fluid permeability, low resistivity, frictional attenuation, brittle fracture, and low frictional resistance. A limiting depth may be determined by the temperature at which rock-forming minerals flow and at which cavities would be eliminated. This depth will probably not exceed 50–100 km.

For some time the mechanical effects of pore fluids have concerned structural geologists, and more recently they have concerned seismologists as well. An important and far-reaching theory of overthrusting, proposed by *Hubbert and Rubey* [1959], was based on high pore pressure in the vicinity of thrust faults; this theory has achieved wide acceptance, particularly with the discovery of various sources of high fluid pressure (summarized by *Bredehoeft and Hanshaw* [1968]). More recently, fluids under pressure in rock have been held responsible for triggering earthquakes near Denver [*Healy et al.*, 1968] and Rangeley, Colorado [*Raleigh*, 1971]. In the application both to faulting and to earthquakes, mechanical behavior was influenced by fluid pressures locally higher than normal. In this paper we show that fluids under pressure in rocks can influence a wide range of properties, including some not normally considered mechanical, and that pore pressure can therefore figure in a variety of geophysical phenomena.

Aqueous solutions are probably the more common pore fluid found in crustal rock, and these solutions are responsible for many geochemical effects [e.g., *Yoder*, 1955; *Hill and Boettcher*, 1970; *Wyllie*, 1971; *Lambert and Wyllie*, 1970; *Scholz*, 1972]. Here we are not concerned with chemical changes due to the interaction of pore fluid and rock. In most cases the following discussion would apply equally well if the pore fluid were an aqueous solution, an inert gas, an organic fluid, or even a partial melt.

The cavities found in rock have been termed cracks or pores depending on their shape [*Walsh and Brace*, 1966]. The expression pore pressure as used here simply refers to the fluid pressure developed in any sort of cavity in a rock. We here review the evidence for a continuous network of cavities in crustal rocks. We argue that some of these cavities have the form of cracks and that cracks can change their shape, owing to changes in pore pressure. We then review the various phenomena that have been found to depend on crack shape, including those noted by *Simmons and Nur* [1968].

EFFECTIVE PRESSURE

Terzaghi [1923] was the first to recognize the role of pore pressure in the deformation of soil. He showed that deformation depended on effective stress, defined as the difference between total stress and pore pressure. This principle has been widely accepted in soil mechanics [e.g., *Scott*, 1963], and, more recently, in rock mechanics as well (see the extended discussion by *Hubbert and Rubey* [1959, pp. 129–142]). A number of investigations of rocks and concrete [e.g., *McHenry*, 1948; *Robinson*, 1959; *Handin et al.*, 1963; *Byerlee*, 1966; *Brace and Martin*, 1968] have shown that fracture strength, frictional resistance, and increase of strength with pressure depend on effective stress or effective pressure. Actually, somewhat different functions of total stress and pore pressure may determine volumetric elastic strain [*Skempton*, 1961; *Nur and Byerlee*, 1971], seismic velocity [*Banthia et al.*, 1965], and perhaps other properties as well [*Robin*, 1972]. However, we are not concerned here with this refinement; in all the examples above, the key feature is the same, namely, that the value of seismic velocity, or frictional resistance, depends not only on total pressure or normal stress but also on pore pressure. We take the effective pressure here to be simply the total pressure minus the pore pressure and the effective stress to be the total normal stress minus the pore pressure. Compressive stress is taken to be positive.

CAVITIES IN ROCKS

It is difficult to determine the shape and the abundance of cavities in deeply buried rocks. The very act of sampling by drilling, for example, tends to introduce flaws not originally present [*Obert*, 1962; *Nur and Simmons*, 1970]. Surface outcrops are probably not representative of rock in its undisturbed state at, say, a depth of 10 km, owing to the well-known development of sheeting and other joints as one approaches a free surface [e.g., *Kieslinger*, 1958; *Snow*, 1968]. To obtain an idea of cavity shape and distribution in buried rock, we must turn to indirect observations, particularly those of electrical resistivity.

Cavity continuity. Deep electrical soundings provide important information about cavity distribution in buried rocks. If a rock is conductive, one of two conditions is required: the minerals in the rock must themselves be conductive, or the rock must contain a continuous network of electrolyte-filled cavities. Minerals in rocks like granite only become conductive at temperatures above about 500°C. Some observations suggest that, at much lower temperatures, crustal rocks are conductive and that they therefore contain a network of electrolyte-filled cavities.

Electrical conductivity of typical rocks under crustal conditions can be estimated with some accuracy for comparison with field observations. These estimates [*Brace*, 1971a] contain the effects of high pressure, high temperature, varying pore solutions, and a wide range of pore pressure. Conductivity based on these estimates should, over this depth range for water-saturated rocks, nowhere be less than about 10^{-5} mhos/m. This figure is within about an order of magnitude of field measurements made in areas underlain by highly resistive rock [*Keller et al.*, 1966]. It would thus appear that, unless the field areas studied are highly atypical, a network of solution-filled cavities exists in typical crustal rocks.

A recent direct comparison of field and laboratory electrical conductivity [*Hermance et al.*, 1972] gave the same result. Magnetotelluric measurements in an area underlaid by a thick sequence of recent volcanic rocks in Iceland gave estimates of conductivities that were within a factor of 2 of those measured on small samples in the laboratory by using pore solutions of comparable conductivity to those found in the field.

Laboratory observations provide further indirect evidence that pore space remains connected in typical rocks in spite of deep burial. It was found that an effective pressure of 10 kb was not sufficient to eliminate water-filled pore spaces in rocks like granite, slate, limestone, and dunite [*Brace and Orange*, 1968b]. On the basis of the effect of pressure on resistivity, it was estimated that an effective pressure in excess of about 60–80 kb would have been required to raise the resistivity of such rocks to values appropriate for the dry minerals themselves. At this point, presumably connected pore space might have been eliminated.

One of the striking features of the earthquakes near Denver [*Healy et al.*, 1968] was the large volume of fluid pumped into what was described as granitic basement. This volume amounted to nearly 10^6 m^3 and suggests that, at least in areas like Denver, even tight basement rocks contain a fairly extensive network of cavities.

A number of arguments can be advanced in support of the view that regional metamorphism of rocks occurs at a very low effective pressure [*Yoder*, 1955]. For the Franciscan sequence, as one example, observations include the occurrence of metamorphic minerals in vuggy veins, the appearance of certain diagnostic minerals, and the consistency in oxygen isotope ratio for specific minerals over large distances [*Ernst*, 1971]. Not only high fluid pressure but also a high degree of connectivity of the pore space is required during metamorphism.

There are also observations that argue against the continuity of pore space in deeply buried rocks, e.g., the observation that a pore pressure considerably in excess of hydrostatic pressure seems to have persisted in some deeply buried rocks for long periods of time [*Hubbert and Rubey*, 1959]. However, such observations are often difficult to assess critically, because a great many hydrologic and geologic factors combine to determine the pore pressure at a particular time and place in a given sequence of rocks [*Bredehoeft and Hanshaw*, 1968]. Most importantly, almost no measurements of pore pressure are available below a few kilometers in the earth, so that it is not known whether these observations can be considered general for greater depths.

Two laboratory observations suggest circumstances under which connectivity of pore space may be destroyed or reduced. The first of these circumstances is due to flow, the second to melting. In measurements of resistivity under stress, *Brace and Orange* [1968a] observed that marble did not show the normal reduction of resistivity at high stress characteristic of silicate rocks. This lack of reduction was attributed to crack closure due to the flow of calcite on the basis of the well-known low resolved shear stress for flow of this mineral [*Paterson*, 1959]. One might infer from this finding that, for any rocks capable of flow, connectivity of pore space might also be reduced. Presumably, this reduction would occur for typical silicate rocks at sufficiently high temperature and pressure. Another effect has been observed in unpublished studies of the melting of granite (A. Arzi, personal communication, 1971). Apparently, normal permeability was retained by a granite until the onset of melting in the presence of water and an effective pressure of several kilobars. However, at melting, and with the appearance of scattered melt films, it became

increasingly difficult for water to move freely through the system. Although a precise estimate is not yet possible, it would appear that permeability was lowered by several orders of magnitude. This reduction in magnitude could also be viewed as a reduction in the connectivity of pore space and presumably would characterize regions in the earth where melting has begun.

Cavity shape. Many of these observations suggest that cavities form a continuous network in deeply buried rocks. We now consider the shape of these cavities. On the basis of their behavior under pressure, cavities in rocks have been classified [*Walsh and Brace*, 1966] as pores or cracks; a pore has a cross section that is more or less equant, whereas a crack has one dimension many times less than the others. A further subdivision of cracks is based on size; microcracks are of the same order as the grain size. Larger cracks include joints, faults, and other fractures.

Cavities in the form of cracks are indirectly suggested by some of the arguments presented above. Thus a continuous network of cavities in rock of low porosity requires that the cavities have a high aspect ratio. If at the same time the cavities must contain or transport large quantities of fluid, as in the Denver situation, a cracklike rather than a tubelike form seems intuitively more likely.

In nearly all deep tunnels, drill holes, and other deep excavations, joints and other cracklike fractures are encountered. *Birch* [1964] reviewed observations that have been made in tunnels and mines and noted that the average fracture spacing was about 5 meters. He showed that this dimension was comparable with the number obtained from the size of scattering 'grains' needed to explain the observed damping of near-earthquake body waves. *Snow* [1968] studied the spacing of fractures in the relatively sound rock beneath 35 damsites. Although the spacing was ≤1 meter near the surface, it did not exceed about 6 meters in the deepest drill holes (130 meters). At depth, fractures such as these, which subdivide the rock, may be largely closed, but they are available and presumably accessible to fluids under pressure.

On the basis of their probable origin [*Adams and Williamson*, 1923; *Birch*, 1961; *Brace and Byerlee*, 1966; *Nur and Simmons*, 1970], microcracks as well as joints may be present at considerable depth. Most rock-forming minerals are brittle and have low crystallographic symmetry. During formation the grains in a typical rock probably fit together perfectly, but, as soon as pressure and temperature change, internal stresses are set up. It is easy to visualize the source of these stresses if one considers the effect of temperature change, for example. For quartz, volumetric thermal expansion is 2–3 times that for other common minerals. An even more extreme case is calcite, in which thermal expansion varies in sign with direction in the crystal. Not only thermal expansion but also the elastic moduli may vary considerably with direction in common minerals. Quartz undergoes a pronounced volume change at the α-β transition. All these characteristics may lead to internal stresses within grains and across grain boundaries in a rock when pressure and temperature change.

Observations of *Simmons and Nur* [1968] were used to suggest that the microcracks found in laboratory samples may actually be absent in rock in situ. They compared the resistivity and the velocity of rocks in a drill hole with samples taken to the laboratory and found that the differences noted in these two cases could be explained by the absence of microcracks in situ. However, *Orange* [1969] noted several difficulties with their comparisons, and thus their conclusion is somewhat uncertain. In any event, at least for velocity, the observed differences could just as well be explained by the presence or the absence of water in microcracks, as the authors themselves have pointed out.

Apparently, grain boundaries in rocks have a relatively low tensile strength [*Brace*, 1964; *Hoek and Bieniawski*, 1965], and, in addition, many minerals cleave readily. As a result of internal stresses, cracks form both along cleavage planes and across grain boundaries. These cracks are particularly evident in rocks in which quartz occurs with other minerals [*Adams and Williamson*, 1923; *Nur and Simmons*, 1970]. If microcracks form because of pressure and temperature changes, it seems unlikely that they are only surface features. Much of the cooling history of granite, for example, occurs well before it is encountered in the typical continental areas of the normal geothermal gradient. Although microcracks may remain closed in deeply buried rocks, they are probably accessible to fluids under pressure. As they do under surface conditions, where effective pressure is 0, cracks in deeply buried rocks probably open if fluid pressures are sufficiently high.

CRACK CLOSURE AND ITS EFFECTS

Assuming that both large- and small-scale cracks are ubiquitous in deeply buried rocks, we next consider the effects cracks have on various physical properties of rocks. These effects arise from a number of causes. First, an open crack can provide a passageway for the transport of material throughout otherwise solid rock. Cracks, then, will strongly influence transport properties such as fluid permeability and electrical conductivity. Also, because cracks form a more or less continuous network in rocks, fluids under pressure have access to rock. As a result, cracks may change shape, and many properties will be drastically affected. We review these many 'crack effects' after noting how crack shape is related to pressure.

Closure of cracks will be affected both by fluid pressure within the crack and by the total pressure applied to the boundaries of the rock containing the crack. *Walsh* [1965a] investigated crack closure under external pressure, approximating the crack by a penny-shaped elliptical slit. For internal fluid pressure as well, it can be shown by superposition that crack closure depends almost entirely on the difference between the external pressure and the internal pore pressure, i.e., on the effective pressure.

Any physical property that depends on crack geometry will, if fluid pressure exists in the crack and the cracks form a continuous network, depend on effective pressure. A good example is fluid permeability [*Brace et al.*, 1968]. As is shown in Figure 1, permeability decreased very markedly with effective pressure up to a pressure of 1 kb and less markedly beyond that. The dramatic changes at low pressure reflect the closure of microcracks. In these experiments, permeability clearly varied with effective pressure. For example, at an effective pressure of 250 bars, permeability was the same, even though pore pressure was 400 bars of water in one case and 50 bars of argon in another. As we show below, quite a number of different properties depend in one way or another on whether cracks are open or closed. By inference, all these properties depend on effective pressure in ap-

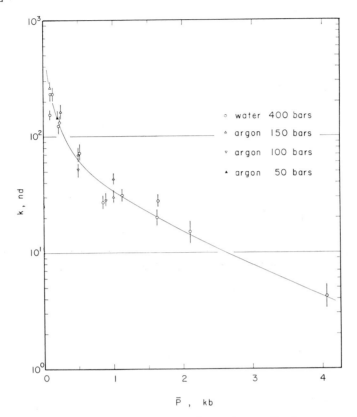

Fig. 1.   Dependence of permeability k of Westerly granite on effective pressure $\bar{P}$. The unit of permeability is the nanodarcy, and the pore pressure during the measurement is given by the different symbols. [After *Brace et al.*, 1968].

proximately the same way as permeability. As we noted earlier, the exact dependence may vary somewhat among the different properties [*Skempton*, 1961; *Nur and Byerlee*, 1971], but the key feature is that all of them are strong functions of both pore and total pressure.

Crack-dependent properties. Rock properties in which pore pressure plays a role through its interaction with cracks can be grouped as elastic, inelastic, and transport properties. Among the elastic properties, compressibility is known to depend on whether confining fluid can penetrate microcracks [*Adams and Williamson*, 1923]. This dependence is seen (Figure 2) in a comparison of compressibility measurements made on jacketed and unjacketed samples. In a jacketed sample, pore pressure is 0, whereas, in an unjacketed sample, fluid at a pressure equal to the confining pressure fills the microcracks and prevents crack closure. Walsh subsequently analyzed the exact way that open

cracks increase compressibility [*Walsh*, 1965a] and Young's modulus [*Walsh*, 1965b] in relation to the value for the solid mineral substance in the rock (Figure 3). Poisson's ratio is also affected by crack closure [*Walsh*, 1965c], as are the dynamic elastic properties [*Zisman*, 1933b; *Birch*, 1961; *Gordon and Davis*, 1968]. For example, jacketed and unjacketed measurements of V_P were compared by *Nur and Simmons* [1969b] for a granite; as is seen in Figure 4, penetration of the confining medium in the unjacketed case prevented the rapid velocity increase at low pressure seen in the jacketed case. The fact that the unjacketed curve is not parallel to the pressure axis reveals that velocity does not depend strictly on the difference between pore and confining pressure [*Nur and Byerlee*, 1971; *Banthia et al.*, 1965].

Certain inelastic phenomena are related to cracks. Attenuation of elastic waves may occur, owing to friction across the walls of cracks that

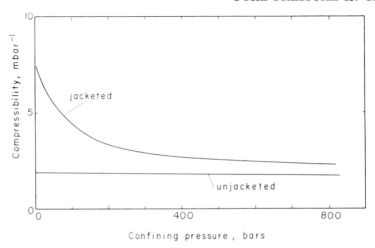

Fig. 2.   Compressibility of Quincy granite as a function of confining pressure for jacketed
and unjacketed samples. [After *Zisman,* 1933a].

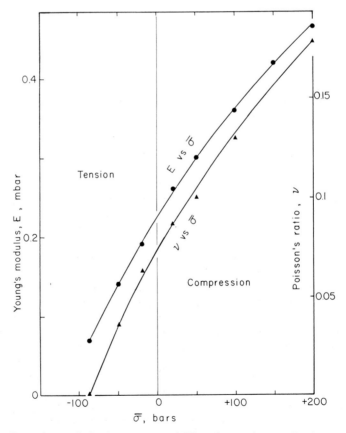

Fig. 3.   Dependence of elastic constants of Westerly granite on effective normal stress $\bar{\sigma}$ near
zero stress. [Reprinted from *Brace,* 1971b].

are nearly closed [*Walsh*, 1966; *Gordon and Davis*, 1968]. Near melting, attenuation in rocks may be caused by thin films of melt; *Walsh* [1968, 1969] suggested that these films could be treated as cracks. Attenuation in this case also depends on the aspect ratio of the 'cracks' and thus on whether cracks are closed. Although this dependency has not yet been observed in partially melted rock, Nur and Simmons demonstrated the general features at room temperature in a rock in which a viscous fluid filled the microcracks [1969*b*].

Frictional sliding on faults and saw cuts and brittle faulting, although they are not strictly crack phenomena in the sense used here, depend on effective pressure [*Robinson*, 1959; *Handin et al.*, 1963; *Brace and Martin*, 1968]. The presence or absence of stick slip during sliding depends on, among other things, the value of the effective normal stress on the fault [*Byerlee and Brace*, 1972].

Among the transport phenomena, fluid permeability has already been noted above. Also to be included (Figure 5) are the thermal conductivity of dry rocks [*Walsh and Decker*, 1966] and the electrical conductivity of water-saturated rocks (Figure 6) [*Brace and Orange*, 1968]. All three of these phenomena have been shown to depend on whether cracks are open or closed. By analogy, high-temperature electrical conductivity might also be affected by cracks,

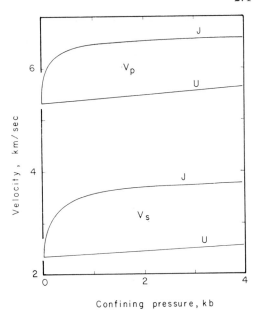

Fig. 4. Dependence of velocity of Casco granite on confining pressure for jacketed (J) and unjacketed (U) samples. The shear wave velocity is V_s, and the compressional wave velocity is V_p. [After *Nur and Simmons*, 1969*b*].

inasmuch as electrical conduction, like thermal conduction, depends on the quality of contact between the grains. The upper curve in Figure 6 illustrates this effect near room temperature for a rock with conductive minerals. If cracks

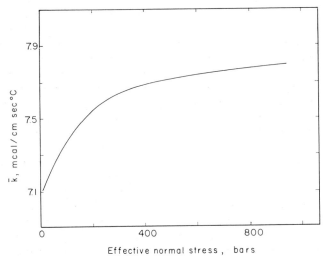

Fig. 5. Dependence of thermal conductivity of dry Casco granite on effective normal stress in the measurement direction. [After *Walsh and Decker*, 1966].

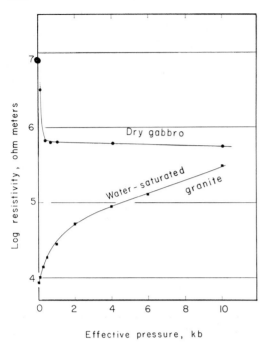

Fig. 6. Dependence of electrical resistivity of two rocks on effective pressure. For the dry gabbro the conduction is primarily through the minerals; for the water-saturated granite it is primarily through the water in the pore spaces. Resistivity of the gabbro has been multiplied by 10^5 in this plot. [After *Brace and Orange*, 1968].

are forced open by fluid pressure, poor contact and poor conduction will result.

APPLICATIONS

Pressure due to the weight of overlying rock increases downward in the earth, and, in general, one determines a pressure-dependent physical property in the earth simply by noting the thickness and the density of overlying rock. Clearly, for properties dependent on effective pressure, this procedure is inadequate. Total pressure, equivalent here to overburden pressure, is not the only parameter. In addition, one must know the pore pressure at the depth in question. If the pore pressure happens to be 0, the value of the property in question would be simply that value appropriate to the overburden pressure at that depth. In general, pore pressure may have to be considered, and, in that case, the appropriate pressure is the effective pressure, which could have any value

from 0 up to the full value of the overburden pressure.

One important consequence of this situation is that many 'low-pressure' phenomena may not be restricted to shallow depth. In the properties reviewed above, abrupt changes or anomalous values characteristically occur at a low effective pressure. For example, the velocity increased rapidly from an anomalously low value, or the permeability dropped rapidly from an anomalously high value, as the pressure increased. At a low effective pressure, brittle fracture is preferred over ductile behavior, and fault creep over stick slip. Frictional resistance is low, and compressibility high. All these characteristics are normally associated with shallow depth; however, they could occur over a wide range of depth if the pore pressure were high.

The occurrence of low-pressure phenomena at depth in rock with a high pore pressure is not a new idea. In one sense it forms the basis of the *Hubbert and Rubey* [1959] theory of overthrusting in that low sliding resistance (associated with low effective normal stress) is postulated to occur at considerable depth beneath certain large overthrusts, where the pore pressure is locally high. Anomalously low strength was observed near certain hydrous to anhydrous phase transitions by *Heard and Rubey* [1966] and *Raleigh and Paterson* [1965]. At the transition in these experiments, the pore pressure was high. These transitions might occur at a depth where the resulting low strength would be entirely inappropriate if one considered only the overburden pressure. Raleigh and Paterson also reported brittle phenomena in serpentinite at 700° and 5-kb confining pressure; presumably this behavior, typical of rocks near the surface, might also be found deep in the earth in regions where the pore pressure attains the high values of their experiments. Finally, seismic velocity in sedimentary formations that is often either anomalously low or increases with depth at a rate far less than typical is usually attributed to the high pore pressure that exists in these formations [e.g., *Musgrave and Hicks*, 1966].

The association of typically shallow phenomena with great depth may have other applications. Thus some regions of low velocity in the lower crust and the upper mantle could result from high fluid pressures rather than from high temperature alone, as is usually assumed. Frictional attenuation or high permeability may have to be considered in deeply buried rocks for which high pore pressure is reasonable and in which, therefore, open cracks may exist. Earthquakes due to brittle phenomena or associated with frictional sliding are often assumed to be restricted to a shallow depth [e.g., *Orowan*, 1960]; in regions of high pore pressure this restriction may not hold. Perhaps some intermediate-focus earthquakes result from the same type of brittle or frictional instability as shallow-focus earthquakes and are therefore simply indicative of regions of high pore pressure.

It is difficult to place exact limits on the depths to which typically shallow phenomena, as discussed above, might extend, although it is likely that high temperature will set a limit. As was noted above, connectivity of pore space is impaired when permanent deformation of minerals occurs. Once pore space is eliminated or pore connectivity reduced, many of the crack-dependent effects noted above disappear.

Whether fluid pressure remained high in isolated cavities in deeply buried rock would be governed by a number of geochemical and other factors [*Yoder*, 1955]. In any event, there is probably no single temperature at which permanent deformation of all minerals begins. To judge from the behavior of gabbro and granite in hot creep tests at high pressure [*Goetze*, 1971], permanent strains first become appreciable near 500°C for silicate rock-forming minerals. This temperature would limit the depth to ≤ 50 km. A higher temperature is possibly more appropriate for rocks found in the upper mantle, particularly at mantle pressures; however, a limiting depth much greater than twice this value seems unlikely, even in regions of exceptionally low geothermal gradient.

Acknowledgments. Discussion of some of these ideas with J. B. Walsh, P. Y. Robin, and S. Kaufmann was particularly helpful, although they should in no way be held responsible for the views stated.

Much of our quoted experimental work was supported by the National Science Foundation.

Faulting and Crustal Stress at Rangely, Colorado

C. B. Raleigh and J. H. Healy

U.S. Geological Survey, Menlo Park, California 94025

J. D. Bredehoeft

U.S. Geological Survey, Washington, D.C. 20242

In Rangely, Colorado, the magnitudes and directions of the principal stresses have been determined from hydraulic fracturing pressure data in Weber sandstone at the depth of the earthquake foci. The total principal stresses are compressive: $S_1 = 590$ bars (8550 psi), $S_2 = 430$ bars (6200 psi), and $S_3 = 315$ bars (4550 psi). The direction of the maximum compressive stress, N 70°E, is consistent with the orientations of the maximum horizontal compressive stress measured by overcoring surface exposures of sandstone of the Mesa Verde formation. The earthquakes at Rangely occur on a pre-existing fault. The orientation of the fault plane and the slip direction are known from focal plane solutions. Resolving the stresses onto the fault plane in the direction of slip gives a shear stress of 80 bars (1120 psi) and a total normal stress of 350 bars (5030 psi). Laboratory data on the initiation of frictional sliding on saw cuts in samples of Weber sandstone and the computed shear and normal stresses on the fault at Rangely indicate that a pore pressure of 260 bars (3730 psi) is required to reduce the effective normal stress sufficiently to induce slip. Pore pressures in the seismically active part of the reservoir at a time of frequent earthquakes were 275 bars (4000 psi), in accord with the calculated pressure.

David Griggs first suggested during the deliberations of the ad hoc Panel on Earthquake Prediction [Press, 1965] that earthquakes might be controlled by fluid injection into a fault zone. This radical hypothesis was not included in the panel's final report. However, without the knowledge of the panel, the U.S. Army had unintentionally been conducting the first experiment in earthquake control with very promising results.

Earthquakes were triggered by injection of waste fluid into the basement rock beneath the Rocky Mountain Arsenal of the Army near Denver, Colorado. Evans [1966] showed convincingly that the fluid injection was linked to the earthquakes and suggested that fluid injection might be used to control earthquakes. The basis for the hypothesis of Griggs and Evans regarding earthquake control came from the work of Hubbert and Rubey [1959], who showed that the fracture strength of rock is proportional to the difference between the total normal stress across the fracture and the pressure of the pore fluids within the rock. Healy et al. [1968] presented evidence that the Hubbert-Rubey effect provided a satisfactory explanation for the triggering of the Denver earthquakes by the injection of fluid at high pressure into rock already stressed to a point near its breaking strength.

Laboratory experiments on the fracture of rocks have shown that failure occurs when the shear stress on some potential shear surface exceeds the product of the normal stress across it (less the fluid pressure) and a frictional coefficient plus the cohesive strength. This simple law has been used successfully by Dieterich [1971] in modeling two-dimensional faults by finite element codes in digital computers. The coefficient of friction and cohesive strength can be determined adequately in the laboratory, but the naturally occurring shear and normal stresses can only be determined by in situ measurements. Of the several questions left unresolved by the incidence of the Denver earthquakes, one of the most important and experimentally most difficult was that of the state of stress in the rocks.

The Denver experiment has had two important implications aside from confirmation of the Hubbert-Rubey theory. Naturally occurring earthquakes might be modified beneficially by fluid injection, but earthquakes might be triggered inadvertently by fluid injection in places where no hazardous seismic activity would normally be anticipated. If earthquakes are ever to be controlled or the seismic effects associated with subsurface fluid injection predicted, the stresses within the rock mass must be known. This paper discusses measurements of stress at the Rangely oil field and their bearing on the seismicity associated with fluid pressures within the reservoir rock.

EARTHQUAKES AT RANGELY

Rangely, Colorado, is the site of a detailed experiment by the U.S. Geological Survey on the relationship between fluid pressure and earthquakes. Water has been injected by the field's producers into the reservoir rock, the Weber sandstone, for secondary recovery of oil since 1958. The fluid pressures in the periphery of the field have risen to 240–275 bars, about 50% in excess of the original field pressure of 170 bars, normal hydrostatic pressure for the reservoir depth of 1.8 km. Earthquakes have been occurring in the vicinity of Rangely at least since 1962, when the Uinta Basin Seismological Observatory was installed and began recording [*Munson*, 1970]. Since detailed recording of the seismic activity was begun by the U.S. Geological Survey in 1969, over 1000 earthquakes of magnitude M_L greater than −0.5 have been located (Figure 1) (C. B. Raleigh, J. H. Healy, J. D. Bredehoeft, and J. Bohn, unpublished data, 1971).

The earthquakes occur along the southwestern extension of a fault mapped through the field by Chevron Oil Company geologists (T. Larson, personal communication, 1969) from well-log data. Displacements of the upper boundary of the Weber sandstone along the fault are small, about 15 meters. As is shown in Figure 1, the earthquakes occur only along that part of the fault at which pore pressures are greater than about 200 bars. Most occur at or somewhat below the depth of the injection horizon, the Weber sandstone. The experiment is designed to lower and raise fluid pressures in

the seismically active zone to determine to what extent the pore pressure controls the occurrence of the earthquakes. The experiment should provide the first adequate field test of the Hubbert-Rubey hypothesis if the fluid pressure distribution away from the pumping-injection wells is known and the rock stresses are measured.

When the orientation and the magnitudes of the principal stresses in the Weber sandstone are given, the orientation of the fault plane and the slip direction still must be determined to calculate the shear and normal stresses on the active fault surface. The direction of the principal stresses in the rock should be consistent with the known sense of shear (right lateral) along the fault. Fault plane solutions have been determined from the sense of the first motion of P waves for Rangely events recorded on 10 or more stations. The first motions are distributed in a quadrantal pattern giving, for most earthquakes (Figure 2), strike slip on vertical, northwest-trending, or northeast-trending nodal planes. The nodal planes showing right lateral strike slip are nearly parallel to the mean trend of the epicenters and thus are likely to represent the active fault planes. This trend of faulting is consistent with the orientation of fracturing in the surface outcrops of Mancos shale at Rangely [*Thomas*, 1945] and the surface faults in exposures to the south of the field.

STRESS MEASUREMENTS

Two modes of stress measurement were used at Rangely. Near-surface stresses were measured by strain relief techniques in the Mesa Verde sandstone that rims the Rangely anticline [*de la Cruz and Raleigh*, 1972]. Overcoring of strain gages cemented to the rock surface and of the U.S. Bureau of Mines borehole deformation gage planted in shallow holes gave similar results. The magnitudes of the stresses, all compressive, were small, 10 bars or so [*de la Cruz and Raleigh*, 1972], because they were apparently relieved by surficial weathering and jointing. The directions of the principal horizontal stress components were reasonably consistent from measurements at three different outcrops. The principal directions at the three sites are shown in Figure 3. The maximum horizontal compressive stress is within 20° of east-west from averages of 6–10 measurements at

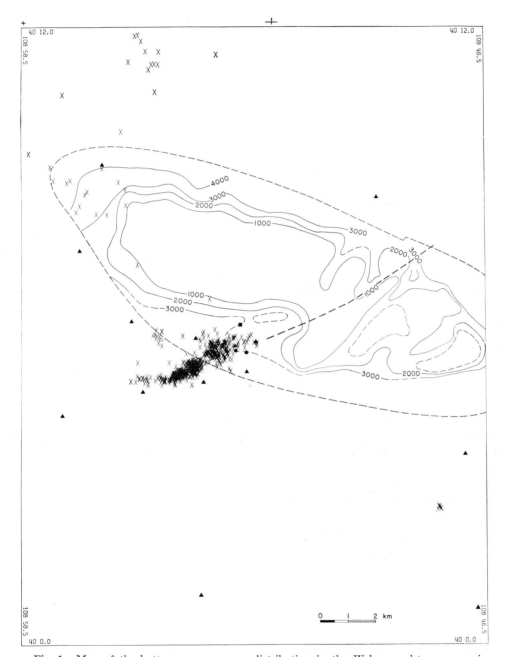

Fig. 1. Map of the bottom pore pressure distribution in the Weber sandstone reservoir in the Rangely oil field as of September 1969. A fault is shown as mapped from the structure contours on top of the Weber sandstone (T. C. Larson, personal communication, 1969). Crosses, epicenters of earthquakes of $M_L > -0.5$ from October 1969 to October 1970; circles, experimental wells; square, hydraulically fractured well UP 72X31.

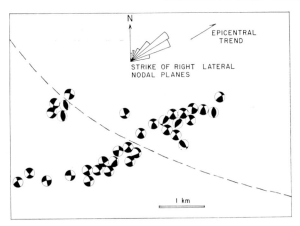

Fig. 2. Well-determined focal plane solutions for Rangely earthquakes for October 1969 to October 1970. Compressional first motions are shown in black, and dilatational first motions in white. The rose diagram gives the azimuths of the right lateral strike slip nodal planes from each solution. The dashed line is the field boundary.

each site. This direction is consistent with both the sense of shear on the fault determined from the focal plane solutions and the greatest compressive stress determined from the orientation of an induced hydraulic fracture within the Weber sandstone (N 70°E) (Figure 3).

The other technique for measurement of in situ stress employed at Rangely is hydraulic fracturing of a borehole into the Weber sandstone. The method, analyzed by *Hubbert and Willis* [1957], *Kehle* [1964], and *Haimson and Fairhurst* [1970], permits a determination of

the magnitude of the maximum and least horizontal stresses, provided that the rock is not already fractured and its tensile strength is known. As performed by *Haimson* [1972] at Rangely, the measurements involved packing off a section of the open hole and raising the fluid pressure in the packed-off section until the rock fractured in tension. When an unfractured column of core 5 meters long was removed during drilling, a packer was set 5 meters above the bottom of the hole so that the fracturing could be accomplished in intact rock. The fluid

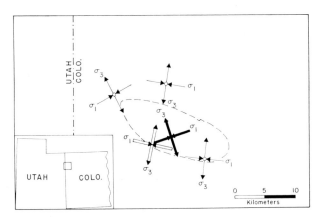

Fig. 3. Map showing orientations of horizontal principal stresses determined from overcoring (single-stemmed arrows) in surface exposures of the sandstone of the Mesa Verde formation, from earthquake focal plane solutions (double-stemmed open arrows), and from the orientation of an induced hydraulic fracture in the Weber sandstone (heavy solid arrows).

pressure was raised by pumping from the surface until a sudden pressure drop occurred, corresponding to the opening of a tensile fracture in the rock. The bottom hole pressure P_F at which the breakdown occurred was 328 bars (Figure 4).

After breakdown a sand and gel mixture was pumped into the fracture at 5 bbl/min to prop open the fracture. After pumps were shut off, the pressure dropped (Figure 4) suddenly to 314 bars, the instantaneous shut-in pressure (ISIP). This pressure held nearly constant until the surface pressure was released. Later a soft rubber packer was inflated in the fractured section, and the orientation of the fracture was determined from its impression on the packer to be vertical, striking N 70°E.

According to theory [Hubbert and Willis, 1957; Haimson and Fairhurst, 1970] the hydraulically induced fractures are opened when the fluid pressure P_F in a borehole in impermeable rock having an interstitial fluid pressure P_0 is

$$P_F = T + 3S_3 - S_1 - P_0 \qquad (1)$$

where T is the tensile strength and S_1 and S_3 are the greatest and least total principal stresses (compression taken as positive). The tensile fracture surface is perpendicular to S_3. The least principal stress S_3 is given by the value of the ISIP [Kehle, 1964]. The tensile strengths of rocks from the borehole are determined from laboratory measurement of the pore pressure required for hydraulically fracturing unconfined 6¼-cm diameter cores. A 7.5-mm diameter hole, coaxial with the core axis, is pressurized until the core ruptures in tension along a fracture parallel to the core axis. The tensile strength is 138 bars, nearly twice that determined from Brazilian tests on the same rock [Haimson, 1972]. The tensile strength derived from pressurization of such very thick-walled hollow cylinders is probably the appropriate value for use in calculation of in situ stresses from (1).

Other conditions for the use of (1) in calculating the maximum principal stress are that the fracturing fluid should not have permeated into the wall rock and that the tensile fracture be produced at the time when the hole is pressurized. In this nonproductive part of the section the Weber sandstone typically has permeability ranging from 0.1 to 1 millidarcy (Chevron Oil Company, personal communication, 1970). Although direct measurement of the permeability in the fractured section has not been determined, core from this section is quite similar to samples from which permeability measurements in this range are available. Because of the low permeability only negligible penetration by the pressurized fluid into the

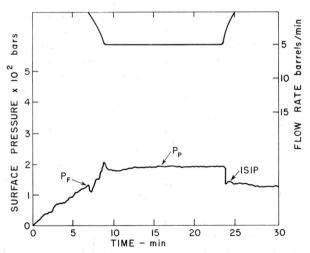

Fig. 4. Surface pressure and flow rate as recorded by a service company during hydraulic fracturing of well UP 72X31. The bottom hole pressure equals the surface pressure plus 190 bars. P_F, breakdown pressure; P_P, pumping pressure; ISIP, instantaneous shut-in pressure.

well rock would have been achieved over the interval of less than ½ hour during which fluid pressure was raised from normal hydrostatic to the breakdown pressure. With regard to the second condition the fracture of the wall of the borehole was recorded over the full length on both sides of the soft inflatable rubber packer following the experiment. Careful examination of the core from that section of the hole showed no such fracturing; in fact, the core was intact except for a few horizontal partings along shaley layers. The fracture recorded by the packer was, therefore, produced by pressurization of the borehole.

The total principal stresses calculated from (1) are:

$S_1 = 590$ bars   N 70°E horizontal
$S_2 = 427$ bars   vertical (assuming the lithostatic pressure to be 0.23 b/m depth)
$S_3 = 314$ bars   N 160°E horizontal

The orientations are accurate to within ±5°; the fluid pressure measurements during fracturing, from which the magnitudes of the stress are derived, are believed accurate to within ±15 bars on the basis of observations from two surface pressure transducers and an Amerada gage recording at the bottom of the hole.

The total least principal stress, being equal to the ISIP, is very well documented from numerous other tests in the field. During routine hydraulic fracturing for well stimulation at Rangely both the ISIP and the breakdown pressure are recorded. The initial breakdown is achieved through two opposing rows of vertical perforations in the casing, and the near-in fracture orientation may be more controlled by the perforations than by the far-field stresses. Whatever the explanation may be, it is observed that the breakdown pressure is quite variable in such tests. However, provided that the fracture attains an orientation normal to S_3 at short distances from the hole, the ISIP should reflect the discontinuous decrease in transmissibility expected when the crack closes as the fluid pressure falls to the magnitude of the normal stress S_3. For 45 wells in the central and eastern parts of the field the ISIP in these more recently treated, cased holes averaged 290 bars [Haimson, 1972]. Of these, 35 wells gave an

ISIP within 10% of that value. The ISIP in the western end of the field is less, averaging 220 bars for 12 measurements [Haimson, 1972]. Therefore, although additional measurements of the breakdown pressure in such controlled tests are desirable, determinations of S_3 from the ISIP over the oil field are consistent and in good agreement with the value used in the following analysis.

EFFECTIVE STRESS DURING FAULTING

If the above values for the principal stresses and the orientation of the fault plane and slip direction are given, the shear and normal stresses acting in the slip direction on the fault plane in the vicinity of well UP 72X31 (Figure 1) can be calculated. The nodal plane and slip direction orientations vary by ±20° for different earthquakes along different parts of the fault zone. Furthermore, the measurements of the orientations of the principal horizontal stresses vary by as much as ±20°. The following calculations are therefore best applied only to the near vicinity of well UP 72X31, where the detailed stress measurements were conducted. The orientations of the nodal planes of the nearest earthquakes, 0.8 km to the south, are plotted in Figure 5 with the orientations of the principal stresses. The right lateral nodal plane parallel to the trend of the epicenters strikes N 50°E, dipping 80° NW. The slip direction plunges 20° toward N 234°E. The normal to the fault and the slip direction in the fault are taken as the coordinate axes x_1' and x_2', respectively, and the normal to the slip direction in the fault plane is taken as x_3'. Rotation of the principal stresses S_{ij} to this new coordinate system gives the stresses S_{kl}' in tensor notation:

$$S_{kl}' = a_{ki}a_{lj}S_{ij} \qquad (2)$$

The direction cosines a_{ij} for the transformation are determined from angles between the coordinate axes measured from the equal-area projection (Figure 5).

	x_1	x_2	x_3
x_1'	cos 70.5	cos 80.5	cos 22
x_2'	cos 154	cos 70	cos 75
x_3'	cos 74	cos 23	cos 106

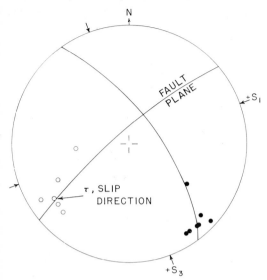

Fig. 5. Equal-area projection showing orientations of principal stress from hydraulic fracturing and mean nodal plane orientations from earthquakes nearest to the fractured well. The slip direction on the fault plane is τ.

The normal stress S_n across the fault plane is

$$S_n = S_{33}' = S_1 a_{11}{}^2 + S_2 a_{22}{}^2 + S_3 a_{33}{}^2$$

$$S_n = 347 \text{ bars} \tag{3}$$

The shear stress τ along the slip direction is

$$\tau = \tau_{12}' = a_{11} a_{21} S_1 + a_{12} a_{22} S_2 + a_{13} a_{23} S_3$$

$$\tau = 77 \text{ bars} \tag{4}$$

Laboratory experiments on the strength of Weber sandstone cores have been conducted by *Byerlee* [1971]. The data in Figure 6 show the shear stress and effective normal stresses at failure on pre-existing fracture or saw cut surfaces. The reported stresses are the peak stress encountered where sliding begins for each specimen. The ratio of the shear stress to the effective normal stress is $\tau/\sigma_n = 0.81$ and defines the coefficient of friction for the initiation of slip on pre-existing faults in Weber sandstone.

If the laboratory data are applied to faulting in the Weber sandstone at Rangely, slip should take place when

$$\tau/(S_n - p) = 0.81 \tag{5}$$

With the τ and S_n values from hydraulic fracturing and the known orientation of the faulting, a pore pressure p of 257 bars satisfies (5). This pressure is very near the bottom hole pressure

in the experimental wells during the injection phase of the experiment when earthquakes were most frequent and the pore pressure was 275 bars. After 1 month of backflowing the experimental wells the bottom hole pressure dropped by 35 bars, and the earthquakes in the vicinity of the wells ceased.

It is useful to examine the effect of the variability in the relative orientations of the fault and the stresses. For a vertical fault plane that is slipping in the horizontal direction and inclined at angle α to S_3, the pore pressure required for faulting can readily be calculated. Figure 7 shows α plotted against the pore pressure p for $S_1 = 590$ bars, $S_3 = 314$ bars, and $\tau/(S_n - p) = 0.81$ at failure. The variation of the pore pressure required to initiate faulting varies between about 235 and 290 bars, where α is between 45° and 85°. The second curve shows the pore pressure variation for initial stresses $S_1 = 570$ bars and $S_3 = 330$ bars. Provided that the magnitudes of the stresses measured are correct to within about 15 bars, the pore pressures existing along the seismically active part of the fault at Rangely are quite close to pressures calculated by *Hubbert and Rubey's* [1959] effective stress law to initiate faulting in previously fractured rock.

CONCLUSIONS

To control earthquakes or to avoid triggering earthquakes inadvertently, we need to know the

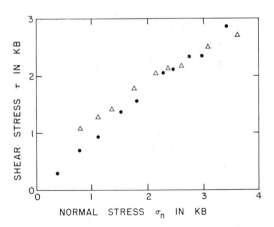

Fig. 6. Shear versus effective normal stresses for fracturing of intact rock and sliding on saw cuts in Weber sandstone [*Byerlee*, 1971]. Triangles, fracture of intact rock; circles, maximum friction for sliding on saw cuts.

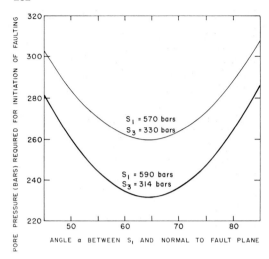

Fig. 7. Pore pressure p required to initiate shear failure in Weber sandstone reservoir as a function of α, the angle between the normal to the fault plane, and S, the maximum principal stress. S_1 and S_3 are measured as 590 and 314 bars, respectively; values of 570 and 330 bars are assumed to show effect of a variation in the state of stress.

state of stress in the rocks and a criterion for failure that takes into account the effect of pore fluids on rock strength and behavior. At Rangely, measurement of the state of stress in situ has been carried out by hydraulic fracturing at the depth of the earthquake foci. In the laboratory the shear strength of previously fractured Weber sandstone has been determined by *Byerlee* [1971] to be the product of a coefficient of friction and the effective normal stress (the total normal stress less the pore pressure of the ambient fluid). At Rangely, if the earthquakes are assumed to occur when slip is initiated on a pre-existing fault, the resolution of the measured stresses onto the fault plane in the direction of slip indicates that a pore pressure of at least 255 bars is required for shear failure in the fault. Earthquakes occur near the bottom of wells at which the pore pressure is well known to be around 275 bars. The seismicity ceased after the pressure in the wells was dropped by 35 bars. The observed pore pressure and that calculated from a simple laboratory-derived criterion for failure combined with stresses measured in situ are thus in good agreement.

The results reported here suggest that the

measurements of stress in situ at considerable depth can be made and that such measurements can be used, in conjunction with the laboratory-determined physical properties, to predict the pore pressures at which earthquakes may be triggered or propagating fractures possibly arrested. Although, at present, little is known about the state of stress, the temperatures, the pore pressures, or the strength of the fault zone material in active faults, it appears that these quantities are measurable. Moreover, even though adequate failure criteria that include the effect of temperature are not completely determined, conventional laboratory methods of obtaining failure criteria by experimentation can be applied to shear failure in active faults with reasonable hope of success. As a result of the Rangely experiment, Griggs's speculation on earthquake control can now be tested by an experiment on an active fault.

APPENDIX

Stress directions and focal plane solutions. One interesting result of this study is the observation that the directions of the principal horizontal stresses from in situ measurements agree with those derived from the focal plane solutions. The radiation pattern of P waves from each of these earthquakes defines a double-couple solution for which the nodal planes represent, in effect, mutually perpendicular conjugate shear planes. The maximum and least principal stresses giving the maximum shearing stress on these planes in the proper direction bisect the dilatational and compressional quadrants, respectively, of the focal sphere [*Scheidegger*, 1964]. For the unambiguous solutions for Rangely earthquakes the maximum and least compressive stresses are derived accordingly and shown in Figure 8. The maximum and least horizontal stresses from the in situ measurements plotted on the same projection agree quite well with those from the earthquakes.

McKenzie [1969a] has pointed out that pre-existing faults may slip even at quite low resolved shear stresses before new fractures in intact rock will develop [also, cf. *Handin*, 1969]. His analysis indicates that the P direction, or S_1, the maximum principal stress in the present terminology, may lie anywhere within the dilatational quadrant of the focal sphere. If this analysis is correct, stress directions derived

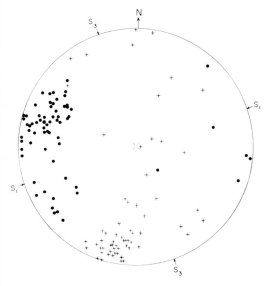

Fig. 8. P (maximum principal stress, compression) directions (circles) and T (least principal stress, tension) directions (plus signs) from fault plane solutions of Rangely earthquakes. The arrows on the circle of projection are the greatest and least horizontal principal stresses from in situ stress measurements.

from focal plane solutions would clearly be of only limited tectonic significance.

However, even intact homogeneous rock has finite strength, so that new faults can be generated even though existing faults are under finite shear stress.

The shear strength τ_c of a fracture in rock is given by the product of the effective normal stress times a coefficient of friction, i.e., for normal stresses corresponding to shallow earthquakes,

$$\tau_c = \sigma_n \mu \qquad (A1)$$

Likewise, the shear strength τ_c' of intact rock is given by the normal stress times the coefficient of internal friction $\tan \phi$ plus a cohesive strength τ_0:

$$\tau_c' = \tau_0 + \sigma_n \tan \phi \qquad (A2)$$

These equations are represented by straight lines on the Mohr diagram, a plot of τ versus σ_n, the shear and normal stresses. With, for example, the data for Weber B sandstone from *Byerlee* [1971] in kilobars,

$$\tau_c = 0.81\sigma_n \qquad (A3)$$

$$\tau_c' = 0.7 + 0.6\sigma_n' \qquad (A4)$$

Failure by faulting in the intact Weber sandstone will occur when a Mohr circle (Figure 9) is tangent to the envelope for τ_c', σ_n'. The circle has its center at $(\sigma_1 + \sigma_3)/2$ and is the locus of points where τ and σ_n are given by

$$\tau = \frac{\sigma_1 - \sigma_3}{2} \sin 2\alpha \qquad (A5)$$

$$\sigma_n = \frac{\sigma_1 + \sigma_3}{2} + \frac{\sigma_1 - \sigma_3}{2} \cos 2\alpha \qquad (A6)$$

Thus, for stresses σ_1 and σ_3 required for fracturing intact material, there will be a range of α over which slip on a pre-existing fracture will take place (Figure 9). The values of 2α at the intercepts of the Mohr circle on the envelope for sliding on existing faults are given by

$$\frac{\sin 2\alpha}{\mu} - \cos 2\alpha = \frac{1}{\sin \phi} - \frac{2\tau_0}{(\sigma_1 - \sigma_3) \tan \phi} \qquad (A7)$$

Two values for α, α_1, and α_2 satisfy the above equation and

$$\mu = -\cot (\alpha_1 + \alpha_2) \qquad (A8)$$

The angles α_1 and α_2 can also be solved graphically in the Mohr construction (Figure 9). Table 1 gives the range of α over which faulting will take place on existing faults in preference to fracture in the intact Weber sandstone as a function of the stresses (for simplicity, $\sigma_1 > \sigma_2$

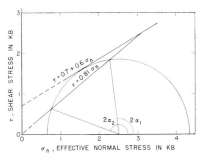

Fig. 9. Mohr diagram showing envelopes for failure in intact Weber sandstone and for slip on pre-existing faults in Weber sandstone and a Mohr circle tangent to the envelope for failure in the intact rock.

TABLE 1.   Range of Angles, α, $\alpha_1 < \alpha < \alpha_2$, between σ and the Normal to the Fault Plane at Stresses over Which Slip Will Occur on a Pre-existing Fault Rather Than on a New Shear Fracture in Intact Rock*

Stress, kb			Range, deg		
$(\sigma_1 + \sigma_3)/2$	σ_1	σ_3	α_1	α_2	$\alpha_2 - \alpha_1$
1.6	2.95	0.25	44	85	41
2.0	3.60	0.40	46	84	38
2.5	4.38	0.62	48	81	33
3.0	5.12	0.88	51	78	27
3.5	5.88	1.12	54	76	22
4.0	6.61	1.39	56	72	16
4.5	7.38	1.62	59	70	11

* Calculated for Weber B sandstone.

$= \sigma_3$). The range is greatest at low stress. From Table 1, σ_1 (or the P axis in *McKenzie's* [1969a] terminology) may lie anywhere between 44° and 85° to the fault plane normal for slip to occur preferentially along the fault at stresses $\sigma_1 = 2.95$ kb and $\sigma_3 = 0.25$ kb. The mean stress is 1.15 kb, corresponding approximately to a depth of 5 km. At greater depths the higher stresses required for sliding on existing fractures approach those required for fracturing in intact rock because $\mu > \tan \phi$.

A more extreme example is the Westerly granite, which has a very high fracture strength but a frictional sliding coefficient very close to that of the Weber sandstone [*Byerlee*, 1967]. When σ_1 lies less than 40° or greater than 82° to a pre-existing fracture, faulting will occur in the intact rock in preference to frictional sliding. These angles hold over a wide range of stress.

The maximum principal stress is taken to be 45° from the fault plane when derived from focal plane solutions. This angle could be in error by as much as 40° where sliding on an existing fault in an otherwise intact rock produces the earthquake. If the nodal plane corresponding to the fault is known, σ_1 should be plotted at 60° to the normal to the fault plane and 30° to the slip direction. In this case the orientation of σ_1 would not be in error by more than 20°.

Acknowledgments. We are grateful to our colleagues, J. D. Byerlee, R. de la Cruz, and J. Dieterich, and to B. Haimson for discussion and criticisms. Publication of this paper was authorized by the Director of the U.S. Geological Survey.

The research reported here was supported by the Advanced Research Projects Agency of the Department of Defense under ARPA orders 1469 and 1684.

Plate Tectonics, the Analogy with Glacier Flow, and Isostasy

F. C. FRANK

Department of Physics, University of Bristol
Bristol, England

A diagrammatic method is presented for calculating resultant horizontal forces on parts of the lithosphere attributable to gravity. These forces give rise to stresses in the lithosphere mainly dependent on changes of large-scale topographic height. Spreading pressures of the ocean floors, sliding away from mid-ocean ridges, approximately balance the spreading pressures of continental mountain masses, in isostatic balance for vertical forces. Shear stresses transmitted to plates from below (by 'conveyor belt' mantle convection) cannot be significantly larger than the gravitational sliding stresses deducible from surface topography and may be small or absent.

In the discussion period following a recent colloquium by D. H. Matthews I expressed the view that the motion of the ocean floors away from mid-ocean ridges should be considered downhill sliding under the force of gravity like that of a glacier and that, as for a glacier, 'downhill' is to be interpreted with reference to the form of the upper surface (in this case, the ocean floor, descending from about 2 km below sea level at the ridge to about 5 km below sea level before reaching the continental shelf, or 10 km below in a trench). As for a glacier, the form of the lower surface is of secondary importance, and this situation is fortunate, because we know comparatively little about the form of the lower surface. Objection was taken to this view on the grounds that it would imply a particular pattern of gravitational anomalies in association with ocean ridges, whereas no strong correlation of gravitational anomalies with ocean ridges is observed. This paper clarifies this issue and shows, among other things, that even a fully isostatic situation, free from gravitational anomaly, is not incompatible with downhill sliding, according to the glacier analogy. I prefer, incidentally, to make the analogy to a glacier rather than to an ice sheet, because the ocean floor, like the glacier, is principally fed with new material in a localized mountaintop region, from which it flows unidirectionally and can flow uniformly, whereas the ice sheet, if

persisting in a steady state, is fed with new material and therefore is undergoing extension (often two dimensionally) over a large part of its area; thus the ocean floor is more analogous to a very wide glacier rather than to an ice sheet. (An illness has both delayed the submission of this paper for publication by about a year and prevented a thorough search for previous work. I am indebted to referees for directing my attention to the work of *Benioff* [1949], *Weertman* [1962, 1963], *Temple* [1968], *Pollock* [1969], and *Hales* [1969]. I believe that the method of treating the problem given here is original. I am advised that what I have referred to as pressure-compensated stresses are most appropriately called Benioff stresses.)

Let us start with an extremely simplified model, Figure 1a, with a flat earth and the flat oceanic plate ABCD, the ocean floor AB sloping down from the ridge at A and the plate sliding on its lower surface CD and being renewed by the solidification of melt fluids injected from below into the opening gap AC. An arbitrary vertical plane BD passes through the plate, where it approaches either the seismic zone commencing with an oceanic trench or a continent receding from the ridge. We calculate forces acting on the plate, which is assumed to be rigid, by first considering that the forces acting on all its surfaces are hydrostatic. This assumption is clearly correct on AB (the drag

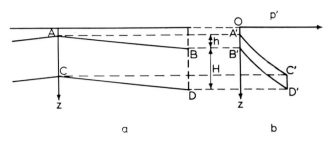

Fig. 1. (a) Idealized model of the downhill sliding of the oceanic floor away from the ridge A. (b) Pressure gradient in the plate.

due to ocean currents is negligible), and, as will be shown later, it is a reasonable one to make with respect to the horizontal forces on AC, where new melt fluid is injected. The resultant force thus calculated must be balanced by the nonhydrostatic forces acting on BD and CD. We take the pressure everywhere to be given by the vertical overburden pressure

$$p(z) = \int_0^z g\rho \, d\zeta \qquad (1)$$

where g is the gravitational acceleration, ρ is the density at depth ζ, which is an integration variable, and z is the integration limit, which is measured downward from the surface of the sea, and the integration path is vertical. The true pressure at a point can deviate from this value by amounts of the order of magnitude of the yield stress of the material by 'arching' or where there is too rapid horizontal change of density, but the mean value of the compressive vertical stress $\langle \sigma_{zz} \rangle$ over any horizontal plane of dimensions that are large as compared with depth must necessarily equal the mean value $\langle p \rangle$ of pressure thus calculated, so that this assumption is well-justified where the vertical distribution of density changes slowly with horizontal position, as in our case. This assumption ensures that vertical forces on the plate (gravity included) are totally balanced except for possible vertical forces on its end. The horizontal force F_H (per unit length normal to the plane of the figure) calculated with these hydrostatic boundary conditions is

$$F_H = \int_{ACD} p \, dz - \int_{ABD} p \, dz \qquad (2)$$

the integrals being taken with respect to depth z but along the left and right bounding surfaces

ACD and ABD of the plate. Since both integrals have the same range in z, we can also introduce $p_w = \int_0^z g\rho_w \, d\zeta$, define $p' = p - p_w$, and write

$$F_H = \int_{ACD} p' \, dz - \int_{ABD} p' \, dz \qquad (3)$$

where ρ_w is the density of sea water (with fictive existence to any depth). Thus the force F_H is given by the area A′C′D′B′ in Figure 1b. The slope dp'/dz of either line A′C′ or line B′D′ is proportional to the density of the plate material (less that of water) at the corresponding depth. Taking densities as being uniform makes A′C′ and B′D′ parallel straight lines. Taking densities as increasing with depth, according to the same law at either end of the plate, makes them similar curves, and in both cases we have

$$F_H = (\langle g\rho \rangle - \langle g\rho_w \rangle) h H \qquad (4)$$

where g is the gravitational acceleration, ρ is the density of the plate material, the angle brackets signify an average, h is the height of the ridge A above B, and H is the thickness of the plate. With little error, g can be treated as a constant, and so can ρ_w, which is a fictive quantity for most of the depth concerned.

If there is no shear traction on CD, there is a mean horizontal nonhydrostatic compressive stress on BD:

$$\langle -\sigma_{zz} \rangle = (\langle g\rho \rangle - \langle g\rho_w \rangle) h \qquad (5)$$

Insertion of values $g = 10$ m sec^{-2}, $(\langle \rho \rangle - \langle \rho_w \rangle)$ $= 2.3 \times 10^3$ kg m^{-2}, and $h = 3$ km gives a stress of 6.9×10^7 N m$^{-2} = 0.69$ kb. A zero nonhydrostatic stress on BD would imply a mean shear traction on CD, smaller in the ratio of H/L, where L is the horizontal length CD, and directed toward the ridge. A shear traction from below directed away from the ridge would act to increase the comprehensive stress across BD, with amplification by a factor of L/H. The stress deduced from free sliding on CD appears to be of a reasonable order of magnitude, being about 10 times greater than the typical average stress release in a large earthquake.

Conditions at AC, where the plates separate to admit melt fluid from below, which progressively solidifies to renew the plates, deserve a little more thought. To conceive of AC as a slot filled with fluid in hydrostatic equilibrium does not make a satisfactory model. Its density ρ_e is less than ρ, that of the plate material it forms on solidification, so that the pressure gradient in it is less than the gradient of the overburden pressure on either side. If its pressure just suffices to reach the top of the ridge, the edges of the plate will be squeezed tightly together lower down, and separation will not occur; if the fluid pressure at the bottom of the plate, or, for that matter, the mean pressure of fluid over the depth of the slot, suffices to open the gap, the fluid will emerge with a large excess pressure (the consequence of which will be to build the ridge to greater height). A more acceptable model is that the fluid flow is resisted by viscous drag in a narrow gap (or, more reasonably, in many narrow channels in a porous and plastic medium, perhaps of substantial width), so that, despite the lower density, the pressure gradient in the fluid remains substantially equal to the overburden pressure gradient on either side. Then the horizontal forces on the edge of the plate correspond to the 'hydrostatic' boundary condition proposed. The viscous drag of fluid flow makes an upward force (per unit length of ridge) on the edge of each plate equal to $\frac{1}{2}gV(\rho - \rho_e)$, if g is taken as constant and V is the volume of fluid in the gap (per unit length of ridge). This force is of the right sign to maintain the tilted attitude of the plate, but it would be quite inadequate in magnitude if there were a 'true' hydrostatic boundary condition on CD, i.e., one with the material below CD behaving as an inviscid fluid. In that case, the fluid density being ρ, the water-compensated pressure p on the base, minus its mean, is $g(\rho - \rho_w)x \sin \alpha$, where x is distance from the midpoint of CD and α is the angle of tilt of the plate. If the moment of these forces about the point $x = 0$ were balanced by that of the upward force at AC, we should have

$$\tfrac{1}{4}gV(\rho - \rho_e)L \cos \alpha$$
$$= (1/12)g(\rho - \rho_w)L^3 \sin \alpha$$

and thus

$$V = \tfrac{1}{3}[(\rho - \rho_w)/(\rho - \rho_e)]Lh$$

about 42,000 km³ of fluid per kilometer of ridge, if $(\rho - \rho_w) \simeq 2.1$ Mg m⁻³, $(\rho - \rho_e) \simeq 0.1$ Mg m⁻³, $L = 2000$ km, and $h = 3$ km. This volume of fluid is unacceptably large. An escape from the difficulty is provided as before by supposing that the water-compensated pressure gradient parallel to CD in material just below is practically annulled by viscous drag arising from its downhill creep on more rigid material beneath. This assumption implies that the plastic zone below CD does not reach to great depths comparable to the 1000-km order of magnitude of the length of a plate. On the other hand, of course, if there are plastic motions at such depths, driven by density gradients there, there may be other mechanisms by which these motions transmit the requisite tilting torque to the plate.

Figure 2 shows the effect of modifying the

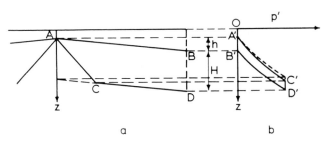

a   b

Fig. 2. (a) Idealized sliding model from Figure 1 with the modified accretion surface AC.
(b) Pressure gradient in the modified model.

form of the bottom surface of the plate. We
have the same sliding surface CD, but we have
a sloping accretion surface AC. The effect in
Figure 2b is to lower the point C', reducing the
area A'C'D'B'. This change is not large. In the
extreme of bringing C into coincidence with D
and C' into coincidence with D', the horizontal
force F_H and the stresses deduced from it would
be halved. The most reasonable assumption is
probably a curved lower boundary to the plate,
to which the sloping lines AC and CD make a
first approximation, with perhaps a resulting
reduction of 10% from the force and stress
values calculated according to Figure 1. Of
course, it is an idealization to treat this boundary as a sharply defined surface at all: it should
rather be a zone of considerable thickness in
which both plastic yield and plate accretion by
melt solidification occur.

We have so far taken the density distribution
in the plate to be the same at AC as at BD.
However, since at AC it is freshly made from
melt, it should be hotter and may still contain
a finite proportion of melt fluid (which is not
necessarily incompatible with rigid behavior),
and its density should therefore be less. The
corresponding modification of diagrams is shown
in Figure 3 (in which, for simplicity, we revert
to a vertical boundary AC, as in Figure 1,
rather than the somewhat more realistic sloping
line AC of Figure 2). No longer is CD parallel
to AB, the length AC contracting thermally to
BD during the progress of the plate. We assume
that the column height changes inversely as the
density, on the grounds that the upper surface
cools at an early stage, becoming most rigid and
preventing contraction in area. The integration
diagram, Figure 3b, is very nearly similar to
Figure 2b, except that the line C'D' is no longer
vertical. The water-compensated overburden

pressure at C is less than that at D by an
amount $\langle g\rho_w \rangle H \langle \rho - \rho' \rangle / \langle \rho \rangle$, where ρ' is the mean
density of a column of plate at AC and ρ is the
mean density of such a column at BD. Thus a
relatively trivial reduction is made in F_H, the
more important reduction coming from the decrease of slope dp'/dz in A'C' in proportion
to the decrease in density from $\langle \rho \rangle$ to $\langle \rho' \rangle$. As
a result we have, very nearly,

$$F_H = \langle g(\rho - \rho_w) \rangle H h$$
$$\cdot \left[1 - \frac{\langle \rho - \rho' \rangle}{2\langle \rho \rangle} \left(\frac{H}{h} + \frac{\langle \rho_w \rangle}{\langle \rho - \rho_w \rangle} \right) \right] \quad (6)$$

If $\langle \rho - \rho' \rangle / \langle \rho \rangle = 1\%$ and $H/h = 20$, the first
correction term makes a reduction of 10%, and
the second is negligible (less than ¼%); the
height of C above D is 20% less than h.

We now turn to the subject of isostasy; first,
we consider it in relation to the more familiar
example of an isostatic continent, since consideration of the horizontal forces is often omitted
from presentations of this subject. Figure 4a
shows a continent, of uniform density ρ_1, supported isostatically by a material of greater
density ρ_2, covered by a layer of thickness w
(5 km, say) of water of density ρ_w. Pressures
are assumed everywhere equal to overburden
pressures, constant across boundaries and constant on horizontal planes in the water and the
medium of density ρ_2. It follows, for the parts
of the continent that are not water immersed,
that

$$(h + d)\rho_1 = d\rho_2 \quad (7)$$

and therefore

$$(h + d)/h = \rho_2/(\rho_2 - \rho_1) \quad (8)$$

where h is the height and d the depth of the
upper and lower continental surfaces measured

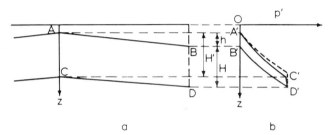

Fig. 3. (a) Idealized model from Figure 1 incorporating effects of nonuniform density and
thickness. (b) Pressure gradient in the modified model.

from an effective flotation level E, which lies $w\rho_w/\rho_2$ above the ocean floor, i.e., $w[1 - (\rho_w/\rho_2)]$, say 3.3 km, below sea level. For the water-immersed margins of the continent we have instead

$$(h' + d')(\rho_1 - \rho_w) = d'(\rho_2 - \rho_w) \quad (9)$$

h' and d' being measured from the ocean floor. The difference of (9) from (8) makes a trivial correction that is of no further importance to us. These conditions being satisfied, vertical forces are balanced on every vertical column of the continental mass. Furthermore, since the integrated mass of matter per unit area below every external point is the same, then, if it is given that the vertical distribution of densities is sufficiently slow in horizontal variation in comparison with the total depth in which there is density variation (it should be considered, of course, that the vertical scale of Figure 4a, as in the previous figures, is greatly exaggerated), it follows that g has the same value everywhere on an external surface at constant height above sea level, after correction for the rotation and the asphericity of the earth.

That is, there is no gravitational anomaly under 'free-air' reduction or the several 'isostatic' reductions to uniform level, which attempt in their various ways to measure departure from this condition and succeed, in agreement with each other, when the horizontal variation of the vertical density distribution is low enough. The Bouguer reduction gives negative anomalies in mountainous areas and positive anomalies in the ocean under the same isostatic conditions.

As was done before, we calculate the horizontal force F_H on the part of the continent lying to the right of the vertical boundary AC, which passes through its highest point, under the initial presumption of hydrostatic stress at all its boundaries, thus deducing the resultant force on boundaries AC and CD, where departures from this condition can occur. Since the system is not entirely water immersed, it is more convenient to use (2) here instead of (3). The integration diagram, Figure 4b, is simple, being essentially the triangle between straight lines A'C' and E'C' with slopes dp'/dz proportional to ρ_1 and ρ_2, respectively, with a small correction from the triangle B'D'E', in which the slope of B'D' is proportional to ρ_w.

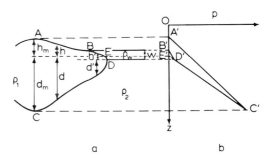

Fig. 4. (a) Model of the continent of uniform density in isostatic equilibrium. (b) Pressure gradient in the continent.

The result is

$$F_H = \tfrac{1}{2}g\rho_1 h_m(h_m + d_m)$$
$$- \tfrac{1}{2}g\rho_w w^2[1 - (\rho_w/\rho_2)] \quad (10)$$
$$= \tfrac{1}{2}g\rho_1 h_m{}^2\rho_2/(\rho_2 - \rho_1)$$
$$- \tfrac{1}{2}g\rho_w w^2[1 - (\rho_w/\rho_2)] \quad (11)$$

where h_m and d_m are the greatest height and the greatest depth of the continent, respectively, above the effective flotation level, which is about 3.3 km below sea level.

With ρ_1 and ρ_2 as 2.8 and 3.1 Mg m^{-3}, respectively, the second term is about 1% and is now neglected. Thus $(h_m + d_m)/h_m = \rho_2/(\rho_2 - \rho_1)$ and is approximately 10. For mountain ranges rising 2, 3.5, and 5 km above sea level the corresponding maximum continental thicknesses are 55, 70, and 85 km, respectively. The horizontal spreading forces F_H are 3.9, 6.5, and 9.6 TN m^{-1}, sufficient to make a stress of 10^8 N m^{-2} = 1 kb on slabs of thickness 39, 65, and 96 km, respectively. If local peaks are disregarded, 2 km represents the height above sea level of the mountain ranges in most continents, 3.5–4 km may represent the Andes, and 5 km the Himalaya and Tibetan Plateau (continuing to the Pamirs at 4 km).

This force can be sustained by tensile stress across AC (of mean magnitude $\tfrac{1}{2}g\rho_1 h_m$, equal to 0.74, 0.95, and 1.16 kb in the three cases considered), by shear stress on the quasi-horizontal part of CD representing the lower surface of the continental mass, or by compressive stress at the edge of the continent. A fourth possibility, that the continent is not isostatic, is not acceptable except as a moderate correction. To annul the spreading force, the continental

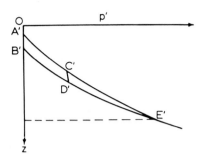

Fig. 5. Pressure gradient in the crustal plate required to produce zero gravity anomaly at depth.

roots would have to reach down to an additional depth $[d(d + h)]^{1/2}$ below the depth required for isostasy, another 52 km under a mountainous region rising to 2 km above sea level, and would produce a negative free-air gravitational anomaly of $2\pi G\rho_1[d(d + h)]^{1/2}$ (where G is the gravitational constant) amounting in the same case to 6.5×10^{-3} m sec^{-2} = 650 mgal (and a still larger negative Bouguer anomaly), which is far larger than anything observed. I prefer to reject the first of these possibilities, that the continent holds together by its own tensile strength, on the assumption that the crust of the earth is 'wet' at all depths below the water table, pore fluids being mobile, if there is enough time, at pressures generally close to the overburden pressure. It thus follows that, on the large scale and in the long term, little or no tensile deviation from lithostatic stress can persist. It is interesting to observe that the horizontal spreading pressure of the continents, except that arising from the Tibetan plateau, is just balanced by that of the oceanic plates if these plates are presumed to have thicknesses in the range 50–100 km, which are reasonable depths for cooling from the surface in a time of the order of 10^8 years to have produced rigidity. The plate thickness of 140 km required to balance the Tibetan plateau seems excessive, but here our simple two-dimensional analysis is probably at fault. We have no obligation to assume 'undertows' of mantle movement dragging the lower surfaces of oceanic plates and continents. Such drags, if present, should generally be oppositely directed for the oceanic plate and the adjacent continent. In the terms of the resultant force produced they can be smaller than or similar in magnitude to the forces we have calculated, but they can hardly be permitted to be larger by an order of magnitude without generating intolerably large stresses.

We come now to isostasy in connection with the oceanic ridges. For zero free-air anomaly with low horizontal variation of the vertical distribution of density, we require the integrated density below any point at the surface of the sea to be the same. When g is taken as a constant (a good approximation), this situation implies the equality of the integral (1) that we used to calculate pressure, which must now be evaluated down to some 'compensation depth' below which truly horizontal stratification of density is presumed to exist. Thus the requirement for zero free-air anomaly is that the pressure-versus-depth lines $A'C'$ and $B'D'$ in, e.g., Figures $1b$, $2b$, and $3b$, continued to greater depth ultimately merge, as is shown in Figure 5. In Figure $3b$, where we presumed a mean density 1% smaller at AC than at BD, the two curves had already approached 20% closer at a depth of about 80 km. Continuation of the same density difference to a greater depth would cause them to come together (and cross) at a depth of 400 km, but continuation of the same fractional density difference, with increasing density, would cause them to come together at a lesser depth. Allowance of 1% may also be ungenerous for the density difference. With a 5% density difference, doubtless an overgenerous one, we produce the situation shown in Figure 6, with a horizontal plate bottom that can itself be taken as a level of compensation. The essential point, however, is as revealed in Figure 5, that the ultimate merging of lines $A'C'$ and $B'D'$ to produce zero gravity anomaly (or their crossing to reverse its sign) in no way conflicts with having a finite area $A'C'D'B'$ that measures the downhill sliding force. In simplest terms, if we see a broad hill with no free-air gravitational anomaly we are entitled to infer that there is relatively light matter beneath it: but this fact should make no change in the direction in which a landslide will go (namely, downhill) nor any appreciable change in its force. And sea level is an adequate reference surface with which to define downhill; if g increases by 0.05 m sec^{-2} (=500 mgal) from A to B as it does from the equator to the poles ('anomalies' are smaller by an order of magni-

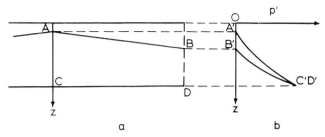

Fig. 6. (a) Idealized model of the compensated crustal plate with a horizontal base.
(b) Pressure gradient.

tude), two level surfaces separated by 5 km at A are 25 meters closer together at B.

Figure 6 represents an improbable extreme. A more likely situation is one somewhat similar to Figure 5. In this case the material below the plate has its own horizontal spreading force F_H, measured by the area C'D'E', which can be of similar order of magnitude as A'C'D'B', which measures the force on the plate, either greater or lesser. There are various ways in which this force can be sustained. The greater part of the volume on which it acts may be too rigid to move at speeds comparable to those of the plate sliding. If some of it spreads under the influence of this force, the larger frictional drag resisting it may be on its lower boundary. In likelihood, only a small part of this force, if any, becomes transmitted to the plate. The boundary CD (as in Figures 1, 2, and 3) represents a zone of relatively high plasticity; motions below CD are to a large extent decoupled from motions above.

An objection can be made to the flat uniform slope of ocean floor represented in Figures 1a, 2a, and 3a, but this objection is readily removed. The required form of the line AB for maintenance of a steady state is one in which each point follows the line in its motion. The general rigid surface permitting such motion is a surface of revolution, and an acceptable form

for the flat earth model, providing a ridge rising more steeply to the enter, is a cylinder with an elevated axis about which it rotates, as shown in Figure 7, which is a revised version of Figure 1a. No change at all is required in Figure 1b.

The corresponding modified form for the plate surface on a spherical earth is adequately represented by an ellipsoid having revolutional symmetry about its own axis of rotation, which is offset from the center of the earth (Figure 8). To level out to the horizontal at a depth δ below the ridge at an angular distance θ from it, the axis must be offset from the center of the earth by

$$p = \delta/(1 - \cos \theta) \qquad (12)$$

and the radius R' must exceed that of the earth taken to the ridge R by

$$R' - R = p - \delta \qquad (13)$$

these distances being 30 and 26 km, respectively, for $\delta = 4$ km and $\theta = 30°$.

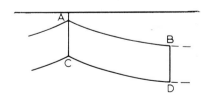

Fig. 7. Model of a plate similar to Figures 1a, 2a, and 3a, modified with curved upper and lower boundaries.

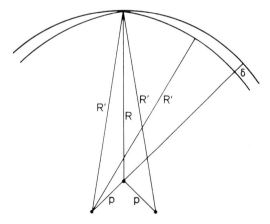

Fig. 8. Geometric model of the plate surface on a spherical earth.

In the conventional theory of the motion of spherical plates on a spherical earth, the velocity at a point $\mathbf{r}$ of the ith plate, $\mathbf{r}$ being the position vector from the center of the earth, is

$$\mathbf{v}_i = \boldsymbol{\omega}_i \times \mathbf{r} \qquad (14)$$

where $\boldsymbol{\omega}_i$ is the rotation vector for the ith plate, and the relative motion at a point where two plates i and j are adjacent is

$$\mathbf{v}_{ij} = \mathbf{v}_j - \mathbf{v}_i = \boldsymbol{\omega}_{ij} \times \mathbf{r} \qquad (15)$$

where $\boldsymbol{\omega}_{ij} = \boldsymbol{\omega}_j - \boldsymbol{\omega}_i$. (It is, in fact, only the relative velocities $\mathbf{v}_{ij}$ and relative rotation vectors $\boldsymbol{\omega}_{ij}$ that can be determined.) The modification made when the rotation vector $\boldsymbol{\omega}_i$ is offset from the center of the earth by $\mathbf{p}_i$ is to make

$$\mathbf{v}_i = \boldsymbol{\omega}_i \times (\mathbf{r} - p_i) = \boldsymbol{\omega}_i \times \mathbf{r} - \boldsymbol{\omega}_i \times \mathbf{p}_i \qquad (16)$$

$$\mathbf{v}_{ij} = \boldsymbol{\omega}_{ij} \times \mathbf{r} + \boldsymbol{\omega}_i \times \mathbf{p}_i - \boldsymbol{\omega}_j \times \mathbf{p}_j \qquad (17)$$

thus adding to the result (15) a small correction term independent of $\mathbf{r}$.

Having thus satisfied the kinematic require-ments for motion on a spherical earth, we can take over the flat-earth calculations of forces unchanged. Lateral dimensions did not enter into these calculations except by the require-ment of minimal variation in these dimensions. All integrations were taken with respect to z, which can simply be interpreted as the direction in which gravitational potential ϕ decreases (at rate g), replacing the elementary interval of integration $g\,dz$ by $d\phi$. Horizontal forces are then reinterpreted to be torques about the rotation axis.

The glaciological analogy for plate tectonics has thus been successfully defended. Moreover, glaciology can be regarded as that branch of tectonics in which, because the motions are most rapid, the basic principles are most vividly illustrated. The essential feature of both plate tectonics and glaciology is thermal mobilization of matter via a fluid phase to produce a state of high gravitational potential energy and the action of the consequent mechanical processes to reduce this energy, thus maintaining an approxi-mately steady state of motion.

Some Shock Effects in Granodiorite to 270 Kilobars at the Piledriver Site

I. Y. BORG

Lawrence Livermore Laboratory, University of California
Livermore, California   94550

Postshot exploration around the 61 ± 10 kT Piledriver underground nuclear explosion in granodiorite indicates that the cavity radius r_c is 40.1 meters. The shape of the vertical chimney, which extends 277 meters above the shot level, was influenced by pre-existing joints and fractures and is asymmetric. The limit of detectable shock-induced microfracturing is $2.7 \pm 0.2 \, r_c$, at which point rocks have been subjected to peak radial pressures of 6–8 kb. Extensive fracturing occurs at distances to the shot point of $<1.3 \pm 0.2 \, r_c$, corresponding to pressures exceeding the granodiorite Hugoniot elastic limit of 45 kb. The onset of slip and twinning in mineral constituents is correlated with measured shock pressures at estimated strain rates of $\leq 10^4$–10^5 sec^{-1}, ambient temperatures of 30°C, and calculated Hugoniot temperatures for granodiorite of <300°C. For quartz, planar lamellas are detectable in some grains subjected to pressures of 75–78 kb and in all grains subjected to a pressure of 205 kb. Mechanical ($\bar{1}01$) twinning in hornblende and sphene ⟨110⟩ is evident in rock that has experienced pressures of 24–40 and 14–18 kb, respectively. Some kinking in biotite is associated with shock pressures as low as 15–16 kb; above 75 kb all biotite contains kink bands. At ≤ 270 kb no shock-induced twinning or planar lamellar structure was noted in either the orthoclase or the albite-oligoclase component of the granodiorite, although there was a noticeable loss of birefringence in both. Glass occurs within the chimney rubble and in distant fractures within the surrounding granodiorite where it was injected by expanding gases. No diaplectic glass was noted in rock forming the cavity walls (270 kb). Dissociation of the hydrous phases, biotite and hornblende, in wall rock surrounding the cavity is attributed to the permeation of hot gases along fractures.

The Piledriver event, a nuclear explosion in granodiorite at the Nevada Test Site, provided an opportunity to study shock effects, since it occurred adjacent to a highly instrumented tunnel drift system and was accompanied by an extensive re-entry and postshot drilling program. Close-in stress gages, instruments measuring particle velocities and accelerations, and in situ cameras recording displacements operated satisfactorily at the time of the shock propagation and provided an unusually complete set of data, which was used to check preshot predictions. Under these circumstances, the specific behavior of the rock surrounding the explosion (vaporization, melting, plasticity, and fracture) can be associated with particular pressures with a minimum of uncertainty.

Postshot exploration resulted in the recovery of three 3-inch cores, one of which penetrated the lower portions of the cavity produced by the explosion. This paper summarizes the mode of failure of the granodiorite and the mineral constituents within these cores as functions of radial distance and peak radial pressure.

Rubble and glass recovered within the cavity chimney are largely ignored in this account. The position of such material at the time of the explosion, and hence the magnitude of the pressure pulse that they saw, cannot be known precisely. The maximum radial stress experienced by any recovered core outside the cavity is 270 kb.

PILEDRIVER EXPERIMENT

The Piledriver site, which is located within the Climax stock of area 15 of the Nevada Test Site, is about 0.4 km from the 5-kT Hardhat experiment, which was conducted in 1962. In both instances a well-instrumented underground tunnel drift complex existed at shot time (Figures 1 and 2). The area of the stock exposed at the surface is 3.4 km²; however, the

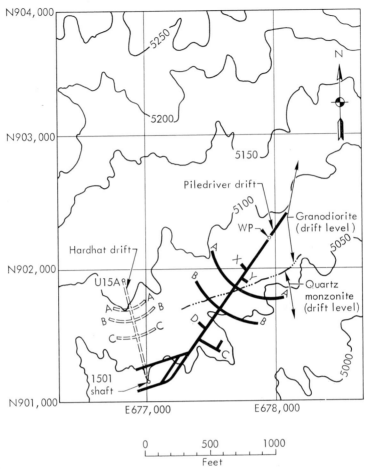

Fig. 1. Piledriver and Hardhat sites in Area 15, NTS. Contours show the surface elevation in feet above sea level. WP marks the shot point. (C. J. DiStephano, D. R. Williams,
and H. A. Jack, personal communication, 1969.)

rock mass broadens to a diameter of 9.7 km
at a depth of 396 meters [*Allingham and Zietz*,
1962]. The stock consists of a granodiorite and
a younger porphyritic quartz monzonite. Although most of the instrumentation surrounding
both Hardhat and Piledriver was placed in the
quartz monzonite part of the intrusive, the
shot points were located well within the granodiorite (Figure 1). Similarly, the three postshot
Piledriver holes were cored within the granodiorite. The device was emplaced below the
water table encountered at 1372 meters mean
sea level (msl); however, perched water was
encountered at variable depths (24–112 meters)
during preshot drilling and tunneling operations.
These findings suggest that it is localized by
major fracture and shear zones [*Houser and*

Poole, 1961]. Additional data concerning the
event are summarized in the appendix with
previously published results and postshot observations.

*In situ stress measurements and displace-
ments.* In addition to instrumentation in the
drifts and associated slant holes, five close-in
radial boreholes were fitted with stress gages
(Figure 2, holes A–F) by contracting agencies.
Data from reliable gages are plotted in Figure 3 as a function of distance from the shot
point. Also included are peak radial stresses calculated from peak particle velocities. The curve
is an arbitrary straight-line fit of the data,
which theoretically can deviate from linearity
[*Cherry and Rapp*, 1968].

The expansion and heave of the cavity follow-

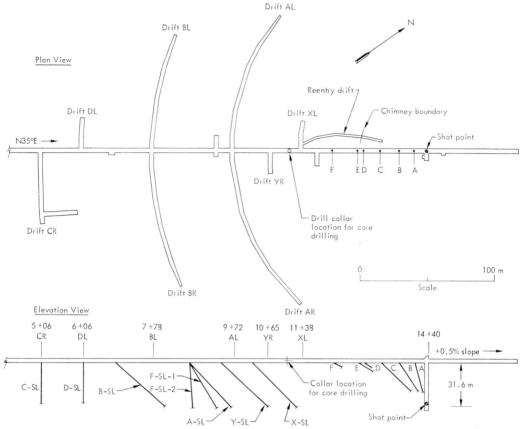

Fig. 2. Piledriver tunnel and drift complex showing locations of instrumented boreholes at shot time. Stations indicated on elevation (e.g., 5 + 06) are referenced to the 15.01 shaft (0 + 00), which is off the diagram. (Modified from drawing by K. Willits of Fenix and Scisson, Inc., Project Engineer, 1968, Piledriver project station 15.01, drawing MER-192 14d-1501, Nevada Operations Office, U.S. Atomic Energy Commission.)

ing the passage of the main shock wave result in displacements of the surrounding rock that diminish in magnitude with distance from the shot point. In order to correlate data such as peak pressures measured within 100 msec of the time of detonation (Figure 3) with observed shock effects in postshot cores and drifts, real displacements must be estimated. Comparing preshot and postshot positions of markers in access tunnels and drifts is somewhat unsatisfactory, because movements along pre-existing joints and faults were facilitated by the tunnel and the drifts. Another measure of displacements is the residual radial displacement calculated by the integration of particle velocity or the double integration of acceleration curves. Such calculated displacements are strongly affected by base line errors in the initial data and

characteristically show great scatter. Data of these types are plotted as a function of preshot distance in Figure 4 with displacements calculated from the SOC code [*Cherry and Peterson*, 1970]. Agreement between measured and calculated values is reasonably good when they can be compared, and thus it is indicated that the SOC calculation can be used fairly accurately over the whole range. In the linear portions of the curve, underground displacements are proportional to the inverse of the square of the radial distance. A similar relationship was established from permanent particle displacements at depth in two earlier granite events at the Nevada Test Site (Hardhat and Shoal), despite greater scatter in the data (J. L. Merritt, personal communication, 1969).

Strain rate of shock loading in field experi-

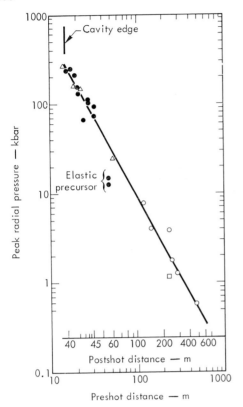

Fig. 3. Measured peak radial stress versus preshot and postshot distance from the Piledriver shot point. Triangles, data from manganin gages (C. T. Vincent and J. Rempel, personal communication, 1968); solid circles, data from electrolytic gages (P. Liebermann, personal communication, 1967); open circles, data calculated from particle velocity and from acceleration data [*Perrett*, 1968]; squares, datum from the quartz piezoelectric gage [*Perrett*, 1968].

ments. In pressure regimes of >100 kb the strain rate associated with shock loading in the field (nuclear) experiment is assumed to be slightly less than that for the flying-plate type of laboratory shock experiment (10^5–10^6 sec^{-1}). From free-field measurements of strain ϵ or particle velocities u_p and pressures σ_1 and from records of rise times it is possible to calculate strain rates $\dot{\epsilon}$ below about 14-kb pressure, a limit set by satisfactory operation of the gages. From stress and strain gages, strain rates of 10^{-1} sec^{-1} can be calculated for peak pressures of 1–1.5 kb. With the relation

$$\epsilon = \rho_0 u_p^2 / \sigma_1$$

derived from shock equations, strain rates of 7 to 7×10^2 sec^{-1} can be calculated in the interval 7–14 kb. The measured rise time to maximum velocity or strain normally includes the time for cables and instruments to transmit and record the signal. The lag time, estimated for velocity gages to be of the order of 3×10^{-3} sec (C. J. Sisemore, personal communication, 1971), has been subtracted from the measured rise times for the calculations. Nonetheless, uncertainties in the exact lag time correction limit the accuracy of the calculated strain rates. They are probably accurate to within a factor of 10. The important observation is that the strain rate falls off with distance from the shock source to low values, in relation to the values associated with shock loading in the >100-kb regime.

Vaporization and melting in the cavity. Also plotted in Figure 4 is a calculated point (inverted triangle) based on the difference between the measured cavity radius and the calculated radius of combined vaporized and melted rock. Although cavity development is a continuous process, it is convenient to describe it as being discontinuous, involving shock vaporization, shock melting, and melting resulting from solid-vapor interactions. The solid-vapor interactions involve thermal conduction and exothermic condensation of ionized and vaporized elements on solid-rock interfaces exposed by slumping of the shock-melted zones to lower portions of the cavity. *Butkovich* [1967, 1968] has estimated the amount of granitic rock in grams per kiloton involved in the three types of transformations:

Shock vaporized	70×10^6
Shock melted	350×10^6
Melted by solid-vapor interactions	250 to 300×10^6

From these estimates the volume of rock affected can be calculated, and the radii of concentric zones can be determined. The difference between the measured cavity radii (40.1 meters) and the calculated outermost zone of melted rock (15.6 $\pm$ 1.0 meters) represents the distance that the rock at the present cavity wall has been displaced by gas expansion (24.5 meters).

Temperature associated with the shock wave. The presence of glass along fractures and the dissociation of the hydrous mineral phases (e.g., biotite and hornblende) indicate that the gran-

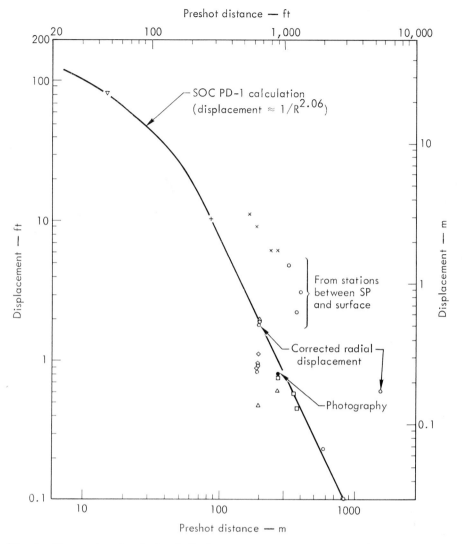

Fig. 4. Measured and calculated displacements as functions of preshot distance from the shot point. Open circles, residual radial displacements [*Perrett*, 1968; *Hoffman and Sauer*, 1969]; crosses, postshot survey in the access tunnel; triangles, average permanent displacement of a test section (C. J. DiStephano and L. J. Asbaugh, personal communication, 1969); squares, horizontal movement of bench marks (C. J. DiStephano, D. R. Williams, and H. A. Jack, personal communication, 1969); diamonds, average packing displacement (H. H. Holmes, H. M. Hanson, and A. O. Bracket, personal communication, 1969); pluses, displacement in the tunnel (D. Rabb, personal communication, 1970); inverted triangles, difference between final cavity radius and the radius of melted and vaporized rock; solid circle, photographic techniques [*Smith*, 1969].

odiorite has been heated to temperatures of >700°C. In order to determine whether the heat was derived from the passage of the shock wave, temperatures were calculated along the Hugoniot at a series of volumes with the method of *Walsh and Christian* [1955]. The equation

describing the conservation of energy in a shock wave,

$$dE = \frac{v_0 - v}{2} \, dp - \frac{p + p_0}{2} \, dv$$

where v_0, p_0 and v, p are initial states and final states, respectively, when combined with the first law of thermodynamics, yields the expression

$$TdS = \frac{p - p_0}{2}\, dv + \frac{v_0 - v}{2}\, dp \qquad (1)$$

Using the identity $dS = C_v(dT/T) + (\partial p/\partial T)_v dv$, in which C_v is the specific heat at constant volume, and equating $(\partial p/\partial T)_v$ by the chain rule to α/β, where α is the thermal expansion and β the isothermal compressibility, we obtain the expression

$$TdS = C_v\, dT + (\alpha/\beta)T\, dv \qquad (2)$$

Entropies of the phase transformations of quartz and feldspar that produce anomalous compressions above 140–150 kb have been neglected. That portion of the total energy of the system is difficult to estimate, because the transformations are incomplete below 300–400 kb and mixed phases exist up to those pressures [Ahrens and Rosenberg, 1968; Ahrens, et al., 1969a]. Omission of transformation energies leads to calculated Hugoniot temperatures that are higher than they would be if the energies were taken into account. Inasmuch as the maximum Hugoniot temperature in the interval 0–270 kb is of particular interest here, the simplification is defensible.

When (1) and (2) are combined with the definition of the Grüneisen constant,

$$\gamma_0 = V(\partial p/\partial E)_v = \alpha V/C_v \beta$$

where V is the specific volume at standard temperature and pressure, the temperature difference between the ambient and the shocked condition of the material is

$$dT = \frac{p - p_0}{2C_v}\, dv + \frac{v_0 - v}{2C_v}\, dp - \frac{\gamma_0 T}{v_0}\, dv \qquad (3)$$

Hence,

phase, and, from a calculated $\gamma = 0.645$ and $\rho = 2.68$ g/cm^3, $\gamma/V = 2.46$ g/cm^3 for the low-pressure phase under ambient conditions. The value of C_v was approximated by the equations given by Birch et al. [1942] for C_p as a function of temperature. The equation-of-state data were compiled from shock experiments on the Climax granodiorite [Van Thiel, 1966]. Data for $P - V$ below 40 kb were taken from isothermal compressibilities [Stephens et al., 1970].

The results of the calculations are plotted in Figure 5 with comparable calculations for single-crystal quartz [Wackerle, 1962], oligoclase [Ahrens et al., 1969a], and Coconino sandstone, 24% porosity [Ahrens and Gregson, 1964]. As might be expected, below 400 kb the granodiorite data closely parallel the quartz and oligoclase data. Above 400 kb, conversion to high-pressure phases of quartz and oligoclase is almost complete, and all the temperature calculations are subject to uncertainties arising from assumed values of ρ_0 (high), γ, C_v, and the entropies of the transformations, if they were taken into account. Thus the differences apparent between the single-crystal and granodiorite curves do not merit close examination. It is sufficient to note that at 270 kb the shock temperature is $<300°$C and not high enough to account for the high-temperature effects noted in granodiorite adjacent to the cavity wall. However, the shape of the release adiabat curves for both quartz and plagioclase [Ahrens and Rosenberg, 1968] indicates that a certain amount of stored energy is released as heat during unloading. Nevertheless, as was demonstrated for oligoclase [Ahrens et al., 1969a], it is unlikely that the temperature of the granodiorite during release from 270 kb rises as high as 700°–900°C; thus the decomposition temperature of biotite and the observed thermal breakdown of the hydrous phases are more reason-

$$T_i = \frac{T_{i-1}[1 - (\gamma_0 v/2V)] + \{[(\langle p \rangle - p_0)\, \Delta v + (v_0 - \langle v \rangle)\, \Delta p]/2C_{v_{i-1}}\}}{1 + (\gamma\, \Delta V/2V)}$$

where $\langle p \rangle$ and $\langle v \rangle$ are mean values of points i and $i - 1$ on the Hugoniot. In the calculations, (γ/V) for the low and high (>370 kb) phases was taken to be a constant. From $\gamma = 0.62$ and $\rho = 3.96$ g/cm^3 for granite [Ahrens et al., 1969b], $\gamma/V = 1.73$ g/cm^3 for the high-pressure

ably attributed to the permeation of hot gases into shock-induced fractures.

Measurements and predictions of cavity size. Before the postshot exploration of the chimney cavity, various estimates of the expected cavity size were made. In Table 1, nine calculations are

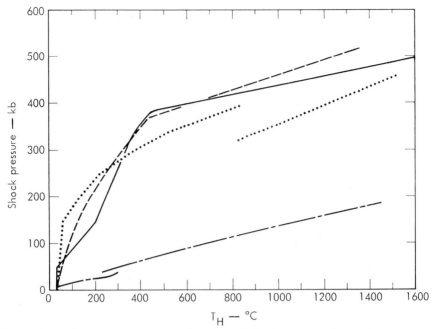

Fig. 5. Hugoniot temperature as a function of shock pressure in granite, quartz, and sandstone. The oligoclase data above 400 kb are based on $\gamma = 1.00$, $\Delta E_{\iota-h} = 0.97$, and $\rho = 3.69$ g/cm³ for the high-pressure phase. Dashed line, granodiorite; solid line, quartz [*Wackerle*, 1962]; dotted line, oligoclase [*Ahrens et al.*, 1969a]; long dashed line, Coconino sandstone [*Ahrens and Gregson*, 1964].

compared. All the calculations are within 18% of the measured radius, and most are within 5%. Thus, if appropriate physical parameters describing the granodiorite are given, the cavity radius can be predicted accurately.

POSTSHOT CORES

Location of cores and chimney definition. The locations of four postshot drill holes are indicated in Figure 6. The drilling demonstrated that the southwest wall of the chimney is canted about 13° from the vertical [*Sterrett*, 1969] and controlled by one or two of the four well-developed joint fracture systems in the stock. The attitude of other parts of the chimney wall is not known. However, the one explored section through the adjacent Hardhat chimney demonstrated that its north-northwest and south-southeast walls are nearly vertical. Since the major joint fracture systems in the stock are pervasive, the north-northwest and south-southeast walls of the Piledriver chimney may also be vertical. By similar reasoning, the unexplored southwest wall of the Hardhat chimney may depart from the vertical.

Description of cores. Figure 7 contains photographs of recovered 3-inch-diameter core from U 15.01 PS-3. Caliper and γ logs in the hole indicate that the granodiorite cavity debris contact occurs at a slant hole distance (SHD) of 71.0 meters (233 feet) [*Sterrett*, 1969]. Recovered core from that depth (Figure 7c) contains a very sharp interface between the highly deformed granodiorite and the black vesicular melted counterpart. Figure 7a shows a core recovered at 70.7 meters (232 feet) SHD, which is coherent although highly fractured, plastically deformed, and partially decomposed (hydrous phases). Figure 7b was recovered from a slant hole distance of 68.9 meters (226 feet). Glass seams filling reopened joints are conspicuous in these samples and others in the SHD of 50.0–71.0 meters (164–233 feet) of PS-3. Injected siliceous glasses can occur along gross fractures and pre-existing shear zones as far as 42.7 meters horizontally from the cavity chimney edge; however, glassy pseudomorphs of minerals constituting the granodiorite, or devitrified counterparts, do not occur within core collected up to the cavity wall.

TABLE 1.   Predictions and Measurements of Piledriver Cavity Radius

Predicted Radius		Prediction Method
meters	feet	
32.7	108	SOC calculation assuming constant overburden and no free surface [*Cherry and Rapp*, 1968]
36.4	119	$r_c = 21.0\ W^{0.306}E^{0.514}/\rho^{0.244}\mu^{0.576}h^{0.161}$ [*Closmann*, 1969]
40.1	131.5	$r_c = CW^{1/3}/(\rho h)^{\alpha}$, where $C = 103$ (granodiorite) and $\alpha = 0.32$ (2% H_2O) [*Higgins and Butkovich*, 1967]
40.1	131.5	Measured value, accepted as most probable cavity radius on basis of γ and caliper logs of core hole PS-3, which intersected cavity below level of shot point [*Sterrett*, 1969]
40.1+	131.7	Assuming adiabatic expansion of cavity gas to twice overburden pressure [*Chapin*, 1969, Figure 2, granite]
40.9	134	$r_c = C\ W^{1/3}/(\rho h)^{1/4}$, where $C = 61.6$ (granite) [*Boardman*, 1967]
41.2	135	$r_c = 16.3\ W^{0.29}E^{0.62}/h^{0.11}\rho^{0.24}\mu^{0.67}$ (H. C. Heard and F. J. Ackerman, personal communication, 1967)
41.6	136	$r_c = 52\ \alpha^{1/3}W^{1/3}/(\rho gh + C_s)^{1/3}\gamma$, where $C_s = 30$ bars (granite), $\gamma = 1.05$ (granite with 2.5% H_2O), [*Michaud*, 1968]
44.5	146	Calculated from void volume of the chimney determined from chimney pressurization tests [*Boardman*, 1967]
45.1	148	$r_c = (3R^2\Delta\sigma)^{1/3}$ from average corrected residual particle displacements at two slant holes (204- and 470-meters radial distance) [*Perrett*, 1968]
45.7	>150	SOC calculation assuming variable overburden and rarefaction at a free surface [*Cherry and Rapp*, 1968]

r_c = cavity radius.
W = yield (kT).
E = Young's modulus (Mb).
ρ = density of rock (g/cm^3).
μ = shear modulus (Mb).
h = depth of burst (meters).
α [*Higgins and Butkovich*, 1967] = $1/3\gamma$.

γ = coefficient of adiabatic expansion.
α [*Michaud*, 1968] = 1.0 (coefficient related to size of the shot chamber).
C_s = coefficient related to the 'contraints of the structure.'
R = radial range (meters).
Δs = residual displacement (meters).

Orientation of core fractures. It is customary to assume that at least three types of fracture systems symmetric about the point source develop around an underground explosion: a set inclined at an acute angle to the peak radial stresses (i.e., shear fractures), a radial set, and a set tangential to concentric spheres about the source. To test the assumption, the attitude of several thousand macroscopic fractures was measured in three postshot cores. Because only the bearing of the core axis is known, such fractures cannot be oriented uniquely in space. However, at any particular position along the core length it is possible to test whether the attitude of the fracture with respect to the core axis is consistent with a radial, a tangential, a shear, or a random orientation. Results of this study [*Borg*, 1970a] are that the fractures are demonstrably nonrandomly oriented. Their positions are consistent with failure along pre-existing regional joints, fractures, and shear zones. With the exception of a short interval of core near a pre-existing tunnel wall where rarefaction of the shock front was probable, the orientation of the fractures is not consistent with radial or tangential positions in any core interval. With the core information available the test for shear failure is not sufficiently precise to determine whether the fractures are also consistent with this possibility.

Fracturing radius. Microfracturing in the Piledriver cores was assessed as a function of distance from the shot point. Approximately 75 thin sections were examined; the results are shown in Figure 8. In general, core PS-3 is slightly more fractured than PS-1 at comparable postshot distances. This finding is in keeping with a lower over-all core recovery record (89 versus 61%) [*Sterrett*, 1969], which reflects different drilling techniques.

Samples were assigned to one of five categories on the basis of the over-all amount of fracturing present: intensely fractured (crushed), highly fractured, moderately fractured, slightly fractured, and unfractured samples. Assignment to one of the five categories was not difficult; however, the usefulness of such a qualitative assess-

ment of fracturing is limited to comparisons made by a single observer. On this account, fracturing in the Hardhat cores was also as-

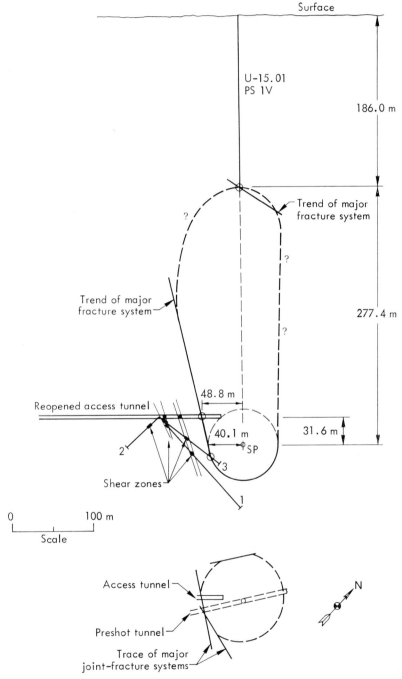

Fig. 6. Vertical cross section (top) and horizontal cross section at the tunnel level (bottom) of the chimney shape resulting from the Piledriver event. Open circles, points established by postshot exploration; solid circles, injected radioactive glass in the tunnel.

a

b

c

Fig. 7. Sections from hole PS-3 showing glass seams filling fractures (top and bottom left) and contact between granodiorite and cavity debris (right).

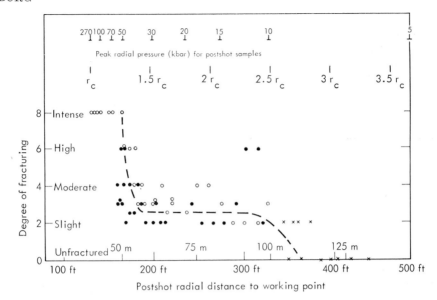

Fig. 8. Degree of microfracturing in the Piledriver postshot cores as a function of postshot radial distance and peak radial pressure. Solid circles, postshot hole 1; crosses, postshot hole 2; open circles, postshot hole 3.

sessed, and the new data are included in Table 2, which summarizes the fracture radii r_f. It was not possible to duplicate the semiquantitative fracturing indices computed by *Short* [1966] for the Hardhat material, but the results of both studies are consistent. When reduced to multiples of the cavity radii (r_f/r_c, Table 2) and associated with peak radial stresses, the limits given in Table 2 for the two events are seen to

be closely related. Comparable limits of crushed and fractured zones (boundary between 600- and 60-millidarcy fracture zones) in French Hoggar granite underground tests are 1.4 and 2.8 r_c, respectively [*Delort and Supiot*, 1970].

The limits of extensive and detectable shock-induced microfracturing occur between peak radial stresses of 42–45 and 6–8 kb, respectively. The first stress, 42–45 kb, corresponds to the

TABLE 2. Microfracturing in Piledriver and Hardhat Postshot Cores

	Piledriver			Hardhat		
		r_f/r_c	Peak radial stress, kb		r_f/r_c	Peak radial stress, kb
Limit of intensely and highly fractured rock	51.8 ± 3.0 meters (170 ± 10 feet)	1.3 ± 0.1	45	24.4 ± 3.0 meters (80 ± 10 feet)	1.3 ± 0.2	42
Limit of detectable microfracturing	109.8 ± 6.1 meters (360 ± 20 feet)	2.7 ± 0.2	8	54.9 ± 7.6 meters (180 ± 25 feet)	2.9 ± 0.4	6
Cavity radius, r_c	40.1 meters (131.5 feet)			19.2 meters (63 feet)		
Chimney height, h_c	277.4 meters (910 feet)			85.7 meters (281 feet)		

measured Hugoniot elastic limit [*Cherry and Peterson*, 1970]. The second, 6–8 kb, corresponds to the normal pressure associated with the expected onset of appreciable fracturing under shock loading [*Cherry*, 1970]. It is assumed that shock loading at low pressures approximates a one-dimensional rather than two-dimensional strain model, such as that used to describe conventional triaxial laboratory deformations of rock.

PLASTICITY

In connection with research at sites of meteor impact, several attempts have been made to correlate plastic phenomena in particular rock-forming minerals to specific critical shock pressures [*Chao*, 1968; *Engelhardt and Stöffler*, 1968; *Stöffler*, 1971]. By necessity, such estimates have been based on shock data from laboratory tests and to a lesser extent on results from static deformations. The perfection and small size of the sample in shock tests and the comparatively slow strain rate in triaxial tests mitigate against an exact analogy with behavior in either meteoric impact or nuclear detonation in a large structurally inhomogeneous body of rock. On this account, plastic phenomena, meaning in this context intracrystalline slip and twinning, observed in recovered core, are described in some detail and correlated with measured peak radial shock pressures. These data are subject to errors of two kinds: those related to the pressure measurements themselves and those related to the accuracy of locating the sample within the drill hole. The pressures are taken from the curve of Figure 3 and believed to be accurate to 10%; the probable error in the locations of the samples is <1 meter. A detailed description of the phenomena in individual mineral species follows.

Quartz. Microscopic structures associated with the plasticity of quartz in high-pressure regimes have been extensively studied both in the laboratory and in the field. The literature is too extensive to review in toto, and the papers of *Carter* [1968] and *Engelhardt and Bertsch* [1969] are recommended to the reader for résumés.

Following *Hörz* [1968], the distinction drawn by *Carter* [1968] between deformation lamellas and planar features has not been attempted here. Apart from being technically impossible for all

lamellas that can be measured on a universal stage, the significance of the optical phenomena associated with deformation lamellas and planar features is still a matter of conjecture. Furthermore, it has yet to be demonstrated either in the shock laboratory or in the field that recognition of one or the other type permits any profound inferences to be drawn concerning the conditions attending the shock metamorphism of a particular rock. In this context three groups of microstructures have been recognized: (1) widely spaced planar discontinuities parallel to rational planes (cleavage), (2) singular irregular irrational surfaces (fractures), and (3) closely spaced planar or nearly planar structures (lamellas).

Insofar as planar features and/or deformation lamellas mark the onset of plasticity in quartz, the following estimates of critical pressures at ambient temperatures have been made: threshold for 'planar fracture sets' in nuclear shocked rock, 50–75 kb [*Short*, 1966]; 'deformation lamellas' in meteoritic impact sites, 30–100 kb [*Chao*, 1968] (the lower limit is described as 'appearance of a trace amount' [*Chao*, 1968, Figure 1 and p. 222]); and 'planar features' in shock-loading experiments, 100–120 kb [*Hörz*, 1968] and 105–140 kb [*Müller and Défourneaux*, 1968]. The lower limit of the pressure seen by the bulk rock indicated by the present data is 75–78 kb. At this point they are rare, short, and barely detectable with conventional microscopy techniques. At 155 kb, 89% of the quartz contains one or more sets of lamellas; at pressures of >205 kb all grains contain lamellas. At 155, 234, and 263 kb there are 1.6, 2.8, and 3.0 sets per grain, respectively. The maximum number of sets observed in any one grain corresponding to the above three pressures is 5, 7, and 7. On the basis of measurements of approximately 100 lamellas in each sample, the amount of lamellas that form angles of 65°–72° to [0001] is 41%, 53%, and 59%, in order of increasing pressure.

Most of the lamellas in this range appear to be {10$\bar{1}$3}, as has been noted by all observers. When a planar {10$\bar{1}$1} rhombohedral parting is also developed, the lamellas can be shown to occur within or close to the ⟨01.0⟩ zones. However, as a rule, fractures are conchoidal, and true cleavages are rare; as a consequence, indexing is tenuous. Any error in measurement of either the c axis or the lamellas greatly affects the zonal assignment. The fractured nature of the quartz, its optically biaxial and strained character, and the lowered birefringence in the more highly shocked samples make meaningful measurements in any one grain very difficult. On this account, positive recognition of other lamellas was not always possible. Form $\{10\bar{1}1\}$ was positively identified as the next most common one (3, 10, and 11% at 155, 234, and 263 kb, respectively). The frequencies are in keeping with $Hörz$'s [1968] histograms describing planar features in laboratory-shocked quartz as well as $Engelhardt\ and\ Bertsch$'s [1969] and $Carter$'s [1968] absolute frequencies at the Ries meteorite crater and the Clearwater Lake structure, respectively. In a few instances, $\{10\bar{1}2\}$ lamellas were positively identified. However, there was no compelling evidence from measurement of angles between lamella sets that the 4–9% of lamellas recorded as occurring at 56°–60° to the c axis were, in fact, parallel to $\{10\bar{1}2\}$. This finding is somewhat surprising since in shock experiments above 200 kb the form becomes increasingly prominent [$Carter$, 1968]; in addition, the form is recorded as conspicuous in extreme examples of naturally shocked quartz, e.g., type D of $Robertson\ et\ al.$

[1968] and $Engelhardt\ and\ Bertsch$ [1969]. Basal and sub-basal (within 10° of $\{0001\}$) lamellas are sparse; they comprise 1, 4, and 6% of the total observed in the 155-, 234-, and 263-kb samples. Assignment of indices to the remainder of the lamellas ($\sim$30% in each of the three samples) was attended by ambiguity. In the 155-kb sample there was a lack of adequate crystallographic control; on the other hand, in the 234- and 263-kb samples, intense fracturing and subsequent rotation of fragments allowed only the approximate position of the lamellas within the structure to be described. In all three cases, 90–93% of the observed lamellas are at angles of $>$45° to the c axis (namely, pole to lamellas [0001] $<$54°). The only noteworthy aspect of any of the unidentified group was a 12% concentration at 45° $\pm$ 2° to [0001] in the 263-kb sample. Most of these sets are cozonal with c $\{0001\}$ and m $\{10\bar{1}0\}$ and, therefore, are of the form $\{10\bar{1}1\}$.

Well-formed rhombohedral and prismatic cleavages are better developed in nuclear shocked rock than in tectonically deformed quartz but are rare as compared with those formed in laboratory shock experiments (Figure 9) [$Hörz$, 1968; $Müller\ and\ Défourneaux$, 1968].

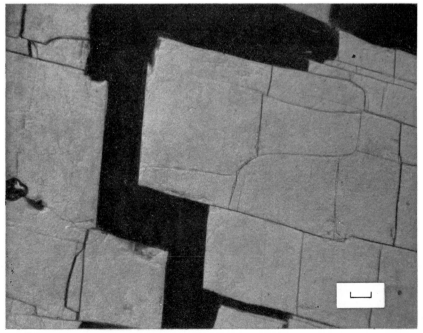

Fig. 9. Photomicrograph of recovered quartz from shock experiment at $\sim$10-kb peak pressure. The scale shown is 1 mm.

From all indications, under shock conditions they are release or tensile phenomena and are best developed on release from low to moderate pressures, i.e., below the Hugoniot elastic limit. Cleavage development has been described from the West Hawk Lake crater, Quebec [*Robertson et al.*, 1968, Figure 4], the Meteor crater, Arizona [*Bunch*, 1968, Figure 2], the Flynn Creek structure, Tennessee [*Short and Bunch*, 1968, Figure 4], and the Ries crater, Germany [*Engelhardt and Bertsch*, 1969].

A weak preferred orientation of lamellas exists in each of the three samples. However, it appears to be related to the pre-existing preferred crystallographic orientation of the grains themselves rather than to a preferential development of certain lamellas in a homogeneous stress field. Occurrence of lamellas at grain boundaries and near fractures suggests that local stress concentrations have been important. A strong preferred orientation is unlikely in view of the rarefaction of the shock wave(s) at discontinuities, which gives rise to a heterogeneous stress history for any rock unit.

Hornblende. The main evidence of plasticity in hornblende is development of polysynthetic twins parallel to $(\bar{1}01)$ or (001), depending on the choice of cell axes. The twin-glide system, as established by *Rooney et al.* [1970], *Buck and Paulitsch* [1969], and *Buck* [1970], is $K_1 = (\bar{1}01)$, $N_1 = [101]$, $K_2 = (100)$, and $N_2 = [001]$ for the $C2/m$ setting, and $K_1 = (001)$, $N_1 = [100]$, $K_2 = (100)$, and $N_2 = [001]$ for the $I2/m$ setting. The twinning is analogous to a twin-glide system recognized for the monoclinic pyroxenes, $C2/c$ [*Griggs et al*, 1960; *Raleigh and Talbot*, 1967]; $K_1 = (001)$, $N_1 = [100]$, $K_2 = (100)$, and $N_2 = [001]$, in keeping with the custom of using the body-centered cell $(I2/m)$ in amphiboles to illustrate morphological and structural similarities between the two groups of chain silicates.

Onset of twinning in the group of shocked rocks is potentially difficult to recognize, since, at low stresses, lamellas are sparse and thin and in many respects resemble $(\bar{1}01)$ exsolution lamellas [e.g., *Jaffe et al.*, 1968; *Ross et al.*, 1968]. The rarity of any lamellas in the unshocked hornblende and the absence of (100) exsolution lamellas in shocked hornblende that commonly accompany $(\bar{1}01)$ exsolution lamellas argue for a deformational origin for all lamellas observed. Occasional simple (100) twins represent the only

unusual feature of the undeformed hornblende. Twin lamellas were first detected in hornblende that had seen a 24- to 40-kb peak radial stress, and the maximum shear stress, $(\sigma_1 - \sigma_3)/2$, on any plane is 4 kb. This stress is considerably lower than the estimated 200-kb pressure associated with development of planar features in amphiboles [*Chao*, 1968; *Engelhardt and Stöffler*, 1968] and higher than the 10-kb axial pressure, $(\sigma_1 - \sigma_3)/2 = 1.3$, associated with the onset of twinning at 20°C and a strain rate of 10^{-4} sec^{-1} [*Buck*, 1970]. At higher stress levels, optics and cleavages within the lamellas can be readily measured. 'Rotation' of the $\{110\}$ cleavage within the lamellas is definitive for twinning since exsolution intergrowths result in a near coincidence of cleavages (Figure 10).

Decomposition of hornblende can be noticed in rock that has been shocked to 140–150 kb. It is probably a thermal effect, reflecting not so much the shock temperature as the proximity to steam and other gases emanating along fractures from the expanding cavity.

Sphene. Polysynthetic twinning in shocked sphene has previously been described from the access drifts at the Piledriver site [*Borg*, 1970b]. The geometry of mechanical twinning in sphene is described by the following elements: $K_1 \simeq \{221\}$, $N_1 = \langle 110 \rangle$, $K_2 = \{\bar{1}31\}$, and $N_2 =$ irrational line. The thinnest twin lamellas are recognized by a 'play' of second- and third-order interference colors when viewed obliquely to the composition plane. However, the paucity of sphene crystals in the granodiorite, $<1\%$, poses a sampling problem in ascertaining the point at which stresses are high enough to produce twinning in a few properly oriented grains. A previous estimate based on rock collected in the drift walls was 5–8 kb and a stress difference of ~ 1 kb. In view of the free surfaces that existed at the time of shock propagation, i.e., the drift walls, and of the results of laboratory deformations of the mineral [*Borg and Heard*, 1972], these values are considered to be too low. Core data indicate that 14–18 kb (from stress gages) and 8–9 kb (from SOC calculations) are more likely values for the peak stress and the stress differential, respectively, under field shock conditions. The 'weakly developed' planar features associated with very high shock pressures (>300 kb) by *Chao* [1968, p. 239] possibly owe their origin to yet another slip mechanism.

Fig. 10. Shock twinning on {$\bar{1}01$} in hornblende (PS-3 core, 67 meters (220 feet) SHD). The scale shown is 0.1 mm.

At 270 kb, sphene is highly twinned and fractured and appears metallic in reflected light, although decomposition to ilmenite recorded by *El Goresy* [1968] was not detected.

Biotite. Kinking in biotite concomitant with bend gliding on (001) has been the subject of numerous investigations both in the laboratory and in the field. *Hörz and Ahrens* [1969] and *Hörz* [1969] in laboratory shock experiments set the lower limit for their formation at >37.5 and ~10 kb for propagation normal and parallel to (001), respectively. *Cummings* [1965, 1968] has investigated biotites from nuclear shocked granodiorite from the Hardhat event and indicates that some kinking occurs as low as 10-kb peak radial stress. It appears, however, that, in using calculated peak pressures as a function of preshot distance from the nuclear explosion, Cummings neglected to consider the 1- to 4-meter displacements seen by the core samples studied. Thus the specimens examined from Hardhat cores have seen pressures higher by 3–7 kb than those suggested by their postshot radial distances from the shot point. Displacements and ΔP given are based on postshot radial distances of 24–43 meters for core samples

studied by *Cummings* [1968, Figure 6]. *Short* [1966] estimated the lower limit for kinking at the Hardhat site at 25–30 kb. The data presented here suggest that incipient kinking is first recognizable in rock that has seen a 15- to 16-kb peak radial stress. Above 75 kb all biotite grains contain kinks. Figure 11 shows the progressive development of kinking as a function of peak radial stress. In all cases the c axis of the grain is nearly normal to the photograph. The examples represent the most highly kinked biotites at that orientation in the sample. At 205 kb the biotites are largely decomposed, magnetic, and opaque. Magnetite, forsterite, and fluor-phlogopite (?) are detectable in Debyé-Scherrer X-ray photographs of single grains. Comparable phases are not recognizable in biotite shocked to 313 kb in the laboratory (sample by courtesy of T. Ahrens), nor are they recognized below an estimated 300 kb at sites of meteor impacts [*Chao*, 1968; *Engelhardt and Stöffler*, 1968]. Decomposition observed in the Piledriver biotite is a thermal effect connected primarily with the heat of the explosion rather than a transient temperature associated with the passage of the shock wave.

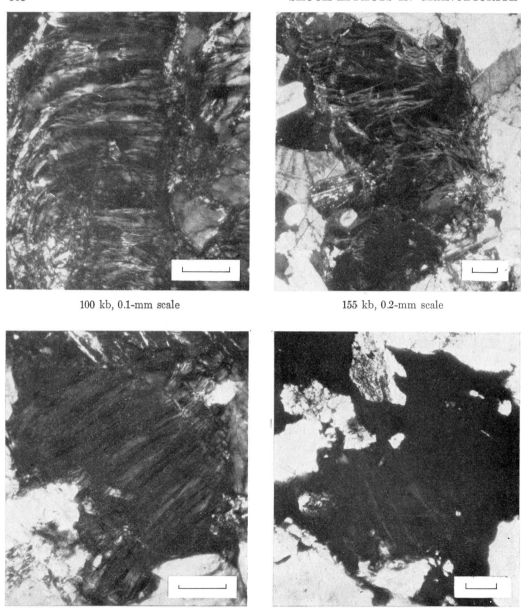

100 kb, 0.1-mm scale   155 kb, 0.2-mm scale

205 kb, 0.1-mm scale   270 kb, 0.1-mm scale

Fig. 11.   Progressive development of kinking with increased peak radial stress in biotite.

There is only a slight preferential develop-ment of kink bands in grains of particular crys-tallographic orientations within the granodiorite. However, kink bands within grains at a single thin section do show a moderate preferred ori-entation. Presumably, the bands tend to be nor-mal to the peak radial stress, although this orientation cannot be tested except in uniquely oriented core. *Cummings* [1965] came to similar conclusions. In regimes above 100 kb, kinking becomes extreme, and the generalizations made above are tenuous. Close-in the stress-time his-tory is particularly complex. The rock has been repeatedly deformed not only by a radiating

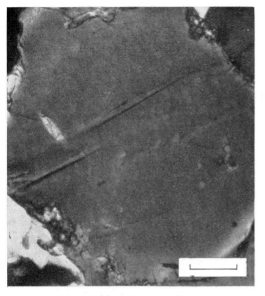

18 kb, 0.1-mm scale

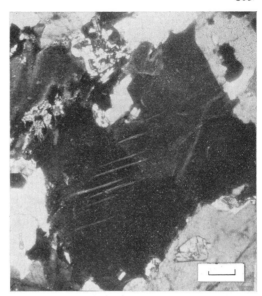

30 kb, 0.2-mm scale

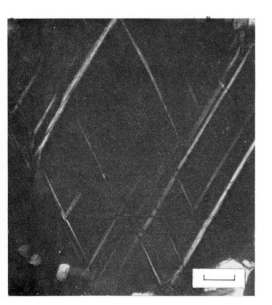

40 kb, 0.1-mm scale

51 kb, 0.2-mm scale

Fig. 11. (continued)

shock pulse but also by strong rarefacted shock waves reflected from pre-existing discontinuities. Locally concentrated and oriented stresses related to the brittle failure of the strong components further complicate the stress picture.

Feldspar. Response of orthoclase and albite-oligoclase to pressures of ≤270 kb is chiefly by fracture and destruction of the structure on a submicroscopic scale. No shock-induced twinning or planar lamellar structure occurs in either feldspar. As was noted by *Short* [1970], failure of orthoclase is associated with a characteristic fracture network or closely spaced irregular 'tears.' Birefringence of both feldspars is grossly lowered to 0.003 at 270 kb; however, neither glass, sandine, nor any dense high-pressure phase was detected.

From equation-of-state data on quartz and plagioclase Stöffler [1967] deduced that such glasses and high-pressure modifications of the feldspars form at pressures in excess of 250 kb at the Ries meteor crater in Bavaria. With an uncertainty of 10% in both Stöffler's estimate and the shock pressures measured at the Piledriver site, the observations are consistent. On the other hand, plasticity recorded by Stöffler [1967] in shocked plagioclase in the 100- to 250-kb range has no counterpart at the Piledriver site.

In view of the intractable nature of acidic plagioclases in static laboratory tests, as contrasted to those whose anorthite content exceeds 30% [Borg and Heard, 1970], it is likely that the observed differences in behavior (between laboratory tests cited by Stöffler and the present test) are related to the basic chemical and structural differences of the plagioclases studied. All the nonisotropic planar features in plagioclase described by Stöffler [1967] as well as Robertson et al. [1968] and Dworak [1969] from Canadian craters occur in plagioclase with an anorthite content of >30%, whereas the present observations are restricted to albites and peristerites. Nonetheless, debris recovered from within both the Hardhat and the Piledriver cavity does contain disordered feldspar with localized lamellar and band development accompanied by patches of glass. The pressures associated with these phenomena cannot be accurately specified, and there is a real possibility that they are products of high temperatures associated with the explosion itself and of only moderate stresses rather than of high temperatures associated with high shock pressures.

RÉSUMÉ AND DISCUSSION

The first part of this paper was a review of the parts of the Piledriver experiment that are relevant to the study of shock deformation of granodiorite and its mineral constituents. Satisfactory operation of a large number of stress gages allowed correlation between shock effects and specific shock pressures. The combined error in the measured stress and the location of the core samples at the shot time results in an estimated error of 10–15% in pressure assigned to any sample (C. T. Vincent and J. Rempel, personal communication, 1968). The highest reliable measurement of peak radial pressure measured was 270 kb at a distance of 14.6 meters from the shot point. The remainder of the stress data (Figure 3) suggest that this pressure may be too low by ~25 kb and that a pressure of 270 kb is appropriate to rock located <1 meter farther from the shot point, i.e., rock constituting the present cavity wall. Inasmuch as recovered debris within the cavity has not been considered here, the maximum pressure seen by any core sample is close to 270 kb.

Dissociation of hydrous minerals (e.g., biotite) in the surviving granodiorite indicates that temperatures have reached as high as 600°–700°C locally [Wones and Eugster, 1965]. Localization of high-temperature effects in areas adjacent to fractures suggests that mobile hot gases have primarily been responsible. In situ melting of the granodiorite was not recognized, although glasses fill in fractures of recovered core <56 meters from the shot point.

Mechanical phenomena, e.g., slip and twinning, noted in the various mineral constituents have been correlated with particular shock pressures. The stress differential $(\sigma_1 - \sigma_3)$ is more difficult to specify. Calculations of the propagating stress field around an explosive source such as the SOC code indicate that in granodiorite below the Hugoniot elastic limit (~45 kb) the average differential is in the 1–10 kb range [Cherry and Rapp, 1968]. However, it is likely that differences in the compressibility of adjacent mineral phases can result in local differentials much greater than these average values. Above the Hugoniot elastic limit, stress differentials can be instantaneously very large, but their magnitudes have not been measured or calculated to date.

Where results of laboratory shock experiments can be compared with results of the present study, e.g., for quartz and biotite, in general, minimum peak shock pressures for the development of lamellas and kinks are lower for the field-shocked samples than for the laboratory-shocked single crystals. The differences may be due to local stress concentrations and locally high stress differentials associated with deformations of heterogeneous materials. They may also reflect the generally lower strain rates in the field test described earlier in this text. *Brace and Jones* [1971] have shown in room temperature experiments on granite at <10-kb confining pressure that brittle failure is not influenced by strain rates over the 10^6–10^{-5} sec^{-1} interval. Nonetheless, it seems likely that the usual strain rate dependence on yielding will hold at higher pressures. Specifically, lower strain rates result in lower yield stresses and lower peak and mean pressures associated with onset of a particular failure mechanism.

Another possible explanation for the differences in the phenomenology observed at given shock pressures in single-crystal shock tests and field shock experiments may lie in a difference in Hugoniot temperatures. The value of T_H is sensitive to void space (pore, fracture, and cleavage porosity), and it is probably locally higher for nuclear explosions and meteoritic impacts than would be indicated by calculations based on equation-of-state measurements of small laboratory samples approaching theoretical density. The effect of increased temperature is to lower the critical resolved shear stress required for the initiation of slip and twinning in mineral components. Thus, owing to higher Hugoniot temperatures, higher local stresses or stress differentials, lower strain rates, or all three factors, it might be anticipated that the initiation of slip and twinning would be associated with lower peak and mean shock pressures for large bodies of shocked rock than for small mineral samples.

APPENDIX: SUMMARY OF PRINCIPAL DATA ON PILEDRIVER EVENT

Shot date: June 2, 1966.

Yield: 61 ± 10 kT, based on determination of the Los Alamos Scientific Laboratory on injected glass encountered in re-entry tunnel 91 meters (300 feet) from the shot point, 14 months after the shot [*Rabb,* 1968; *Boardman,* 1967].

Shot depth: 462.8 meters (1518 feet) below the surface (1089 meters (3572 feet) above msl).

Cavity radius r_c: 40.1 meters (131.5 feet), measured in the drill hole that intersected the cavity wall at a point 10.4 meters (34 feet) below the shot level [*Sterrett,* 1969] (the cavity is a nearly spherical void space formed within seconds of the explosion by the melting and the vaporization of rock and by the thermal expansion of gaseous products).

Chimney height: 277.4 meters (910 feet) above the shot level (the chimney is a vertical column of rubble formed at the collapse of the cavity upon cooling; the void space between the rock debris within the chimney and the cavity is equal to the cavity volume before collapse).

Chimney half-width: 48.8 meters (160 feet) measured in the re-entry tunnel at a point 31.6 meters (103.5 feet) above the shot level.

Chimney volume: 19.0×10^5 m^3 (6.69 × 10^7 ft^3) or $>29.4 \, r_c^3$.

Pressure of overburden at shot point: 121 bars (1721 psi), calculated on the basis of the average density of 2.66 g/cm^3.

Apical void: 0.61 meters (2 feet) high (the apical void is the empty space between the top of the chimney fill material and the chimney ceiling [*Boardman,* 1967]).

Time of cavity collapse: within 14 sec after the shot [*Rabb,* 1968].

Maximum vertical extent of increased air permeability: 314.6 ± 11.0 meters (1032 ± 36 feet) above the shot level [*Boardman,* 1967].

Acknowledgments. Discussions with H. Heard, T. Ahrens, and T. Butkovich on various aspects of shock metamorphism were very profitable. D. Larsen contributed substantially to the development of equations used to calculate Hugoniot temperatures. Comments on the text by D. Stöffler and F. Hörz were appreciated.

This work was performed under the auspices of the U.S. Atomic Energy Commission.

General Shock Conditions

MONTGOMERY H. JOHNSON AND EDWARD TELLER

Lawrence Livermore Laboratory, University of California Livermore, California 94550

The behavior of hydrodynamic shocks has been treated relativistically in the last 25 years (Taub, 1948; de Hoffman and Teller, 1950; Colgate and Johnson, 1960; Johnson and McKee, 1971; Teller, 1971). The present paper gives a simple treatment of the basic relations that also includes the influence of a transverse magnetic field and the change of entropy in the shock. The formulas can be written in a form that is no more complicated than the forms of the formulas used in the nonrelativistic treatment.

SHOCK CONDITIONS

Equations connecting quantities ahead of a shock front to the same quantities behind the shock front are derived by assuming conservation of matter, energy, and momentum. In the relativistic case, conservation of matter is replaced by conservation of the net number of nucleons, i.e., the difference between the number n of nucleons and antinucleons. (In most cases, antinucleons are not produced in the shock, so that the nucleon number is conserved.) Also in the relativistic case, if E is the proper energy density and p the proper pressure, the energy density is $(E + \beta^2 p)(1 - \beta^2)^{-1}$, and the pressure $(p + \beta^2 E)(1 - \beta^2)^{-1}$ for a reference frame in which the material moves with a speed $c\beta$, c being the velocity of light. The flux of energy is $c\beta(p + E)(1 - \beta^2)^{-1}$. These statements follow immediately from the fact that under a Lorentz transformation the stress energy momentum tensor transforms as the direct product of 2 four vectors. This tensor is diagonal in the proper reference frame with the space components equal to p and the time component equal to E. Since the tensor is symmetric, the momentum flux, apart from a factor of $1/c$, is the same as the energy flux.

It is convenient to use a coordinate system fixed in the shocked fluid. Quantities referring to the shocked fluid carry the subscript 1, and those referring to the unshocked fluid carry the subscript 2. Also, $c\beta_r$ is used for the speed of material 2 relative to that of the shocked mate-

rial 1. Of course, the speed of 1 relative to that of 2 is equal to $-c\beta_r$. Let the speed of the shock front be $c\beta_1$ (relative to that of the shocked material 1 as measured by an observer at rest in 1) and let n_1 be the proper nucleon number density in material 1.

Consider nucleon conservation in a cylinder of unit cross section with one end fixed in the shocked material and the other open to the unshocked material and suppose that the shock is planar. Unshocked material enters at the rate of $c\beta_r n_2 (1 - \beta_r^2)^{-1/2}$. The factor $(1 - \beta_r^2)^{-1/2}$ arises from the Lorentz contraction that increases the density of the moving material. At the same time the amount of material in the cylinder increases by $c\beta_1 n_1$ as material of density n_1 appears behind the shock and decreases by $c\beta_1 n_2 (1 - \beta_r^2)^{-1/2}$ as material of density $n_2 (1 - \beta_r^2)^{-1/2}$ disappears into the shock. Nucleon conservation requires that

$$c\beta_r n_2 (1 - \beta_r^2)^{-1/2}$$
$$= c\beta_1 \{ n_1 - [n_2 (1 - \beta_r^2)^{-1/2}] \}$$

or

$$\beta_1 n_1 = (\beta_r + \beta_1) n_2 (1 - \beta_r^2)^{-1/2} = F_1/c \quad (1)$$

The quantities equated in (1) have been designated F_1/c, where F_1 is the number of nucleons passing through 1 cm² of the shock front per second as seen by an observer at rest in the shocked material. The left-hand side in (1) is evidently the flux obtained from the flow behind the shock, whereas the right-hand side is the flux obtained from the flow ahead of the shock.

Next consider energy conservation. The energy entering the tube via the unshocked material is $c\beta_r(p_2 + E_2)(1 - \beta_r^2)^{-1}$. As the shock advances, the energy $c\beta_1 E_1$ appears behind the shock, and the energy $c\beta_1(E_2 + \beta_r^2 p_2)(1 - \beta_r^2)^{-1}$ disappears into the shock. Energy conservation requires that

$$c\beta_r(p_2 + E_2)(1 - \beta_r^2)^{-1}$$
$$= c\beta_1\{E_1 - [(E_2 + \beta_r^2 p_2)(1 - \beta_r^2)^{-1}]\}$$

or

$$\beta_1(p_2 + E_1)$$
$$= (\beta_r + \beta_1)(p_2 + E_2)(1 - \beta_r^2)^{-1} \quad (2)$$

Finally, consider momentum conservation. The pressure difference $p_1 - [(p_2 + \beta_r^2 E_2)(1 - \beta_r^2)^{-1}]$ is balanced by the disappearance of momentum in the cylinder at the rate $\beta_1\beta_r(p_2 + E_2)(1 - \beta_r^2)^{-1}$. Then conservation requires that

$$p_1 - (p_2 + \beta_r^2 E_2)(1 - \beta_r^2)^{-1}$$
$$= \beta_1\beta_r(p_2 + E_2)(1 - \beta_r^2)^{-1}$$

or

$$p_1 - p_2 = \beta_r(\beta_r + \beta_1)(p_2 + E_2)(1 - \beta_r^2)^{-1} \quad (3)$$

$$p_1 - p_2 = \beta_r\beta_1(p_2 + E_1) \quad (4)$$

Equation 4 follows from (3) when (2) is used. (A recent work by one of the present authors [*Teller*, 1971] failed to note the term p_2 on the right-hand side of (4).)

SHOCK AND FLUID SPEEDS

The shock speed can be eliminated by inserting the value of β_1 from (4) into (2).

$$\frac{p_1 - p_2}{\beta_r} = \frac{(p_2 + E_2)}{1 - \beta_r^2}\left[\beta_r + \frac{p_1 - p_2}{\beta_r(p_2 + E_1)}\right]$$

The solution for β_r^2 is

$$\beta_r^2 = [(p_1 - p_2)(E_1 - E_2)]$$
$$\div [(p_2 + E_1)(p_1 + E_2)] \quad (5)$$

Dividing (4) by (5) gives

$$\beta_1/\beta_r = (p_1 + E_2)/(E_1 - E_2) \quad (6)$$

Multiplying the square of (6) by (5) gives

$$\beta_1^2 = [(p_1 - p_2)(p_1 + E_2)]$$
$$\div [(E_1 - E_2)(p_2 + E_1)] \quad (7)$$

Equations 5 and 7 determine the material speed and the shock speed in terms of pressure and energy densities before and behind the shock.

The shock speed relative to the unshocked material $c\beta_2$ can be obtained by the observation that all the equations given above will apply in the reference frame of the unshocked material if the subscripts 1 and 2 are interchanged. Hence (7) changes into

$$\beta_2^2 = [(p_2 - p_1)(p_2 + E_1)]$$
$$\div [(E_2 - E_1)(p_1 + E_2)] \quad (8)$$

According to (7) and (8), the shock speeds satisfy the following simple relations:

$$\beta_1\beta_2 = (p_1 - p_2)/(E_1 - E_2) \quad (9)$$

$$\beta_1/\beta_2 = (p_1 + E_2)/(p_2 + E_1) \quad (10)$$

Sound speed $c\beta_a$ can be obtained from (7) by letting $p_1 \to p_2$ and $E_1 \to E_2$:

$$\beta_a^2 = (dp/dE)_a \quad (11)$$

where the subscript a on the right side of (11) indicates the adiabatic derivative.

It is interesting to calculate explicitly the flux of material through the shock. In the frame of the shocked material the flux is, according to (1),

$$F_1^2 = c^2\beta_1(\beta_1 + \beta_r)n_1 n_2(1 - \beta_r^2)^{-1/2}$$

This expression can be put into a symmetric form by transforming to the frame in which the shock is at rest by multiplying F_1 by $\gamma_1 = (1 - \beta_1^2)^{-1/2}$. With the help of (6) and (7) the equation given above for F_1^2 becomes

$$F^2 = \gamma_1^2 F_1^2$$
$$= \gamma_r \frac{n_1 n_2(p_1 - p_2)c^2}{(E_1 - E_2) - (p_1 - p_2)} \quad (12)$$

Since $\gamma_r = (1 - \beta_r^2)^{-1/2}$, (12) is indeed symmetric.

LIMITING CASES

In the nonrelativistic limit, E can be set equal to the mass energy ρc^2, where ρ is the mass density, and p can be neglected as compared with E. Then (5), (7), (11), and (12) become the well-known relations

$$c^2\beta_r^2 = (p_1 - p_2)[(1/\rho_2) - (1/\rho_1)] \qquad (5')$$

$$c^2\beta_1^2 = p_1 - p_2/[(\rho_1/\rho_2) - 1]\rho_1 \qquad (7')$$

$$c^2\beta_a^2 = (dp/d\rho)_a \qquad (11')$$

$$F^2 = (p_1 - p_2)[(1/\rho_2) - (1/\rho_1)]^{-1}(n/\rho)^2 \qquad (12')$$

In the last equation, ρ/n is the mass of the nucleon and the same on both sides of the shock.

In the relativistic limit the interesting cases are strong shocks in which p_2 and E_2 are negligible in comparison with E_1 and p_1. Furthermore, p_2 is small in comparison with E_2 because E_2 contains the nucleon rest energy. Then (5) and (7) become

$$\beta_r^2 \approx 1 \qquad (5'')$$

$$\beta_1^2 \approx (p_1/E_1)^2 \qquad (7'')$$

For the extreme relativistic case, $p_1 = \frac{1}{3}E_1$, so that $\beta_1 = \frac{1}{3}$ and $\beta_a = 1/(3)^{1/2}$. In the relativistic regime the energy factor $\gamma_r = (1 - \beta_r^2)^{-1/2}$ is a more convenient quantity than β_r. The energy factor, when (5) is used, is

$$\gamma_r^2 = [(p_2 + E_1)(p_1 + E_2)]$$
$$\div [(p_2 + E_2)(p_1 + E_2)] \qquad (13)$$

In the extreme relativistic limit, γ_r^2 increases as $E_1(4E_2)^{-1}$, so that γ_r increases as the square root of the energy density behind the shock. Then, according to (12), F is given by

$$F \approx (c/2)(n_1 n_2)^{1/2}(E_1/E_2)^{1/4} \qquad (14)$$

The energy factor for β_2, when (8) is used, is

$$\gamma_2^2 = \frac{(E_1 - E_2)(p_1 + E_2)}{[(E_1 - E_2) - (p_1 - p_2)](E_2 + p_2)}$$

$$\approx \frac{E_1}{2E_2} = 2\gamma_r^2 \qquad (15)$$

Thus in the extreme relativistic limit γ_2 also increases as the square root of the energy density behind the shock.

In the case of the strong shock ($p = 0$) the division of (2) by (1) gives

$$E_1/n_1 = \gamma_r E_2/n_2$$

The left-hand side is the energy per particle in the shocked material. The right-hand side is the energy per incoming particle. This simple conservation law is well known in the nonrelativistic case and is seen here to be generally valid.

COMPRESSION

The compression of the shocked fluid is determined by eliminating the velocities from (1). The result of inserting (5), (7), and (13) into the square of (1) is

$$\frac{n_1^2}{n_2^2} = \frac{(p_1 + E_1)(p_2 + E_1)}{(p_1 + E_2)(p_2 + E_2)} \qquad (16)$$

In general, if the conditions ahead of the shock, E_2, p_2, and n_2, are given and the equation of state connecting E_1, p_1, and n_1 behind the shock is known, (16) can be solved simultaneously with the equation of state to determine the compression as a function of the shock pressure.

To examine the nonrelativistic limit, put

$$E = nmc^2 + \eta \qquad (17)$$

where η is the internal energy density. Equation 16 becomes

$$1 = \frac{\{1 + [(p_1 + n_1)/n_1 mc^2]\}\{1 + [(p_2 + n_1)/n_1 mc^2]\}}{\{1 + [(p_1 + n_2)/n_2 mc^2]\}\{1 + [(p_2 + n_2)/n_2 mc^2]\}} \qquad (18)$$

Expansion of the right side of (15) in inverse powers of c^2 to the order of $1/c^2$ gives the familiar result

$$(p_1 + p_2)\left(\frac{1}{\rho_2} - \frac{1}{\rho_1}\right) = 2\left(\frac{n_1}{\rho_1} - \frac{n_2}{\rho_2}\right) \qquad (19)$$

For a strong relativistic shock ($p_2 = 0$ and $E_2 \ll p_1$), (16) reduces to

$$n_1/n_2 = (E_1/n_1)(E_2/n_2)^{-1}[(p_1 + E_1)/p_1] \qquad (20)$$

The quantity $(p_1 + E_1)/p_1$ approaches 4 in the extreme relativistic case, so that n_1/n_2 increases in proportion to the specific energy density behind the shock, in contrast to the nonrelativistic case in which the compression of an ideal gas at infinite shock pressure approaches a definite value fixed by the ratio of the specific heats. However, the apparent compression as seen by an observer in the frame of reference of the shocked fluid $n_1(\gamma_r n_2)^{-1}$ obtained from (1) and (6) is

$$n_1(\gamma_r n_2)^{-1} = (p_1 + E_1)/(p_1 + E_2) \quad (21)$$

and approaches the definite limit $(p_1 + E_1)/p_1$ of approximately 4 as the shock becomes infinitely strong.

HYDROMAGNETIC SHOCKS

The results presented above are easily extended to include a magnetic field H perpendicular to the flow direction. (A more complete and much more complicated discussion of this topic has been given by *de Hoffmann and Teller* [1950]. Our present results are not only simpler but also go beyond the previous paper by discussing the change of entropy in the relativistic case.) The relevant diagonal components of the stress/energy momentum tensor in the proper frame of reference are increased by the term $H^2/8\pi$, corresponding to the magnetic pressure and magnetic energy density. Consequently, all the previous formulas remain valid if p and E are replaced by $\bar{p}$ and $\bar{E}$ where

$$\bar{p} = p + (H^2/8\pi) \quad (22)$$
$$\bar{E} = E + (H^2/8\pi)$$

An additional condition on H is needed to complete the set of equations. In astronomical applications the conductivity is large enough to freeze the lines of force in the material. For a transverse field this condition means that, in any material element, H/n remains constant as the material passes through the shock front:

$$H_2/n_2 = H_1/n_1 \quad (23)$$

Consequently, the magnetic pressure and the magnetic energy density behind the shock increase as the square of the compression.

When the changes required by (22) are introduced, the energy factor for the material velocity (13) in the case of a strong shock becomes

$$\gamma_r^2 = \frac{[E_1 + (H_1^2/8\pi)][p_1 + (H_1^2/8\pi)]}{E_2[E_1 + p_1 + (H_1^2/4\pi)]} \quad (24)$$

In (24), H_2^2 has been neglected in comparison with H_1^2 and E_2. In the extreme relativistic limit, γ_r^2 still varies as $\bar{E}_1/E_2$, but the factor of

$\frac{1}{4}$ changes to $\frac{1}{2}$ when $H_1^2/8\pi$ becomes large in comparison with E_1. Then $\bar{E}_1$ becomes $H_1^2/8\pi$, and γ_r is proportional to the field strength behind the shock.

With the same approximations as those in the previous paragraph, the shock speed (7) becomes

$$\beta_1 \approx [p_1 + (H_1^2/8\pi)][E_1 + (H_1^2/8\pi)]^{-1}$$

In the extreme relativistic case, β_1 changes from the value of $\frac{1}{3}$ to that of 1 as $H_1^2/8\pi$ becomes large in comparison with E_1. Evidently β_1 is always larger than p_1/E_1, the shock speed in the absence of a magnetic field.

The expression for sound speed, obtained by letting $p_1 \to p_2$ and $E_1 \to E_2$ in the expression for the shock speed, is

$$\beta_a^2 = \left\{ \frac{d[p + (H^2/8\pi)]}{d[E + (H^2/8\pi)]} \right\}_a$$

$$= \left[\left(\frac{dp}{dE} \right)_a + \frac{1}{8\pi} \left(\frac{dH^2}{dE} \right)_a \right]$$

$$\cdot \left[1 + \frac{1}{8\pi} \left(\frac{dH^2}{dE} \right)_a \right]^{-1} \quad (25)$$

When the adiabatic condition

$$[d(E/n)] + [p d(1/n)]$$
$$= 1/n[dE - (p + E)(dn/n)] = 0 \quad (26)$$

and the flux condition (23) are used, the derivative $(dH^2/dE)_a$ is readily evaluated:

$$(dH^2/dE)_a = H^2/n^2(dn^2/dE)_a = 2H^2/(p + E)$$

Hence the sound speed (25) can be written:

$$\beta_a^2 = \left[\left(\frac{dp}{dE} \right)_a + \frac{H^2}{4\pi(p + E)} \right]$$

$$\cdot \left[1 + \frac{H^2}{4\pi(p + E)} \right]^{-1} \quad (27)$$

Equation 27 expresses the sound speed in terms of $(dp/dE)_a$, the sound speed in the absence of the magnetic field, and the parameter $H^2/[4\pi(p + E)]$. Evidently a transverse magnetic field always increases the sound speed.

The compression in the presence of a magnetic field is given by

$$\frac{n_1^2}{n_2^2} = \frac{H_1^2}{H_2^2} = \frac{(\bar{p}_1 + \bar{E}_1)(\bar{p}_2 + \bar{E}_1)}{(\bar{p}_1 + \bar{E}_2)(\bar{p}_2 + \bar{E}_2)} \quad (16')$$

In the presence of strong magnetic fields where

all other quantities can be neglected as compared to $H^2/8\pi$, (16') becomes an identity. For the interesting relativistic case where $(H_2^2/8\pi) \ll E_2$ and $p_2 = 0$, it follows from (16') that $(H_1^2/8\pi) \ll E_1$. If, furthermore, $E_1 = 3p_1 \gg E_2$ is assumed, (16') becomes

$$H_1^2/H_2^2 \approx 4E_1/E_2 \qquad (28)$$

It can be generally shown from (16') that the magnetic field dominates either in both the shocked and the unshocked material or in neither material. Conversely, the magnetic field can be weak in both regions or in neither region.

CHANGE OF ENTROPY

In this section the notations E and p are used instead of the more general $\bar{E}$ and $\bar{p}$. The results are, however, generally applicable.

Consider the Hugoniot curve where the initial state characterized by E_2, p_2, and n_2 is held fixed while the quantities characterizing the final state E_1, p_1, and n_1 are considered variable. The logarithmic derivative of (16) is

$$\frac{2\,dn_1}{n_1} + \frac{dp_1}{E_2 + p_1} = \frac{dE_1}{E_1 + p_2} + \frac{d(E_1 + p_1)}{E_1 + p_1} \qquad (29)$$

Assume that, on the Hugoniot, higher n_1 values correspond to higher values of E_1 and p_1. Furthermore, assume that the entropy increases as n increases. Thus

$$[dE_1/(E_1 + p_1)] > dn_1/n_1 \qquad (30)$$

The assumption of increasing entropy is demanded by thermodynamics and determines the direction in which the shock proceeds.

The introduction of (30) into (29) leads to

$$\frac{dE_1}{E_1 + p_1} - \frac{dE_1}{E_1 + p_2} > \frac{dp_1}{E_1 + p_1} - \frac{dp_1}{E_2 + p_1}$$

or

$$dE_1 \frac{p_2 - p_1}{(E_1 + p_1)(E_1 + p_2)} > dp_1 \frac{E_2 - E_1}{(E_1 + p_1)(E_2 + p_1)}$$

Since $E_2 - E_1$ is negative, the last inequality can be written as

$$\frac{dp_1}{dE_1} > \frac{(p_1 - p_2)(E_2 + p_1)}{(E_1 - E_2)(E_1 + p_2)} \qquad (31)$$

Note that according to (7) the right-hand side of (31) is β_1^2. However, this fact will not be used.

The inequality (31) can be written in the differential form

$$\frac{dp_1}{(p_1 - p_2)(E_2 + p_1)} > \frac{dE_1}{(E_1 - E_2)(E_1 + p_2)}$$

Assuming that this inequality holds everywhere on the Hugoniot and starts at the point characterized by E_2 and p_2, one can integrate to obtain

$$\ln\left(\frac{p_1 - p_2}{E_2 + p_1}\right) > \ln\left(\frac{E_1 - E_2}{E_1 + p_2}\right) + \ln\left(\frac{dp}{dE}\right)_2 \qquad (32)$$

The last term in (32) is a constant of integration introduced in such a manner as to turn the inequality into an equality at $p_1 \to p_2$ and $E_1 \to E_2$. It is to be noted that $(dp/dE)_2$ is to be taken on an adiabat, because the Hugoniot curve approaches an adiabat for weak shocks. Therefore, $(dp/dE)_2$ is equal to the square of the sound velocity in the unshocked material, $(dp/dE)_2 = \beta_{2S}^2$, where subscript S indicates sound.

Equation 32 can be written in the form

$$[(p_1 - p_2)(E_1 + p_2)]$$
$$\div [(E_1 - E_2)(E_2 + p_1)] > \beta_{2S}^2 \qquad (33)$$

The expression on the left-hand side of (33) is actually β_2^2 as given in (8). Therefore, (33) becomes

$$\beta_2^2 > \beta_{2S}^2 \qquad (34)$$

Thus the shock speed relative to the unshocked material $c\beta_2$ is greater than the sound speed in the unshocked material multiplied by c. Therefore, in the general case the postulate of increasing entropy leads to the conclusion that the shock is supersonic with respect to the unshocked material, a result that is well known in the nonrelativistic regime.

A condition on the shocked material can be obtained by considering a 'Hugoniot' with E_1, p_1, and n_1 held fixed and E_2, p_2, and n_2 variable. Equation 29 then holds with 1 and 2 interchanged, whereas the inequality in (30) with 1 and 2 interchanged is reversed. By the same series of steps, (34) is replaced by

$$\beta_1{}^2 < \beta_{1s}{}^2 \qquad (35)$$

Hence, the condition of increasing entropy along the Hugoniot also requires the shock to be subsonic with respect to the shocked material.

As has been pointed out in the earlier discussion, these results are not changed by the presence of a transverse magnetic field.

Acknowledgment. This work was done under the auspices of the U.S. Atomic Energy Commission.

Lunar Tidal Acceleration on a Rigid Earth

R. A. Broucke,[1] W. E. Zürn,[2] and L. B. Slichter[1]

An improved computer program has been developed for finding the theoretical lunar tidal acceleration at designated points and times on a rigid earth. Older programs may often be in error by 0.5–1.0 μgal. For the present program the maximum absolute error is about 0.01 μgal. This improved precision is needed for evaluating small earth tide residuals due to the yielding of the crust and the mantle caused by ocean tidal loads and for studying the relatively small long-period fortnightly earth tides, for which the geophysically significant residual is only about 3–4 μgal in amplitude. The reference tidal acceleration values computed at 32 test locations provide a convenient means for verifying other tide computer programs.

An accurate and economical computer program for evaluating at any time and place on a rigid earth the lunar and solar tidal accelerations is essential for earth tide investigations. By far the largest part of a measurement of tidal gravity or tidal tilt consists of the corresponding reading that theoretically would have occurred on a rigid earth. The residual signal of geophysical interest that is caused by tidal loads and deformations of the earth frequently represents only 5% of the rigid-earth reading. Obviously, high precision must be maintained to permit even crude estimates of the small remainder.

The notation and the values of the constants for this paper are as follows.

a geocentric radius of the spheroid.
a_1 equatorial radius of the spheroid, 6.378160×10^8 cm.
l flattening parameter, $1/298.25 = 0.003353$.
p horizontal parallax of the moon.
δ declination of the moon.
τ local hour angle of the moon, positive to the west.
ϕ geocentric latitude of the observing station.
$C = 1 - l \sin^2 \phi$.

$a = Ca_1$.
$\alpha = Cp$.
G Newton's constant.
g_h, g_r, g_s, g_E components of tidal acceleration.
M_M mass of the moon.
$K = GM_M a_1^{-2} = 12.0515$.
$B \equiv K p^3 a_1^{-1}$.
$GM_M = 4.90268 \times 10^{18}$.
z zenith angle of the moon.
$\cos z = \sin \delta \sin \phi + \cos \delta \cos \phi \cos \tau$.
$\xi = 1 + \alpha^2 - 2\alpha \cos z$.

PRECISION DESIRED

A feasible and desirable precision in computing the lunar radial tidal acceleration on a rigid earth is suggested to be ± 0.01 μgal, which is a relative precision of 1 in 10,000 in a major lunar tidal amplitude of 100 μgal. The suggested precision can be attained with reasonable convenience and computational economy. It meets the needs of the present most precise types of observations and either avoids corrections or permits the following types of assumptions that might need consideration at higher sensitivity levels: (1) the correction for the tides of Venus, which reach 0.0056 μgal in amplitude at times of the closest approach of Venus, is insignificant, (2) the moon can be regarded as a point mass, (3) field observations with gravimeters and tilt meters necessarily use the plumb line direction as a reference, whereas for usual theoretical calculations the reference frame is geocentric. To first-power terms in the flattening ratio, l, the angle ϵ from the geocentric direc-

[1] School of Engineering, University of California, Los Angeles, California 90024 and Jet Propulsion Laboratory, Pasadena, California 91103.
[2] Institute of Geophysics and Planetary Physics, University of California, Los Angeles, California 90024.

tion to the normal to the spheroid $a = a_1$ $(1 - l \sin^2 \phi)$ is $\epsilon = l \sin 2\phi$. (The normal to the spheroid will serve here as a satisfactory approximation to the plumb line direction.) Let g_R and g_S denote the radial and the south component, respectively, of tidal acceleration in the geocentric system. On the spheroid, let g_r be the component normal to the spheroid and g_s be the south component, perpendicular to g_r in the plane of the meridian. Then

$$g_r = g_R \cos \epsilon - g_S \sin \epsilon$$

$$g_s = g_S \cos \epsilon - g_R \sin \epsilon$$

The ratios of g_r/g_R and g_s/g_S can evidently differ from 1 by approximately ϵ or about 3 parts per thousand. For some earth tide observations of high precision, corrections for the angle ϵ may be desirable in making a comparison with the rigid-earth theoretical values. (4) A small error occurs in the reduction from universal to ephemeris time, owing to variations in the rotation rate of the earth. These variations are known accurately only after astronomical observations have been analyzed. The extrapolated time corrections are now valid within ± 0.10 sec. The corresponding maximum error in (rigid earth) lunar tidal gravity is ± 0.0015 μgal.

The program retains the 71 harmonic constituents in Brown's theory of the moon with amplitudes that exceed $1.0''$ in the moon's latitude, the 61 that exceed $2.0''$ in longitude, and the 30 that exceed $0.1''$ in horizontal parallax [Eckert et al., 1954; Broucke, 1966]. Of the nutation correction terms, 13 terms are included for longitude, and 13 terms are included for obliquity corrections. The program computes the radial (positive upward), south, and east components of lunar tidal acceleration for points of given latitude, longitude, and elevation on the surface of a rigid reference spheroid $a = a_1$ $(1 - 0.003353 \sin^2 \phi)$ at selected intervals of local mean solar time.

EFFECT OF ERRORS IN POSITION OF OBSERVATORY

On the surface of a spherical earth the loci of lunar equitidal accelerations are obviously circles centered on the geocentric line to the moon. Specifically, to the first order, the radial and horizontal accelerations are

$$g_r = 2BaP_2(\cos z) \qquad (1)$$

$$g_h = Ba[\partial P_2(\cos z)/\partial z] \qquad (2)$$

where $P_2 (\cos z)$ is the second-degree zonal harmonic and B (see the notation list) is independent of the coordinates of the observatory. Equations 1 and 2 express the linear relation between the tidal acceleration and the geocentric distance a of the station. Thus

$$\Delta g_r/g_r = \Delta g_h/g_h = \Delta a/a \qquad (3)$$

If these ratios are regarded as relative errors, say of magnitude 10^{-4}, the allowable absolute elevation error Δa is impressively large, 637 meters.

For errors in the geographic position of the observatory resulting in the pertinent change Δz in the zenith angle, the associated acceleration errors are

$$\Delta g_r = \partial g_r/\partial z \, \Delta z = 2Ba(-3/2 \sin 2z) \, \Delta z \quad (4)$$

and

$$\Delta g_h = \partial g_h/\partial z \, \Delta z = Ba(-3 \cos 2z) \, \Delta z \qquad (5)$$

For the unfavorable values $\sin 2z = 1.0$ and $\cos 2z = 1.0$ these acceleration errors are equal. Let $3Ba$ have essentially its maximum value, about 200 μgal. Then for an allowable $|\Delta g_r| = |\Delta g_h| = 0.01$ μgal the allowable $|\Delta z|$ is $(0.01/200) = 5 \times 10^{-5}$ rad or $10''$ of arc, or it is about 320 meters on the earth's surface. If Δz is a change in longitude, the associated time error is $2/3$ sec. In many applications, however, a more realistic present limit on observational precision is ± 0.05 μgal. This limit increases the tolerance to 1.5 km in station location and to 3 sec in time.

TESTS OF PRECISION

The standard closed expressions for the three components of tidal gravity are

$$g_r = Kp^2[(\xi^{-3/2} - 1) \cos z - \alpha\xi^{-3/2}] \qquad (6)$$

$$g_s = Kp^2(\xi^{-3/2} - 1)$$

$$\times (\sin \phi \cos \delta \cos \tau - \sin \delta \cos \phi) \qquad (7)$$

$$g_E = -Kp^2(\xi^{-3/2} - 1) \cos \delta \sin \tau \qquad (8)$$

where g_r is positive upward, g_s is positive to the south, and g_E is positive to the east.

The values of the moon's declination and horizontal parallax, taken from the The Ameri-

TABLE 1. Comparison of Computations of Vertical Lunar Tidal Accelerations on Rigid Earth on November 11, 1970, 22h 26m 18.24s ET for 14 Check Points

Station		Vertical Lunar Tidal Accelerations, μgal			Differences, 10^{-3} μgal	
Latitude	Longitude*	Ephemeris	Longman [1959]	New Results	Longman [1959]	New Results
90°	0°	−48.997	−48.283	−48.997	−714	0
60°56.4′	−0°10.307′	30.357	31.164	30.355	−807	2
15°56.4′	−0°10.307′	127.377	126.634	127.373	+743	4
−29°3.6′	−0°10.307′	30.417	29.342	30.406	+1075	11
−90°	0°	−46.698	−45.971	−46.697	−727	−1
60°56.4′	44°49.693′	− 2.780	−2.305	−2.780	−475	0
15°56.4′	44°49.693′	36.499	35.536	36.499	+963	0
−29°3.6′	44°49.693′	−23.934	−24.834	−23.937	+900	3
60°56.4′	89°49.693′	−52.226	−51.816	−52.226	−410	0
15°56.4′	89°49.693′	−61.319	−61.105	−61.318	−214	−1
−29°3.6′	89°49.693′	−58.056	−57.423	−58.053	−633	−3
60°56.4′	179°49.693′	−51.324	−51.492	−51.322	+168	−2
15°56.4′	179°49.693′	71.338	70.187	71.338	+1151	0
−29°3.6′	179°49.693′	111.760	111.520	111.766	240	−6

The standard deviations are 0.75 μgal for *Longman*'s [1959] results and 0.004 μgal for the results of the new program.
Universal time at this moment is 22h 25m 37.128s.
* West of Greenwich.

TABLE 2. Comparison of Computations of Vertical Lunar Tidal Accelerations on a Rigid Earth on November 5, 1970, 00h 00m 00s ET for 18 Check Points

Station		Vertical Lunar Tidal Accelerations, μgal			Differences, 10^{-3} μgal	
Latitude	Longitude	Ephemeris*	Longman [1959]	New Results	Longman [1959]	New Results
90°	0°	−26.9973	−27.2444	−26.9950	+247.1	−2.2
35°23.4558′	−0°10.307′	−16.4088	−15.9707	−16.4046	−433.9	−4.1
0°	−0°10.307′	−42.4109	−42.4961	−42.4051	+ 91.0	−5.8
−24°36.5442′	−0°10.307′	−57.0776	−57.6728	−57.0723	+600.5	−5.3
−54°36.5442′	−0°10.307′	−54.8131	−55.6261	−54.8101	+816.0	−3.0
−90°	0°	−29.6612	−29.9540	−29.6594	+294.6	−1.8
35°23.4558′	44°49.693′	−58.2743	−59.1140	−58.2704	+843.6	−3.8
0°	44°49.693′	−32.4470	−33.4160	−32.4476	+968.4	+0.7
−24°36.5442′	44°49.693′	− 9.6508	−10.4190	− 9.6535	+765.5	+2.7
−54°36.5442′	44°49.693′	−2.7372	−3.1478	−2.7392	+408.6	+2.0
35°23.4558′	89°49.693′	−23.0645	−23.6097	−23.0638	+545.9	−0.7
0°	89°49.693′	+72.8756	73.2615	72.8683	−393.2	+7.2
−24°36.5442′	89°49.693′	+104.9289	105.7414	104.9180	−823.4	+10.8
−54°36.5442′	89°49.693′	+65.7725	66.4804	65.7666	−663.8	+5.9
35°23.4558′	179°49.693′	−59.1299	−59.9036	−59.1255	+778.1	−4.3
0°	179°49.693′	−44.7004	−44.8495	−44.6996	+149.9	−5.0
−24°36.5442′	179°49.693′	−25.6567	−25.3503	−25.6521	−301.8	−4.6
−54°36.5442′	179°49.693′	−14.0529	−13.7057	−14.0498	−344.1	−3.0

The standard deviations are 0.604 μgal for *Longman*'s [1959] results and 4.85×10^{-3} μgal for the results of the new program.
Universal time at this moment is −40.98 sec from November 5, 1970, 00h 00m 00s ET.
* West of Greenwich.

TABLE 3. Comparison of Computations of Horizontal Lunar Tidal Accelerations on a Rigid Earth on November 5, 1970, 00h 00m 00s ET for 18 Check Points

| Station | | Horizontal Lunar Tidal Accelerations | | | | | |
| Latitude | Longitude* | East Component | | | South Component | | |
		Ephemeris, μgal	New Results, μgal	Difference, 10^{-3} μgal	Ephemeris, μgal	New Results, μgal	Difference, 10^{-3} μgal
90°	0°	63.5688	63.5631	4.7	+21.7461	21.7457	0.4
35°23.4558'	−0°10.307'	73.4410	73.4368	4.2	−14.3166	−14.3148	−1.8
0°	−0°10.307'	46.0205	46.0196	−0.8	−22.2944	−22.2991	−0.3
−24°36.5442'	−0°10.307'	15.9496	15.9515	1.9	−9.3208	−9.3221	1.3
−54°36.5442'	−0°10.307'	−24.3909	−24.3867	4.2	13.7137	13.7118	1.9
−90°	0°	−63.2247	−63.2197	−5.0	21.6284	21.6283	0.1
35°23.4558'	44°49.693'	−10.8552	−10.8519	3.3	7.5658	7.5632	2.6
0°	44°49.693'	−57.4859	−57.4797	6.2	29.2788	29.2752	3.6
−24°36.5442'	44°49.693'	−78.1751	−78.1680	7.1	20.3634	20.3620	1.4
−54°36.5442'	44°49.693'	−83.3295	−83.3229	6.5	−8.4675	−8.4651	2.4
35°23.4558'	89°49.693'	−24.2150	−24.2148	−0.2	68.2894	68.2836	5.8
0°	89°49.693'	−46.5515	−46.5504	−1.1	65.2985	65.2924	6.1
−24°36.5442'	89°49.693'	−51.9540	−51.9524	−1.6	3.6314	3.6314	0.0
−54°36.5442'	89°49.693'	−45.1899	−45.1891	−0.8	−69.9310	−69.9241	−5.9
35°23.4558'	179°49.693'	−1.1681	−1.1658	−2.3	−0.6951	−0.6938	−1.3
0°	179°49.693'	+44.5685	44.5681	0.4	21.5958	21.5957	0.1
−24°36.5442'	179°49.693'	+67.5868	67.5845	4.1	20.0533	20.0521	1.2
−54°36.5442'	179°49.693'	78.2730	78.2689	4.1	−0.0754	−0.0770	2.7

The standard deviations are 4.00×10^{-3} μgal for the east component and 3.00×10^{-3} μgal for the south component.
Universal time at this moment is −40.98 sec from November 5, 1970, 00h 00m 00s ET.
* West of Greenwich.

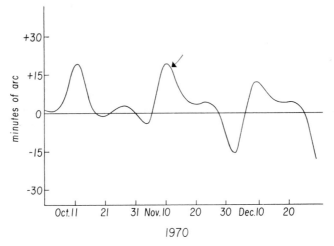

Fig. 1. Difference between the approximate declination of the moon computed by the *Longman* [1959] program (δ Bartels) and the corresponding ephemeris (δ ephemerides) values. The arrow indicates the check point, November 11, 1970, 22h 26m 18.24s ET.

can Ephemeris and Nautical Almanac for the Year 1970 [1968], were inserted into (6) for two chosen ephemeris times, November 11, 1970, 22h 26m 18.24s, and November 5, 1970, 00h 00m 00s. For November 11, g_z was computed for 14 different geographic locations; for November 5, it was computed at 18 locations. The two horizontal components were also computed

for the latter time at the 18 locations. Since the ephemeris values are listed to two more decimal places than were needed in the new computer programs, these direct computations of (6)–(9) are regarded as the exact reference values. For the interpolations involved at the first chosen time, the fourth-degree Bessel interpolation formula described in *The American Ephemeris and*

TABLE 4. Comparison of Computations of Vertical Solar Tidal Accelerations on a Rigid Earth on November 11, 1970, 22h 26m 18.24s ET for 24 Check Points

Station		Vertical Solar Tidal Accelerations, μgal		Difference, μgal
Latitude	Longitude*	Ephemeris	*Longman* [1959]	*Longman* [1959] − Ephemeris
90°	0°	−18.920	−18.927	−0.007
60°56.4′	−0°10.307′	12.184	12.210	0.026
15°56.4′	−0°10.307′	44.078	44.108	0.030
−29°3.6′	−0°10.307′	5.951	5.935	−0.016
−90°	0°	−18.917	−18.927	−0.010
60°56.4′	44°49.693′	−9.291	−9.297	−0.006
15°56.4′	44°49.693′	−8.179	−8.203	−0.024
−29°3.6′	44°49.693′	−22.475	−22.507	−0.032
60°56.4′	89°49.693′	−25.067	−25.089	−0.022
15°56.4′	89°49.693′	−22.176	−22.189	−0.013
−29°3.6′	89°49.693′	−12.020	−12.009	−0.011
60°56.4′	179°49.693′	−23.634	−23.661	−0.027
15°56.4′	179°49.693′	21.747	21.747	0.000
−29°3.6′	179°49.693′	41.837	41.870	0.033

Standard deviation of the differences is 0.022 μgal.
Universal time at this time is 22h 25m 37.128s.
* West of Greenwich.

Nautical Almanac for the Year 1970 [1968] was used.

Table 1 displays comparisons on November 11, 22h 26m 18.24s ET of three computations of the vertical lunar tidal acceleration, the ephemeris exact values, the results of an older commonly used program, and the results of the new program. Of the 14 values by the old program, 9 values have absolute errors between 0.6 and 1.15 μgal. The standard deviation of the 14 errors is 0.750 μgal. The maximum error in the new program is 0.011 μgal, and the standard deviation of its errors is 0.004 μgal. Similarly, for the second 18 stations (Table 2), the respective maximum errors for the old and new programs are 0.968 and 0.011 μgal, and the standard deviations are 0.604 and 0.005 μgal.

In Table 3 the exact horizontal tidal accelerations are compared with those computed by the new program at the second group of 18 stations. For the east component the maximum absolute discrepancy is 0.007 μgal, and the standard deviation of the errors is 0.004 μgal. For the south component the corresponding numbers are 0.006 and 0.003 μgal.

The errors associated with the older computer programs appear to be caused by its imperfect representation of the position of the moon. In Figure 1 the differences between the approximate declination of the moon computed by the old program and the corresponding ephemeris values are plotted for a 3-month period, October–December 1970. These differences periodically rise to about 18 min of arc. Such a discrepancy produces an error in the vertical acceleration approaching 1 μgal when the moon is in the local meridian at zenith angle about 45°. The two chosen times for the comparisons, November 11 and 5, represent times of large and very small discrepancies in the declination, respectively. However, the standard deviations noted earlier for the older program were reduced only from 0.75 to 0.60 μgal by the reduction in the declination error.

For the vertical solar earth tide the commonly used computer programs [*Longman*, 1959] are much more satisfactory, as is illustrated in Table 4 for the first group of 14 locations. The maximum absolute discrepancy is 0.033 μgal, and the standard deviation of the 14 differences is 0.022 μgal. Our present computer program incorporates the older program for computing solar earth tides with the new program for computing lunar tides.

COST

The program computes the three components of the tidal acceleration of the moon and the sun at a given location for 1000 successive times in 12 sec on an IBM 360/91 computer. At the present rates of the University of California at Los Angeles the cost of this computer time is about $6.00.

Oscillating Disk Dynamo and Geomagnetism

E. C. BULLARD AND D. GUBBINS

Department of Geodesy and Geophysics, Cambridge University
Cambridge, England

Higgins and Kennedy (1971) have suggested that the core of the earth may be in stable equilibrium and incapable of supporting large-scale motions with a radial component. This suggestion presents a difficulty for the dynamo theory of the earth's magnetic field. We show that, in a double-disk dynamo, oscillatory motion is able to support a steady electric current and a magnetic field. It seems possible that magnetohydrodynamic waves in the earth's core can support a field having a time scale that is very large in comparison with the period of the waves.

Higgins and Kennedy [1971] have given rather convincing arguments showing that the temperature gradient in the earth's core is less than the adiabatic gradient and that the core is in stable equilibrium. If the core is in stable equilibrium, it can have no large-scale steady motions with a radial component. These arguments present a difficulty for the dynamo theory of the earth's magnetic field, which is usually supposed to require such motions, produced either by thermal convection or by the forces associated with precession [*Malkus*, 1968].

The simplest oscillations of a stable core in the absence of a magnetic field are internal waves. From the results of Higgins and Kennedy it seems that the periods of these waves are likely to be a few hours. Rossby waves would also be possible and would have periods of a few days or possibly of as long as a year. Both these types of waves are probably irrelevant to motions in the earth's core, since the electromagnetic forces will greatly exceed the inertial force and will be comparable with the Coriolis force. *Hide* [1966a, b] has studied the oscillations of a perfectly conducting core in a toroidal magnetic field. He finds rapid eastward-traveling waves with periods of a few days and slow westward-traveling waves with periods of a few hundred years. These oscillations can be considered as magneto-hydrodynamic waves considerably modified by the rotation and the boundaries. All the periods are short in comparison with the decay time for large-scale electric currents in the core (10^4 years) and with the time for which the field is maintained between reversals (10^5–10^6 years).

It might be possible to avoid the difficulties for the dynamo theory by questioning the assumptions of Higgins and Kennedy, for example, their extrapolation of melting-point temperatures to high pressures. Also, it is essential to their argument that the transition from the inner to the outer core be due to melting. Again, it is possible, as von Zeipel's theorem shows [*Wasiutyński*, 1946; *Greenspan*, 1969], that stable stratification in a rotating sphere does not entirely prevent large-scale steady motions. In this paper we do not pursue any of these possibilities but accept the results of Higgins and Kennedy at face value and consider how an oscillatory motion might produce a steady magnetic field.

A number of authors [*Steenbeck et al.*, 1966; *Rädler*, 1968; *Moffatt*, 1970a, b] have shown that turbulence in a conducting fluid can maintain a magnetic field having length and time scales that are much greater than those of the turbulence. *Roberts* [1970] has proved a remarkable theorem that almost all motions periodic in both space and time act as dynamos. He does not discuss the time constants involved, but it seems from his Appendix A that the field has a component constant in time.

To make the nature of the phenomena clear

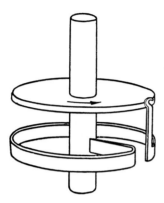

Fig. 1. Disk dynamo.

without the elaborate analysis needed to treat fluid dynamos, we consider a disk dynamo with a prescribed motion. The simplification is that, instead of representing the magnetic field by vectors that are functions of the space coordinates and time, we have one or two scalars, the electric currents in the circuits, which are functions only of time.

SINGLE-DISK DYNAMO

A single-disk dynamo [*Bullard*, 1955] is a disk coupled to a coil as in Figure 1. The disk is here supposed to rotate with a specified velocity (an 'inexorable motion') and not in response to applied forces. It is easily shown that the single disk of Figure 1 will not act as a dynamo if it has an oscillatory motion. The current I must satisfy

$$L\dot{I} + RI = M\Omega I \qquad (1)$$

where L is the inductance of the circuit, R its resistance, $2\pi M$ the mutual inductance between the coil and the periphery of the disk, and Ω the imposed angular velocity. The solution at time t is

$$I = I_0\, e^{-Rt/L} e^{M\theta/L} \qquad (2)$$

where θ is the angle through which the disk has turned since time zero. If θ is a periodic function (i.e., if the disk has no average rotation but only an oscillatory motion), the solution (2) is a periodic function of time, which decays with a time constant L/R. The decay time is thus independent of the motion and the same as that for a stationary disk.

The reason for the failure of the oscillations

to maintain a current is that the long-term average of the right-hand side of (1) is 0. To maintain a current, the product of Ω and the periodic part of I must have a nonzero average. Over a period τ,

$$\int_0^\tau \Omega e^{M\theta/L}\, dt = \int_0^\tau d\theta/dt\; e^{M\theta/L}\, dt = 0$$

if θ is periodic in time.

TWO COUPLED DISK DYNAMOS

If two disk dynamos are coupled, the disk of one feeding the coil of the other [*Rikitake*, 1958; *Allan*, 1962], there is an adjustable parameter, the relative phase of the oscillations of the two disks. This parameter changes the relative phases of the motions and the currents so as to give dynamo action.

For the first disk let the current be I_1 and its angular velocity be Ω_1; for the second disk let them be I_2 and Ω_2. Thus, if the disks and coils are similar,

$$L\dot{I}_1 + RI_1 = M\Omega_2 I_2$$

$$L\dot{I}_2 + RI_2 = M\Omega_1 I_1$$

If both disks perform simple harmonic oscillations with period ω,

$$L\dot{I}_1 + RI_1 = M\Omega_{02} I_2 \cos \omega t$$

$$L\dot{I}_2 + RI_2 = M\Omega_{01} I_1 \cos (\omega t + \phi)$$

where ϕ is the phase difference between the motions of the two disks and Ω_{01} and Ω_{02} are the amplitudes of their angular velocities.

If the variables are changed by the substitutions

$$t = z/\omega$$

$$I_1 = u e^{-Rt/L} / \Omega_{02}^{1/2}$$

$$I_2 = v e^{-Rt/L} / \Omega_{01}^{1/2}$$

$$2q = M\Omega_{01}^{1/2}\Omega_{02}^{1/2}/L\omega$$

the equations become

$$du/dz = 2qv \cos z$$

$$dv/dz = 2qu \cos (z + \phi)$$

For $\phi = 0$ or π the solutions are similar to (2) and give no dynamo action. We therefore take $\phi = -\frac{1}{2}\pi$ and get

$$du/dz = 2qv \cos z$$

$$dv/dz = 2qu \sin z \tag{3}$$

Floquet's theorem [*Whittaker and Watson*, 1927] shows that there are solutions of the form

$$u = e^{\mu z} U(z)$$

$$v = e^{\mu z} V(z) \tag{4}$$

where U and V are periodic in z with period 2π.

$$\frac{iq^2[\mu + i(n + 1)]}{(\mu + in)[(\mu + in)^2 + 1] + 2q^2} \quad 1 \quad \frac{-iq^2[\mu + i(n - 1)]}{(\mu + in)[(\mu + in)^2 + 1] + 2q^2}$$

The value of the constant μ in (4) must be determined from the differential equations (3). Suppose that U and V are expanded in Fourier series:

$$U = \sum_{n=-\infty}^{\infty} a_n e^{inz}$$

$$V = \sum_{n=-\infty}^{\infty} b_n e^{inz}$$

when these are substituted in (3), μ, a_n, and b_n can be determined. If one of the currents is to have a steady or growing component, μ must be positive and $> R/L\omega$, and a_0 or b_0 must be nonzero.

Substitution in (3) gives recurrence relations for a_n and b_n:

$$(\mu + in)a_n = q(b_{n+1} + b_{n-1})$$

$$(\mu + in)b_n = q(a_{n+1} - a_{n-1}) \tag{5}$$

All the coefficients b_n can be eliminated from (5), and thus

$$\frac{iq^2 a_{n-2}}{\mu + i(n - 1)} + \left[\mu + in + \frac{iq^2}{\mu + i(n + 1)}\right.$$

$$\left. - \frac{iq^2}{\mu + i(n - 1)}\right]a_n - \frac{iq^2 a_{n+2}}{\mu + i(n + 1)} = 0 \tag{6}$$

These equations fall into two sets, one for even n and one for odd n. If there is to be a nonzero solution for a_n, μ must be chosen so that one of the sets of coefficients has a zero determinant. When μ and the a_n have been found from (6), the b_n are given by (5). The coefficients b_n for even n are only nonzero if the a_n for odd n are nonzero, and a similar situation exists for the odd b_n and the even a_n. Thus we have two values

of μ and two sets of coefficients, one with all even a_n and odd b_n being 0 and the other with all odd a_n and even b_n being 0. Identical results are obtained by eliminating b_n, and it is not necessary to consider separately the determinants of the coefficients of the b_n.

The determinants are all tridiagonal. If factors are removed to make the diagonal terms 1, then, provided that $\mu \neq 0$, the nonzero terms in row n are

where for one determinant

$$n = \cdots -2, 0, 2 \cdots$$

and for the other

$$n = \cdots -3, -1, 1, 3 \cdots$$

For large n the off-diagonal terms are $O(n^{-2})$, which is sufficient to make the determinants convergent [*Whittaker and Watson*, 1927].

The computation of the values of μ for which the determinants vanish is greatly simplified if q is small. For small q it can be shown that μ is also small and $O(q^2)$. In the physical problem, q and μ are indeed small. If the dynamo is to give a nonzero average current, the ohmic decay $e^{-Rt/L}$ must be balanced by the term $e^{\mu\omega t}$ in (4); that is

$$\mu = R/L\omega$$

Now L/R is the time of ohmic decay, which in the earth's core is about 10^4 years; $2\pi/\omega$ is the period of oscillation of the disks, of which the analogy in the earth is the period of internal waves of perhaps between 10^2 and 10^{-2} years. Thus μ can be expected to be between 10^{-3} and 10^{-7} years, and q^2 can be expected to be of the same order. If the field grows with a time constant that is large in comparison with the period of oscillation, say a few hundred years (10^4–10^5 times the period of oscillation), μ can exceed $R/L\omega$ but will still be much less than 1. It is therefore reasonable to expand the determinants in powers of q^2 and μ and retain only the leading terms.

If the leading term in the determinant for odd a_n is equated to 0, it is found that

$$\mu = 2q^2 + O(q^4) \tag{7}$$

The recurrence relations (5) then give

$$a_1 = a_{-1} = iq b_0 + O(q^3 b_0)$$

The odd a_n and the even b_n each decrease as q^2 and are small in comparison with b_0, a_1, and a_{-1}. These values of μ and the coefficients give as a solution of (3)

$$\mu = 2q b_0 \exp (2q^2 z) \sin z + O(q^3 b_0)$$
$$v = b_0 \exp (2q^2 z) + O(q^2 b_0) \tag{8}$$

The other solution can be obtained from the determinant from the even a_n. It gives

$$\mu = -2q^2 + O(q^4)$$

This solution decays with time and is of no interest for the present purpose. The relevant solution (8) gives for the currents

$$I_1 = b_0 M \Omega_{01}^{1/2} (L\omega)^{-1} e^{\alpha t} \sin \omega t$$
$$I_2 = b_0 \Omega_{01}^{1/2} e^{\alpha t}$$

where

$$\alpha = \tfrac{1}{2} M^2 \Omega_{01} \Omega_{02} [L^2 \omega - (R/L)]$$

We thus have the unexpected result that the current I_1 is oscillatory but the current I_2 is not. Current I_2 and the amplitude of I_1 can be stationary, can decay, or can grow exponentially. The oscillating current is small $O(q)$ in comparison with the steady current; this situation is expected, since the steady current decays very little during one period of oscillation of the disks and a small oscillating current can maintain it. The terms neglected in (8) are harmonics of the periodic term and have amplitudes at most $O(q^2)$ in relation to the leading terms. If the phase difference ϕ had been taken as $+\tfrac{1}{2}\pi$, the behavior of I_1 and I_2 would have been interchanged.

The same method can be used for solving the more general problem with disk motions $\Omega_{01} \cos \omega t$ and $\Omega_{02} \cos (\omega t + \phi)$. Again, growing solutions for the electric current can exist with

$$\mu = 2q^2 |\sin \phi|$$

The problem can also be solved for completely general periodic disk motions. Solutions with a positive μ are again possible, provided that for at least one of the harmonics the motion of one disk has a cosine dependence on time and the motion of the other disk has a sine dependence.

It seems very probable that spherical fluid dynamos with similar properties will exist, and it may be that magnetohydrodynamic waves in the earth's core can maintain a field, even if the material is stably stratified and the period of the waves is small on the time scale of geomagnetism.

We have not considered disk dynamos driven by periodic forces; it would be interesting to know if the field would reverse on a time scale that is long in comparison with the period of the forces. It seems unlikely that the possibility of such a reversal could be shown analytically, and there would be difficulties in treating numerically the important case where the forces oscillate many times between reversals.

Notes on Geyser Temperatures in Iceland and Yellowstone National Park[1]

Francis Birch

Department of Geological Sciences, Harvard University
Cambridge, Massachusetts   02138

George C. Kennedy

Institute of Geophysics and Planetary Physics
University of California, Los Angeles, California   90024

Temperatures, at various depths and at various times before eruption, have been measured at Great Geysir, Iceland, and at Old Faithful Geyser, Lion Geyser, Great Fountain Geyser, and Sapphire Pool, all of Yellowstone Park, Wyoming. Geyser action is initiated when bodies of hot water convectively overturn and cross the boiling-point curve at relatively shallow levels. This process produces a body of steam that displaces the overlying water, reduces pressure on the system, and causes subsequent flash boiling to greater depths. The geysers we have examined all show similar thermal regimes before eruption. An upper regime of cooler water is separated by a mixing zone from a deeper regime of hotter water. Boiling and subsequent eruption is normally initiated in the mixing zone.

Recent interest in geothermal regions has included a revival of curiosity about geysers, notable features of a few of the hot-spring areas of Iceland, New Zealand, and the western United States. Measurements in deep boreholes in these areas have greatly increased our understanding of the underground thermal regimes, and the discussions by *Benseman* [1965], *White* [1967, 1968], *White et al.* [1971], *Rinehart* [1968], and *Rinehart and Murphy* [1969] provide a generally satisfactory theory of geyser action. In view of the current interest, it seems worthwhile to publish measurements made in Iceland in 1947 and in Yellowstone National Park in 1948, hitherto presented only orally and pictorially [*Birch et al.*, 1949; *Graton*, 1949; see also *Barth*, 1950, p. 67]. The measurements made in Yellowstone Park show conditions before the Hebgen Lake earthquake of 1959. They were made with a multielement string of thermocouples with short response times. Temperatures were recorded every few seconds, and the records give a more complete representation of short-term changes in temperature, both in space and in time, than has yet been obtained elsewhere in natural geysers. The measurements in the Great Geysir of Iceland have been repeated and improved by *Sigurgeirsson* [1949] and *Barth* [1950, p. 68] and are of interest because of the unusual accessibility and symmetry of this historic geyser, now dormant for several years.

This work was initiated and largely carried out by the late L. C. Graton as part of his life-long study of the 'ore-forming fluid'. In 1942 Graton and Esper S. Larsen, Jr., plumbed some of the geysers in Yellowstone Park with an electrical resistance thermometer and were impressed by the fluctuations of temperature with time at fixed depths. In 1947 Graton and F. Birch, with the collaboration of the late Steinthor Sigurdsson, logged temperatures in the Great Geysir with the resistance thermometer and with various combinations of maximum thermometers. Again the fluctuations were notable, and it became clear that instruments with shorter time constants were needed to fol-

[1] Publication 954, Institute of Geophysics and Planetary Physics, University of California, Los Angeles, California   90024.

low large variations occurring within seconds. In 1948 L. C. Graton, F. Birch, and G. C. Kennedy undertook to record temperatures in several Yellowstone Park geysers using instruments with fast responses, and their results are presented below.

GREAT GEYSIR OF ICELAND

Great Geysir in Haukadalur, Iceland, the prototype that has given its name to erupting hot springs, has been described fully in the works of *Einarsson* [1937, 1942, 1949, 1967] and in those of *Barth* [1940, 1950], who gives excellent photographs of several stages of the cycle. A mound of siliceous sinter has been built to a height of some 6 meters and a diameter of about 100 meters. At the top is a shallow basin about 16 meters in diameter and about 1 meter deep at the top of the geyser tube; this tube is nearly circular and has a diameter of 3 meters to a depth of about 20 meters [*Einarsson*, 1937]. Great Geysir became dormant in 1915 but was rejuvenated in 1935 by Einarsson, who cut a narrow gate in the rim of the basin. With this gate open the water level, when full, stood about 60 cm below its former level. Normally, in 1947 the gate was kept closed, but, when an eruption was wanted, usually on Sunday afternoons, the level was lowered by removal of the gate, a large quantity of soft soap was introduced, and after a short interval a large eruption emptied the tube and covered the mound with soapsuds. The tube, having a visible volume of about 150 m³, refilled in 10–12 hours with a mean rate of inflow of about 4 l/sec. When the tube was filled only to the level of the open gate, small spontaneous eruptions occurred at intervals of several hours; after each of these eruptions the water level was lowered by 3–6 meters.

Measurements with mercury maximum thermometers and with a Leeds and Northrup nickel resistance thermometer were made during the week of June 29 to July 4, 1947. This procedure allowed observations before and after the eruption on Sunday, June 29, and subsequently during a period in which several small spontaneous eruptions occurred. A line was rigged across a diameter of the geyser tube, and the thermometers were lowered as nearly as possible on the center, where the hottest water is found. The bare mercury ther-

mometers had time constants of about 1 sec, which could be increased to several minutes by enclosing the thermometers in heavy rubber tubing. By lowering bare thermometers and insulated thermometers together and leaving them in position for 5 min it was possible to obtain a rough measure of the amplitude of fluctuations of temperature with short periods; the full amplitude must have been at least twice as great as the observed difference. The resistance thermometer had a time constant of about 1 min and also showed significant fluctuations until the cable was ruptured by a violent surge.

Measurements with the water at a high level are shown in Figure 1. There had been a large induced eruption on June 29 and a smaller one that lowered the water level about 5 meters on July 1. *Barth's* [1940, p. 396; 1950, p. 67] measurements in 1937 show a slightly cooler geyser. The boiling curve is shown for reference, adjusted to a water level at overflow with the gate in place.

Measurements with the water at a low level are shown in Figure 2. The measurements of this work in the figure indicate maximum temperatures recorded on June 29 on repeated lowerings to 10 meters; at 15 meters, repeated soundings showed a virtually unchanging temperature of 124°C. Soap was then added; temperatures remained about the same until an eruption emptied the tube. The water level regained its low level overnight, and another profile was observed with bare maximum thermometers on June 30. Measurements with the combination of the bare and the insulated thermometers made on July 1 are shown. A small eruption occurred spontaneously shortly after these measurements were completed. For comparison, *Einarsson's* [1937] measurements for 1935 are also shown.

Sigurgeirsson [1949] repeated these observations with thermistors with time constants of about ½ sec and found a similar pattern of fluctuation, greatest at depths of about 9 meters, where the temperature changed by as much as 10°–20° in a few seconds.

The accessible portion of Great Geysir thus shows three distinct regions, each about 6 meters in vertical extent. In the uppermost region, which includes the shallow basin, the temperature remains between 80° and 90°C, and little fluctuation occurs. In the lowest region the temperature remains close to 120°C, and little fluctuation occurs. In the central region, violent mixing takes place between the upper cold and lower hot masses, and large fluctuations occur. Boiling and condensation occur when rising hot water reaches the saturation pressure. The presence of variable amounts of steam accounts for the characteristic surging of the water surface. The imminence of an eruption is signaled by an increased displacement and overflow of water and finally by a striking doming of the water surface. Since the mean temperature of the water before eruption is approximately 110°C, there is enough thermal energy in the hot water to vaporize nearly 2% when the pressure is reduced to the atmospheric level. There seems to be no necessity to postulate large invisible storage chambers for water or steam. The initiation of eruption appears to take place at the depth of about 10 meters, where several degrees of superheat have been observed, as is shown in Figure 2, by points above the reference boiling curve. The rapid fluctuations demonstrated in this highly accessible geyser led to the return to Yellowstone Park with more appropriate instruments.

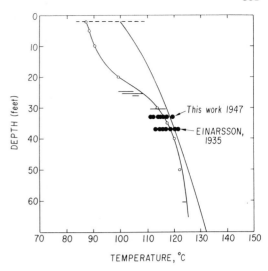

Fig. 2. Temperatures in Great Geysir, Iceland, with water at low level. Open circles, readings with the bare maximum thermometers on June 30; bars, ranges observed with the combination of the bare and the insulated maximum thermometers on July 1; dashed line at top, the water level; solid undotted curve, the reference boiling curve.

TEMPERATURES IN YELLOWSTONE GEYSERS

In 1948 a six-channel 'Speedomax' Leeds and Northrup recording potentiometer became available. It printed one reading every 4 sec. A cable was made with six duplex neoprene-insulated iron-constantan thermocouples with the junctions spaced at 3-meter intervals and a total length of 45 meters. With this combination a temperature profile spanning 15 meters could be recorded every 24 sec. Part of a record for Old Faithful Geyser is reproduced as Figure 3. The numbers on the record, ranging from 1 to 6, are printed automatically and identify the junction being read. The potentiometer was recorded on the Fahrenheit scale, from 0° to 350°F, and was frequently calibrated at the steam point (about 199°F or 93°C at the elevation of Old Faithful).

The Old Faithful and Lion Geysers are surmounted by sinter cones with relatively small apertures, which prevent visual observation of the water level and interfere with free placement of the cable. Before lowering the thermocouple cable into these geysers, we plumbed as far as possible with a small weight and a light line. Various constrictions and obstructions were

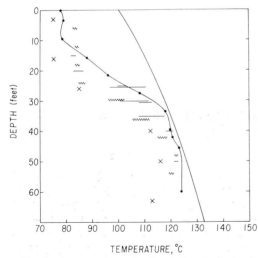

Fig. 1. Temperatures in Great Geysir, Iceland, with water at high level. Solid dots, readings with the bare maximum thermometers on June 29; bars, ranges observed with the combination of the bare and the insulated maximum thermometers on July 2; zigzag pattern, ranges with resistance thermometer on July 3; crosses, measurements by *Barth* [1940] with maximum thermometers; solid undotted curve, the reference boiling curve.

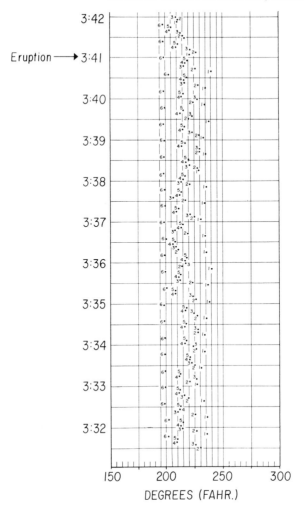

Fig. 3. Speedomax record of temperature versus time for Old Faithful Geyser, Yellowstone Park, August 23, 1948.

found. In Lion Geyser a constriction at 10 meters allowed passage of a lead plummet of 1-inch diameter but not one of 1.25-inch diameter. Past this point the line paid out freely to a maximum depth of 24 meters. Old Faithful Geyser had a constriction or ledge at about 11 meters, but after passing this point, the cable paid out to about 21 meters without difficulty. The entire bundle of cable, with its half-pound weight, was thrown clear several times during eruptions. *Marler* [1953] notes a fissure 10–20 cm wide at about 3 meters below the top of the cone. *Rinehart* [1969] reports single thermistor measurements as deep as 175 meters in Old Faithful.

OLD FAITHFUL GEYSER

Seven complete eruptive cycles were recorded for Old Faithful Geyser. Early in the first cycle the thermocouple junction at the 10-meter level (number 5 on the chart) failed, and consequently the profiles were determined by five measurements per cycle instead of six. Since all seven cycles observed at Old Faithful Geyser had essentially similar features, the details of only one, the cycle of 4:45 P.M. on August 23, 1948, are shown (Figures 4 and 5). The eruption preceding this cycle took place 64 min earlier, and our record of it is shown in Figure 3. The mean interval for 1540 eruptions in 1949 was 63.8 min.

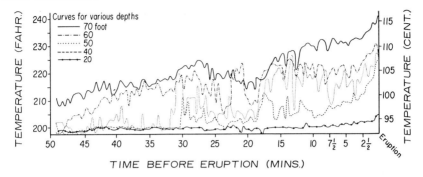

Fig. 4. Eruption of 4:45 P.M., August 23, 1948, Old Faithful Geyser, Yellowstone Park.

Figure 4 shows all the readings during the 50 min preceding the eruption. Rapid fluctuations in temperature at a given junction are noteworthy and could not have been followed with resistance thermometers with long thermal time constants. The cable was emplaced about 15 min after the eruption of 3:41 P.M. (Figure 3); the water then covered only the two lowest junctions, at 18 and 21 meters below the cone. Junction 3 at the 15-meter level was intermittently in steam and water for the first minutes of observations (inferred from temperature fluctuations; temperatures near 94°–95°C are assumed to indicate steam, and temperatures much above 95°C require pressures above the atmospheric level; thus some pres-

sure by liquid water is implied). The rising waters submerged the junction at the 12-meter level (junction 4) approximately 30½ min before the eruption, and junction 3 was continuously submerged about 32 min before eruption. The junction at 6 meters (junction 6) was not continuously submerged until about 6 min before the eruption. Measurements with a wooden float indicated that from 30 to 10 min before the eruption the surface of the water was at a depth of about 11 meters, although rapid fluctuations were noted. The position of the reference 'boiling-point curve' in these geysers, in which the water surface was invisible, is probably uncertain by several meters and certainly fluctuates rapidly; in addition, the curve

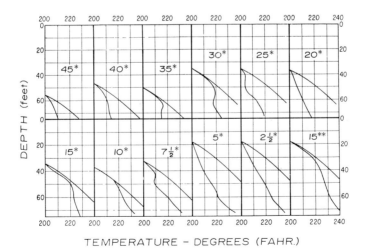

Fig. 5. Profiles of temperature versus depth at Old Faithful Geyser at various times before eruption. The depth to water, the temperature distribution in the water, and the reference boiling curve are shown. One asterisk, minutes before the eruption; two asterisks, seconds before the eruption.

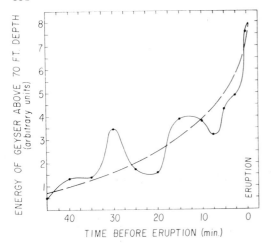

Fig. 6. Estimated available energy in the tube of Old Faithful Geyser at various times before eruption. Dashed curve, the mean increase in the energy of the system; solid curve, data points taken from the profiles shown in Figure 6.

assumes a single liquid phase at all levels, not strictly valid for the top of a boiling and surging mixture of vapor and liquid.

Figure 5 shows temperature versus depth for various times before eruption. Clearly, there is no unique temperature-depth profile, as sometimes assumed. Estimated available energy in the geyser tube as a function of time is shown in Figure 6; we have assumed that the geyser tube is of uniform cross section and that at any moment the observed temperatures sample the entire cross section of a measured depth. Both assumptions obviously are only approximately true. The excess of temperature over the surface boiling point has been integrated over the entire water column. Despite the crudity of the process, the results show a striking general increase in the energy of the system from the end of one eruption to the initiation of the next one. The increase reflects in part the rise in the water level surface and in part the rise of water temperature, particularly during the last few minutes before eruption.

The depth-temperature profiles in Figure 5 show that the eruption was probably initiated in the uppermost 5 meters of the water column. Throughout most of the time between eruptions, temperatures were everywhere below the boiling-point curve. However, in every one of the seven records the temperatures in the top

few meters of water increased rapidly during the last few seconds, and thus the temperature-depth curve and the boiling-point curve were brought practically into coincidence. Boiling and overflow then occurred as masses of hot water were convected above the boiling-point curve. The resultant decrease in pressure as water was displaced from the top of the geyser tube by steam formation precipitated the eruption.

LION GEYSER

Nearly 2 days of recording, including one of the relatively rare eruptions, were taken at Lion Geyser. A sample of the record is shown in Figure 7. The eruption took place at 12:15 P.M., soon after the cable was lowered into the geyser tube. The lowest junction reached an apparent depth of about 24 meters below the top of the vent. Figure 8 shows three temperature-depth profiles and the ideal boiling curve. Curve A represents the temperature distribution approximately 1 min before the eruption; curve C is the last record, eruption having already begun. The cable was replaced, and curve B shows the temperatures at 7:25 P.M.; the

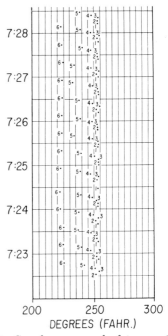

Fig. 7. Speedomax record of temperature versus time in Lion Geyser, Yellowstone Park, at 24 meters.

cable was raised to complete the profile. The tube slowly filled and warmed, and the water level returned to approximately 5 meters below the surface vent 3 hours after the eruption; the temperature at 24 meters had then reached 116°C. The fluctuations of temperature in Lion Geyser are smaller than those in Old Faithful, and the deep temperatures are higher. The relationship between temperature and depth is close to that observed by *Allen and Day* [1935].

GREAT FOUNTAIN GEYSER

Great Fountain Geyser is an open pool about 12 meters deep and roughly 4 × 6 meters in plan. It appears to be fed from a small vent at an estimated depth of 12 meters into which it was not possible to lower the thermocouple cable. The temperature of the pool varied from close to the boiling point at the surface to about 95°C at 12 meters. The temperature of the water at 6 meters fluctuated between 95° and 97°C. In the pool there appears to be convective sinking of the cooled surface water, which flows along the bottom of the pool toward the inlet, where incoming hot water creates an upward current. Unless measurements are made close to the incoming current, the temperatures near the bottom of the pool may be even lower than those at intermediate depths. Reports of profiles showing temperature decreasing slightly with depth in geyser tubes or pools represent only one aspect of a complex pattern of circulation in three dimensions.

At the time of our visit Great Fountain was erupting at intervals of about 12 hours. For several hours before eruption the temperature of the pool remained fairly uniform; in the last few moments before eruption, rapid overflow from the basin took place, and the bottom temperature increased rapidly from 97° to 107°C. This increase was followed by a sudden surge of boiling water, and eruption continued as intermittent fountain play for some 45 min. The eruption probably commences in the narrow orifice somewhere below the open pool; present observations do not fix the point of initiation. The displacement and the rapid rise in temperature of so large a volume of water just before eruption demand the existence of feeding channels and a reservoir. The pool is not emptied during eruption, but its level is lowered by some 3 meters.

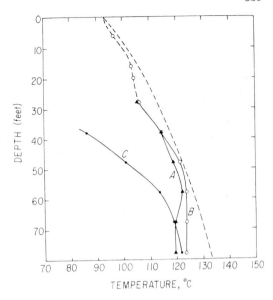

Fig. 8. Depth-temperature profiles in Lion Geyser on August 24, 1948. Curve A, the temperature distribution approximately 1 min before the eruption; curve B, the temperature 7 hours and 10 min after the eruption; curve C, the temperature as the eruption began; dashed curve, the reference boiling curve.

SAPPHIRE POOL

Sapphire Pool shows features similar to those of Great Fountain Geyser on a somewhat smaller scale. Its depth is about 10 meters, and its diameter 8 meters. Small eruptions lasting 1 or 2 min occurred at intervals of about 12 min. Steam bubbles could be seen entering the pool through the feeding vents just before the eruptions. Temperatures at the surface ranged from 93° to 95°C between eruptions, and temperatures at the depth of 6 meters were several degrees higher than those observed near the bottom. A marked overflow occurred just before the eruptions, and during the last few seconds the bottom temperature rose to an observed maximum of 106°C. The initiation of eruption is below the visible region and probably in a relatively small channel.

It should be emphasized that these observations were made before the Hebgen earthquake of 1959. After the earthquake Sapphire Pool became a major geyser and greatly enlarged its tube.

DISCUSSION

Recent papers by R. F. Benseman, D. E. White, and others, together with the monographs by *Allen and Day* [1935] and *Barth* [1950], leave little to be added on the geyser mechanism. The work reported here documents the extremely rapid convective motions that have been inferred by other observers. It is difficult, however, to reconcile our readings of temperature in Old Faithful Geyser with those of *Rinehart* [1969]; his Figure 10, for example, shows a pre-eruption temperature at the 23-meter level of about 97°C, whereas at the same presumed depth we consistently found 112°–116°C (235°–240°F, Figures 4 and 5) for 10 min or more before eruptions. There is little control over the positioning of cables in geysers with narrow apertures, and the simplest explanation may be that Rinehart's thermistor was hung at a shallower depth than 23 meters. The same explanation may account for the low temperatures shown in his Figures 1, 2, 3, and 11.

By contrast, the broad open tube of Great Geysir, with its visible water surface, affords an ideal example of convective behavior in the presence of a mean temperature gradient of roughly 2°C per meter. This gradient is several thousand times greater than the adiabatic gradient in liquid water at 100°C, and vigorous overturning and mixing are to be expected. But water from the deeper levels cannot reach the surface along an adiabat without crossing the saturation curve; whereupon it must either boil or persist as superheated water. The actual regime in Great Geysir shows three remarkably distinct regions, each having a depth of about twice the diameter of the tube: adiabatic (virtually uniform) temperatures in the upper and lower cells, separated by a mixing cell, with large fluctuations in which the whole rise of temperature is concentrated. Similar high gradients in less accessible geysers have been ascribed to unobservable constrictions of the tube, but Great Geysir shows that this interpretation is not necessary. It seems possible that there may be more than one mixing zone, as was suggested by a figure given by *White* [1967, p. 677, Figure 14]; and, indeed, in a tube with a smaller diameter one would expect, with adequate instrumentation, to find numerous mixing cells with smaller individual increments of temperature.

The present work supports the premise that eruptions are initiated by production at relatively shallow levels of a volume of steam sufficiently large to cause overflow, reduction of pressure, and subsequent flash boiling to greater depths. Self-boiling can vaporize only a small fraction of the hot water; what is not ejected returns or remains behind, depleted of its excess thermal content and, possibly with the addition of cooler ground water, it restores the pressure. Many geysers, such as Great Fountain, play in pools that are not emptied but merely are lowered by the eruptions.

Acknowledgments. For the opportunity to work in Iceland we are much indebted to the Icelandic authorities. Steinthor Sigurdsson of the National Research Institute of Iceland took part in all the observations; his participation was indispensable. We had the benefit of discussion with Professors T. Thorkelsson and Trausti Einarsson. The kindness of the U.S. Consul, Mr. William Trimble, and his staff in Reykjavik is not forgotten. For the translation of several papers from Icelandic, we thank Mr. S. Norland. In the Yellowstone National Park, our presence created many problems for the Park officials, and we are especially grateful to the Superintendent, Mr. Edmund B. Rogers, the Chief Park Naturalist, Mr. David Condon, and to Park Ranger George D. Marler for their forbearance as well as active assistance. We are greatly indebted to Mr. G. D. Robinson, U.S. Geological Survey, for the use of a pick-up truck for transporting our equipment and to the Leeds and Northrup Company for the loan of a Speedomax recording potentiometer. We are much indebted to Dr. Donald E. White for constructive comment on this manuscript.

This work was performed under the auspices of the Committee on Experimental Geology and Geophysics at Harvard University.

References

Adams, L. H., and E. D. Williamson, On the compressibility of minerals and rocks at high pressures, *J. Franklin Inst., 195,* 475–529, 1923.

Ahrens, T. J., and V. G. Gregson, Shock compression of crystal rocks: Data for quartz, calcite, and plagioclase rocks, *J. Geophys. Res., 69,* 4839–4874, 1964.

Ahrens, T. J., and J. T. Rosenberg, Shock metamorphism: Experiments on quartz and plagioclase, in *Shock Metamorphism of Natural Materials,* edited by B. M. French and N. M. Short, pp. 58–82, Mono, Baltimore, 1968.

Ahrens, T. J., C. F. Petersen, and J. T. Rosenberg, Shock compression of feldspars, *J. Geophys. Res., 74,* 2727–2746, 1969a.

Ahrens, T. J., D. L. Anderson, and A. E. Ringwood, Equations of state and crystal structures of high-pressure phases of shocked silicates and oxides, *Rev. Geophys. Space Phys., 1,* 667–707, 1969b.

Aki, K., Some problems in statistical seismology, *Jishin, 8,* 205–228, 1956.

Alexander, H., and P. Haasen, Dislocations and plastic flow in the diamond structure, *Solid State Phys., 22,* 27–158, 1968.

Alexsandrov, K. S., and T. V. Ryzhova, The elastic properties of rock-forming minerals, 1, Pyroxenes and amphiboles, *Bull. Acad. Sci. USSR, Geophys. Ser.,* Engl. Transl., no. 9, 871–875, 1961.

Alexsandrov, K. S., T. V. Ryzhova, and B. P. Belikov, The elastic properties of pyroxenes, *Sov. Phys. Crystallogr., 8,* 589–591, 1964.

Allan, D. W., On the behavior of systems of coupled dynamos, *Proc. Cambridge Phil. Soc., 58,* 671–693, 1962.

Allen, C. R., B. Kamb, M. F. Meier, and R. P. Sharp, Structure of the lower Blue Glacier, Washington, *J. Geol., 68,* 601–625, 1960.

Allen, E. T., and A. L. Day, Hot springs of the Yellowstone National Park, *Carnegie Inst. Wash. Publ. 466,* 525 pp., 1935.

Allingham, J. W., and I. Zietz, Geophysical data on the Climax Stock, NTS, Nye Co., Nevada, *Geophysics, 28,* 599, 1962.

Amelinckx, S., Etchpits and dislocations along grain boundaries, sliplines, and polygonization walls, *Acta Met., 2,* 848–853, 1954.

Anderson, D. L., Latest information from seismic observations, in *The Earth's Mantle,* edited by T. F. Gaskell, pp. 354–420, Academic, New York, 1967.

Anderson, D. L., C. Sammis, and T. Jordan, Composition and evolution of the mantle and core, *Science, 171,* 1103–1112, 1971.

Anderson, O. L., and R. Liebermann, Sound velocities in rocks and minerals, *Vesiac Rep. 7885-4-x,* Willow Run Lab., Univ. of Mich., Ann Arbor, p. 138, 1966.

Andreatta, C., Analisi strutturali di rocce metamorfiche, 5, Olivinite, *Period. Mineral., 5,* 237–253, 1934.

Arrhenius, S., *Zur Physik der Salzlagerstatten,* vol. 2, pp. 1–25, Meddelanden Vetenskapsakademiens Nobelinstitut, 1912.

Aschner, J. F., Self-diffusion in sodium and potassium chloride, Ph.D. dissertation, 90, pp., Univ. of Ill., Urbana, 1954.

Avé Lallemant, H. G., Structural and petrofabric analysis of an 'Alpine-type' peridotite, The lherzolite of the French Pyrenees, *Leidse Geol. Meded., 42,* 1–57, 1967.

Avé Lallemant, H. G., and N. L. Carter, Syntectonic recrystallization of olivine and modes of flow in the upper mantle, *Geol. Soc. Amer. Bull., 81,* 2203–2220, 1970.

Bader, H., Introduction to ice petrofabrics, *J. Geol., 59,* 519–536, 1951.

Baëta, R. D., and K. H. G. Ashbee, Mechanical deformation of quartz, 1, Constant strain-rate compression experiments, *Phil. Mag., 22,* 601–623, 1970.

Baker, D. W., On the symmetry of orientation distribution in crystal aggregates, *Advan. X-Ray Anal., 13,* 435–454, 1970.

Baker, D. W., and N. L. Carter, Seismic velocity anisotropy calculated for ultramafic minerals and aggregates, this volume.

Baker, D. W., and H.-R. Wenk, Preferred orientation in a low-symmetry quartz mylonite, *J. Geol., 80,* 81–105, 1972.

Baker, D. W., H.-R. Wenk, and J. M. Christie, X-ray analysis of preferred orientation in fine-grained quartz aggregates, *J. Geol., 77,* 144–172, 1969.

Banthia, B. S., M. S. King, and I. Fatt, Ultrasonic shear-wave velocities in rocks subjected to simulated overburden pressure and internal pore pressure, *Geophysics, 30,* 117–121, 1965.

Barber, D. J., Thin foils of non-metals made for electron microscopy by sputter-etching, *J. Mater. Sci., 5,* 1–8, 1970.

Barr, L. W., I. M. Hoodless, J. A. Morrison, and R. Rudham, Effects of gross imperfections on chloride ion diffusion in crystals of sodium chloride and potassium chloride, *Trans. Faraday Soc., 56,* 697–708, 1960.

Barrett, C. S., and T. B. Massalski, *Structure of Metals,* McGraw-Hill, New York, 654 pp., 1966.

Barth, T. F. W., Geysir in Iceland, *Amer. J. Sci., 238,* 381–407, 1940.

Barth, T. F. W., Volcanic geology, hot springs and geysers of Iceland, *Carnegie Inst. Wash. Publ., 587,* 67, 1950.

Bell, R. T., and J. B. Currie, Photoelastic experiments related to structural geology, 2, *Proc. Geol. Ass. Can., 15,* 33–51, 1964.

Benioff, H., Seismic evidence for the fault origin of oceanic deeps, *Geol. Soc. Amer. Bull., 60,* 1837–1856, 1949.

Bennett, E. M., Lead-zinc-silver and copper deposits of Mount Isa, in *Geology of Australian Ore Deposits,* vol. 1, *8th Commonwealth Mineralogy and Metallurgy Congress,* 2nd ed., edited by J. McAndrew, pp. 233–246, Australia–New Zealand, 1965.

Benseman, R. F., The components of a geyser, *N. Z. J. Sci., 8,* 24–44, 1965.

Berg, C. A., The diffusion of boundary disturbances through a non-Newtonian mantle, *Application of Modern Physics to the Earth and Planetary Interiors,* edited by S. K. Runcorn, Wiley-Interscience New York, pp. 253–272, 1969.

Berner, K., and H. Alexander, Versetzungsdichte und lokale Abgleitung in Germaniumeinkristallen, *Acta Met., 15,* 933–941, 1967.

Biot, M. A., Theory of folding of stratified viscoelastic media and its implications in tectonics and orogenesis, *Geol. Soc. Amer. Bull., 72,* 1595–1620, 1961.

Biot, M. A., Theory of internal buckling of a confined multilayered structure, *Geol. Soc. Amer. Bull., 75,* 563–568, 1964.

Biot, M. A., Theory of similar folding of the first and second kind, *Geol. Soc. Amer. Bull., 76,* 251–258, 1965a.

Biot, M. A., Theory of viscous buckling and gravity instability of multilayers with large deformation, *Geol. Soc. Amer. Bull., 76,* 371–378, 1965b.

Biot, M. A., Further development of the theory of internal buckling of multilayers, *Geol. Soc. Amer. Bull., 76,* 833–840, 1965c.

Biot, M. A., *Mechanics of Incremental Deformations,* John Wiley, New York, 504 pp., 1965d.

Biot, M. A., H. Odé, and W. L. Roever, Experimental verification of the theory of folding of stratified viscoelastic media, *Geol. Soc. Amer. Bull., 72,* 1621–1632, 1961.

Birch, F., The velocity of compressional waves in rocks to 10 kilobars, 2, *J. Geophys. Res., 66,* 2199–2224, 1961.

Birch, F., Megageological considerations in rock mechanics, in *State of Stress at the Earth's Crust,* edited by W. R. Judd, pp. 55–80, Elsevier, New York, 1964.

Birch, F., Compressibility; elastic constants, in *Handbook of Physical Constants, Mem. 97,* rev. ed., edited by S. P. Clark, Jr., pp. 97–173, Geological Society of America, Boulder, Colo., 1966.

Birch, F., and G. C. Kennedy, Notes on geyser temperatures in Iceland and Yellowstone National Park, this volume.

Birch, F., J. F. Schairer, and H. C. Spicer, *Handbook of Physical Constants, Spec. Pap. 36,* 325 pp., Geological Society of America, Boulder, Colo., 1942.

Birch, F., L. C. Graton, and G. C. Kennedy, Temperatures of geysers in Iceland and Yellowstone Park (title only), *Eos Trans. AGU, 30,* 471, 1949.

Blacic, J. D., Hydrolytic weakening of quartz and olivine, Ph.D. thesis, Univ. of Calif. at Los Angeles, 205 pp., 1971.

Blacic, J. D., Effect of water on the experimental deformation of olivine, this volume.

Blum, W., and B. Ilschner, Über das Kriechver-

halten von NaCl-Einkristallen, *Phys. Status Solidi, 20,* 629–642, 1967.

Boardman, C. R., Results of an exploration into the top of the Piledriver chimney, *Rep. UCRL-50385,* Lawrence Livermore Lab., Livermore, Calif., 1967.

Boland, J. N., A. C. McLaren, and B. E. Hobbs, Dislocations associated with optical features in naturally deformed olivines, *Contrib. Mineral. Petrol., 30,* 53–63, 1971.

Borg, I. Y., Survey of the Piledriver event and preliminary interpretation of three postshot cores in and near the cavity, *Rep. UCRL-50865,* Lawrence Livermore Lab., Livermore, Calif., 1970a.

Borg, I. Y., Mechanical ⟨110⟩ twinning in shocked sphene, *Amer. Mineral., 55,* 1876–1887, 1970b.

Borg, I. Y., Some shock effects in granodiorite to 270 kilobars at the Piledriver site, this volume.

Borg, I. Y., and J. Handin, Experimental deformation of crystalline rocks, *Tectonophysics, 3,* 251–367, 1966.

Borg, I. Y., and H. C. Heard, Experimental deformation of plagioclases, in *Experimental and Natural Rock Deformation,* edited by P. Paulitsch, pp. 375–403, Springer-Verlag, Berlin, 1970.

Borg, I. Y., and H. C. Heard, Mechanical twinning in sphene at 8 kbar, 25–500°C, in *The Hess Volume: Studies in Earth and Space Sciences, Mem. 132,* edited by R. Shagam, in press, Geological Society of America, Boulder, Colo., 1972.

Brace, W. F., Orientation of anisotropic minerals in a stress field: Discussion, in *Rock Deformation, Mem. 79,* edited by D. T. Griggs and J. W. Handin, pp. 9–20, Geological Society of America, Boulder, Colo., 1960.

Brace, W. F., Brittle fracture of rocks, in *State of Stress in the Earth's Crust,* edited by W. R. Judd, pp. 110–178, Elsevier, New York, 1964.

Brace, W. F., Resistivity of saturated crystal rocks to 40 km based on laboratory measurements, in *The Structure and Physical Properties of the Earth's Crust, Geophys. Monogr. Ser.,* vol. 14, edited by J. G. Heacock, pp. 243–255, AGU, Washington, D. C., 1971a.

Brace, W. F., Micromechanics in rocks systems, in *Structure, Solid Mechanics and Engineering Design,* edited by M. Te'eni, John Wiley, New York, pp. 187–204, 1971b.

Brace, W. F., Pore pressure in geophysics, this volume.

Brace, W. F., and J. D. Byerlee, Recent experimental studies of brittle fracture of rocks, in *Failure and Breakage of Rock,* American Institute of Mining, New York, pp. 58–81, 1966.

Brace, W. F., and A. H. Jones, Comparison of uniaxial deformation in shock and static loading of three rocks, *J. Geophys. Res., 76,* 4913–4921, 1971.

Brace, W. F., and R. J. Martin III, A test of the law of effective stress for crystalline rocks of low porosity, *Int. J. Rock Mech. Min. Sci., 5,* 415–426, 1968.

Brace, W. F., and A. S. Orange, Electrical resistivity changes in saturated rocks during fracture and fictional sliding, *J. Geophys. Res., 73,* 1433–1445, 1968a.

Brace, W. F., and A. S. Orange, Further studies of the effect of pressure on electrical resistivity of rocks, *J. Geophys. Res., 73,* 5407–5420, 1968b.

Brace, W. F., J. B. Walsh, and W. T. Frangos, Permeability of granite under high pressure, *J. Geophys. Res., 73,* 2225–2236, 1968.

Braun, R. L., J. S. Kahn, and S. Weissmann, X-ray analysis of plastic deformation in the Salmon event, *J. Geophys. Res., 74,* 2103–2117, 1969.

Bredehoeft, J. D., and B. B. Hanshaw, On the maintenance of anomalous fluid pressures, 1, Thick sedimentary sequences, *Geol. Soc. Amer. Bull., 79,* 1097–1106, 1968.

Broucke, R. A., A program to compute the improved lunar ephemeris, technical memorandum, pp. 312–754, Jet Propul. Lab., Pasadena, Calif., Nov. 7, 1966.

Broucke, R. A., W. E. Zürn, and L. B. Slichter, Lunar tidal acceleration on a rigid earth, this volume.

Brunner, G. O., H. Wondratschek, and F. Laves, Ultrarotuntersuchungen über den Einbau von H in naturliechen Quarz, *Z. Elektrochem., 65,* 735–750, 1961.

Buck, P., Verformung von Hornblende-Einkristallen bei Drucken bis 21 kbar, *Contrib. Mineral. Petrol., 28,* 62–71, 1970.

Buck, P., and P. Paulitsch, Experimentelle Verformung von Glimmer- und Hornblende-Einkristallen, *Naturwissenschaften, 9,* 460–461, 1969.

Buerger, M. J., Translation-gliding in crystals, *Amer. Mineral, 15,* 1–20, 1930a.

Buerger, M. J., Translation gliding in crystals of the NaCl structural type, *Amer. Mineral., 15,* 174–187, 1930b.

Bullard, E. C., The stability of a homopolar dynamo, *Proc. Cambridge Phil. Soc., 51,* 744–760, 1955.

Bullard, E. C., and D. Gubbins, Oscillating disk dynamo and geomagnetism, this volume.

Bunch, T. E., Some characteristics of selected minerals from craters, in *Shock Metamorphism of Natural Materials,* edited by B. M. French and N. M. Short, pp. 413–432, Mono, Baltimore, 1968.

Bunge, H. J., Zur Darstellung allgemeiner Texturen, *Z. Metallk., 56,* 872–874, 1965.

Bunge, H. J., Die dreidimensionale Orientierungsverteilungsfunktion und Methoden zu ihrer Bestimmung, *Krist. Tech., 3,* 439–454, 1968a.

Bunge, H. J., The orientation distribution function of the crystallites in cold-rolled and annealed low-carbon steel sheets, *Phys. Status Solidi, 26,* 167–171, 1968b.

Bunge, H. J., *Mathematische Methoden der Texturanalyse,* 300 pp., Akademie Verlag, Berlin, 1969.

Bunge, H. J., and W. Roberts, Orientation distribution, elastic and plastic anisotropy in stabilized steel sheet, *J. Appl. Crystallogr., 2,* 116–128, 1969.

Burke, P. M., High temperature creep of polycrystalline sodium chloride, Ph.D. dissertation, 112 pp., Stanford University, Stanford, Calif., 1968.

Burridge, R., and L. Knopoff, Model and theoretical seismicity, *Bull. Seismol. Soc. Amer., 57,* 341–371, 1967.

Butkovich, T. R., The gas equation of state for natural materials, *Rep. UCRL-14729,* Lawrence Livermore Lab., Livermore, Calif., 1967.

Butkovich, T. R., Rock melted per kiloton energy yield, *Rep. UOPKL 68-10,* Lawrence Livermore Lab., Livermore, Calif., 1968.

Butkovich, T. R., and J. K. Landauer, The flow law for ice, *Publ. 47,* Ass. Int. Hydrol. Sci., 318–325, 1958.

Byerlee, J. D., The frictional characteristics of westerly granite, Ph.D. thesis, Mass. Inst. of Technol., Cambridge, Mass., June 1966.

Byerlee, J. D., Frictional characteristics of granite under high confining pressure, *J. Geophys. Res., 72,* 3639–3648, 1967.

Byerlee, J. D., Mechanical behavior of Weber sandstone (abstract), *Eos Trans. AGU, 52,* 343, 1971.

Byerlee, J. D., and W. F. Brace, Fault stability and pore pressure, *Bull. Seismol. Soc. Amer.,* in press, 1972.

Cahn, R. W., Slip and polygonization in aluminum, *J. Inst. Metals, 79,* 129–158, 1951.

Carter, N. L., Dynamic deformation of quartz, in *Shock Metamorphism of Natural Materials,* edited by B. M. French and N. M. Short, pp. 453–474, Mono, Baltimore, 1968.

Carter, N. L., Static deformation of silica and silicates, *J. Geophys. Res., 76,* 5514–5540, 1971.

Carter, N. L., and H. G. Avé Lallemant, High temperature flow of dunite and peridotite, *Geol. Soc. Amer. Bull., 81,* 2181–2202, 1970.

Carter, N. L., and H. C. Heard, Temperature and rate dependent deformation of halite, *Amer. J. Sci., 269,* 193–249, 1970.

Carter, N. L., and C. B. Raleigh, Principal stress directions from plastic flow in crystals, *Geol. Soc. Amer. Bull., 80,* 1231–1264, 1969.

Carter, N. L., D. T. Griggs, and J. M. Christie, Experimental deformation and recrystallization of quartz, *J. Geol., 72,* 687–733, 1964.

Carter, N. L., D. W. Baker, and R. P. George, Jr., Seismic anisotropy, flow, and constitution of the upper mantle, this volume.

Chao, E. C. T., Pressure and temperature histories of impact metamorphosed rocks based on petrographic observations, in *Shock Metamorphism of Natural Materials,* edited by B. M. French and N. M. Short, pp. 135–158, Mono, Baltimore, 1968.

Chapin, C. E., Cavity pressure history of contained nuclear explosions, *Rep. UCRL-72140,* Lawrence Livermore Lab., Livermore, Calif., 1969.

Chapple, W. M., A mathematical theory of finite-amplitude rock-folding, *Geol. Soc. Amer. Bull.*, *79*, 47–68, 1968*a*.

Chapple, W. M., The equations of elasticity and viscosity and their application to faulting and folding, in *Advanced Science Seminar in Rock Mechanics*, vol. 1, edited by R. E. Riecker, pp. 225–284, Air Force Cambridge Research Laboratory, Bedford, Mass., 1968*b*.

Chapple, W. M., Fold shape and rheology: The folding of an isolated viscous-plastic layer, *Tectonophysics, 7*, 97–116, 1969.

Chapple, W. M., The finite-amplitude instability in the folding of layered rocks, *Can. J. Earth Sci., 7*, 457–466, 1970*a*.

Chapple, W. M., The initiation and spacing of folds in viscous multilayered media, *Geol. Soc. Amer. Abstr. Programs, 2*, 276, 1970*b*.

Chen, H. S., J. J. Gilman, and A. K. Head, Dislocation multipoles and their role in strain hardening, *J. Appl. Phys., 35*, 2502–2514, 1964.

Cherry, J. T., Rock breakage from an explosive source, *Intern. Doc. UOPKA 70-26*, Lawrence Livermore Lab., Livermore, Calif., 1970.

Cherry, J. T., and F. L. Peterson, Numerical simulation of stress wave propagation from underground nuclear explosions, in *Proceedings of the Symposium on Engineering with Nuclear Explosives*, vol. 1, pp. 142–220, American Nuclear Society, Hinsdale, Ill., 1970.

Cherry, J. T., and E. G. Rapp, Calculation of free-field motion for the Piledriver event, *Rep. UCRL-50373*, Lawrence Livermore Lab., Livermore, Calif., 1968.

Christensen, N. I., Elasticity of ultrabasic rocks, *J. Geophys. Res., 71*, 5921–5931, 1966.

Christensen, N. I., Fabric, seismic anisotropy, and tectonic history of the Twin Sisters dunite, *Geol. Soc. Amer. Bull., 82*, 1681–1694, 1971.

Christensen, N. I., and R. I. Crosson, Seismic anisotropy in the upper mantle, *Tectonophysics, 6*, 93–107, 1968.

Christensen, N. I., and R. Ramananantoandro, Elastic moduli and anisotropy of dunite to 10 kilobars, *J. Geophys. Res., 76*, 4003–4010, 1971.

Christie, J. M., Mylonite rocks of the Moine thrust-zone in the Assynt region, northwest Scotland, *Trans. Edinburgh Geol. Soc., 18*, 687–733, 1960.

Christie, J. M., The Moine thrust zone in the Assynt region, northwest Scotland, *Univ. Calif. Berkeley, Publ. Geol. Sci., 40*, 345–440, 1963.

Christie, J. M., and H. W. Green, Several new slip mechanisms in quartz (abstract), *Eos Trans. AGU, 45*, 103, 1964.

Christie, J. M., D. T. Griggs, and N. L. Carter, Experimental evidence of basal slip in quartz, *J. Geol., 72*, 734–756, 1964.

Clabaugh, P. S., Petrofabric study of deformed salt, *Science, 136*, 389–391, 1962.

Clark, S. P., Jr., and A. E. Ringwood, Density distribution and constitution of the mantle, *Rev. Geophys. Space Phys., 2*, 35–88, 1964.

Clauer, A. H., B. A. Wilcox, and J. P. Hirth, Dislocation substructure induced by creep in molybdenum single crystals, *Acta Met., 18*, 381–397, 1970.

Closmann, P. J., On the prediction of cavity radius produced by an underground nuclear explosion, *J. Geophys. Res., 74*, 3935, 1969.

Coble, R. L., and Y. H. Guerard, Creep of polycrystalline aluminum oxide, *J. Amer. Ceram. Soc., 46*, 353–354, 1963.

Coe, R. S., and M. S. Paterson, The α-β inversion in quartz: A coherent phase transition under nonhydrostatic stress, *J. Geophys. Res., 74*, 4921–4948, 1969.

Coleman, B. D., and W. Noll, On certain steady flows of general fluids, *Arch. Ration. Mech. Anal., 3*, 289–303, 1959.

Coleman, B. D., and W. Noll, An approximation theorem for functionals with applications in continuum mechanics, *Arch. Ration. Mech. Anal., 6*, 355–370, 1960.

Coleman, B. D., H. Markovitz, and W. Noll, *Viscometric Flows of Non-Newtonian Fluids*, 130 pp., Springer-Verlag, Berlin, 1966.

Colgate, S. A., and M. H. Johnson, Hydrodynamic origin of cosmic rays, *Phys. Rev. Lett., 5*, 235–238, 1960.

Conel, J. E., Studies of the development of fabrics in naturally deformed limestones, Ph.D. thesis, Calif. Inst. of Technol., Pasadena, 1962.

Cottrell, A. H., *The Mechanical Properties of Matter*, 430 pp., John Wiley, New York, 1964.

Cottrell, A. H. and B. A Bilby, Dislocation theory of yielding and strain ageing of iron, *Proc. Phys. Soc. London, Sect. A, 62*, 49–62, 1949.

Crittenden, M. D., Jr., Viscosity and finite strength of the mantle as determined from water and ice loads, *Geophys. J. Roy. Astron. Soc., 14*, 261–279, 1967.

Crosson, R. S., and N. I. Christensen, Transverse isotropy of the upper mantle in the vicinity of Pacific fracture zones, *Bull. Seismol. Soc. Amer., 59*, 59–72, 1969.

Crosson, R. S., and J. W. Lin, Voigt and Reuss prediction of anisotropic elasticity of dunite, *J. Geophys. Res., 76*, 570–578, 1971.

Cummings, D., Kink-bands: Shock deformation of biotite resulting from a nuclear explosion, *Science, 148*, 950–952, 1965.

Cummings, D., Shock deformation of biotite around a nuclear explosion, in *Shock Metamorphism of Natural Materials*, edited by B. M. French and N. M. Short, pp. 211–218, Mono, Baltimore, 1968.

Currie, J. B., H. W. Patnode, and R. P. Trump, Development of folds in sedimentary strata, *Geol. Soc. Amer. Bull., 73*, 655–674, 1962.

Davis, B. T. C., and J. L. England, The melting of forsterite up to 50 kilobars, *J. Geophys. Res., 69*, 1113–1116, 1964.

Davis, L. A., and R. B. Gordon, Pressure dependence of the plastic flow stress of alkali halide single crystals, *J. Appl. Phys., 39,* 3885–3897, 1968.

de Hoffmann, F., and E. Teller, Magneto-hydrodynamic shocks, *Phys. Rev., 80,* 692–703, 1950.

de la Cruz, R., and C. B. Raleigh, Absolute stress measurements at the Rangely anticline, northwestern Colorado, *Int. J. Rock Mech. Min. Sci., 9,* 625–634, 1972.

Delort, F., and F. Supiot, Nuclear stimulation of oil reservoirs, in *Proceedings of the Symposium on Engineering with Nuclear Explosives,* vol. 2, pp. 649–661, American Nuclear Society, Hinsdale, Ill., 1970.

den Tex, E., Origin of ultramafic rocks, their tectonic setting and history: A contribution to the discussion of the paper 'The origin of ultramafic and ultrabasic rocks' by P. J. Wyllie, *Tectonophysics, 7,* 457–488, 1969.

de Sitter, L. U., *Structural Geology,* 2nd ed., McGraw-Hill, New York, 551 pp., 1964.

Dickinson, G., Geological aspect of abnormal reservoir pressures in Gulf Coast, Louisiana, *Bull. Amer. Ass. Petrol. Geol., 37,* 410–432, 1953.

Dieterich, J. H., Origin of cleavage in folded rocks, *Amer. J. Sci., 267,* 155–165, 1969.

Dieterich, J. H., Computer experiments on mechanics of finite amplitude folds, *Can. J. Earth Sci., 7,* 467–476, 1970.

Dieterich, J. H., Computer modeling of earthquake stimulation by fluid injection (abstract), *Eos Trans. AGU, 52,* 313, 1971.

Dieterich, J. H., and N. L. Carter, Stress-history of folding, *Amer. J. Sci., 267,* 129–154, 1969.

Dieterich, J. H., and E. T. Onat, Slow finite deformation of viscous solids, *J. Geophys. Res., 74,* 2081–2088, 1969.

Dommerich, S., Festigkeitseigenschaften bewässerter Salzkristalle, 4 Richtungsabhängigkeit der Streckgrenze gleichmässig abgelöster Steinsalzstäbchen, *Z. Phys., 90,* 189–196, 1934.

Donath, F. A., The development of kink bands in brittle anisotropic rock, in *Igneous and Metamorphic Geology, Mem. 115,* edited by L. Larsen, M. Prinz, and V. Manson, pp. 453–493, Geological Society of America, Boulder, Colo., 1968.

Donath, F. A., and R. B. Parker, Folds and folding, *Geol. Soc. Amer. Bull., 75,* 45–62, 1964.

Dorn, J. E., Some fundamental experiments on high temperature creep, *J. Mech. Phys. Solids, 3,* 85–116, 1957.

Dworak, U., Stosswellenmetamorphose des Anorthosits vom Manicouagan Krater, Quebec, Canada, *Contrib. Mineral. Petrol., 24,* 306–347, 1969.

Eckert, W. J., R. Jones, and H. K. Clark, *Improved Lunar Ephemeris 1952–1959,* U.S. Government Printing Office, Washington, D.C., 1954.

Einarsson, T., Ueber die neuen Eruptionen des Geysir in Haukadalur, *Visindafélag Islandinga, Greinar I, 2,* 149–166, 1937.

Einarsson, T., Ueber das Wesen der heissen Quel-

len Islands, *Visindafélags Islandinga, 26,* 57–69, 1942.

Einarsson, T., Gos geysis i Haukadal, *Náttúrufraedingurinn, 19,* 20–26, 1949.

Einarsson, T., *The Great Geysir and the Hot-Spring Area of Haukadalur, Iceland,* Geysir Committee, Reykjavik, Iceland, 24 pp., 1967.

El Goresy, A., The opaque minerals in impactite glasses, in *Shock Metamorphism of Natural Materials,* edited by B. M. French and N. M. Short, pp. 531–554, Mono, Baltimore, 1968.

Engelhardt, W. V., and W. Bertsch, Shock induced planar deformation structures in quartz from the Ries crater, Germany, *Contrib. Mineral. Petrol., 20,* 203–234, 1969.

Engelhardt, W. V., and D. Stöffler, Stages of shock metamorphism in crystalline rocks of the Ries Basin, Germany, in *Shock Metamorphism of Natural Materials,* edited by B. M. French and N. M. Short, pp. 159–168, Mono, Baltimore, 1968.

Eringen, A. C., *Nonlinear Theory of Continuous Media,* McGraw-Hill, New York, 477 pp., 1962.

Eringen, A. C., *Mechanics of Continua,* 502 pp., John Wiley, New York, 1967.

Ernst, W. G., Do mineral parageneses reflect unusually high-pressure conditions of Franciscan metamorphism?, *Amer. J. Sci., 270,* 81–108, 1971.

Eshelby, J. D., Elastic inclusions and inhomogeneities, in *Progress in Solid Mechanics,* vol. 2, North-Holland, Amsterdam, 89–140, 1961.

Evans, D., The Denver area earthquakes and the Rocky Mountain arsenal well, *Mt. Geol., 3,* 23–36, 1966.

Explanatory Supplement to the Astronomical Ephemeris and the American Ephemeris and Nautical Almanac, Her Majesty's Stationery Office, London, 1961.

Farquharson, R. B., and C. J. L. Wilson, Rationalization of geochronology and structure at Mount Isa, *Econ. Geol., 66,* 574–582, 1971.

Forman, D. J., The Arltunga Nappe Complex, MacDonnell Ranges, Northern Territory, Australia, *J. Geol. Soc. Aust., 18,* 173–182, 1971.

Forman, D. J., E. N. Milligan, and W. R. McCarthy, Regional geology and structure of the northeast margin, Amadeus Basin, Northern Territory, *Rep. 103,* Bur. Miner. Resour. of Aust., Canberra, 1967.

Francis, T. J. G., Generation of seismic anisotropy in the upper mantle along the mid-oceanic ridges, *Nature, 221,* 161–165, 1969.

Frank, F. C., Plate tectonics, the analogy with glacier flow, and isostasy, this volume.

Frank, W., and A. Seeger, Recent advances in the understanding of work-hardening in alkali-halide crystals, *Comments Solid State Phys., 2,* 133–142, 1969.

Friedman, M., Petrofabric techniques for the determination of principal stress directions in rocks, in *State of Stress in the Earth's Crust,* edited by W. R. Judd, pp. 451–552, Elsevier, New York, 1964.

Friedman, M., and G. M. Sowers, Petrofabrics:

A critical review, *Can. J. Earth Sci., 7*, 477–497, 1970.

Friedman, M., and D. W. Stearns, Relations between stresses derived from calcite twin lamellae and macrofractures, Teton Anticline, Montana, *Geol. Soc. Amer. Abstr. Programs, 2*, 555–556, 1970.

Frisillo, A. L., and G. R. Barsch, The pressure dependence of the elastic coefficients of orthopyroxene (abstract), *Eos Trans. AGU, 52*, 359, 1971.

Frondel, C., Secondary Dauphiné twinning in quartz, *Amer. Mineral., 30*, 447–460, 1945.

Frondel, C., *Dana's The System of Mineralogy*, vol. 3, *Silica Minerals*, John Wiley, New York, 334 pp., 1962.

Fulda, E., Salztektonik, *Z. Deut. Geol. Ges., 79*, 178–196, 1927.

Fung, Y. C., *Foundations of Solid Mechanics*, Prentice-Hall, Englewood Cliffs, N.J., 525 pp., 1965.

Garofalo, F., *Fundamentals of Creep and Creep Rupture in Metals*, MacMillan, New York, 258 pp., 1965.

Ghosh, S. K., Experiments of buckling of multilayers which permit interlayer gliding, *Tectonophysics, 6*, 207–249, 1968.

Gibbs, J. W., On the equilibrium of heterogeneous substances, in *The Scientific Papers of J. Willard Gibbs*, vol. 1, Longmans, Green, Toronto, Ont., 434 pp., 1906.

Gillespie, P., A. C. McLaren, and J. N. Boland, Operating characteristics of an ion-bombardment apparatus for thinning non-metals for transmission electron microscopy, *J. Material Sci., 6*, 87–89, 1971.

Gillis, P. P., and J. J. Gilman, Dynamical dislocation theory of crystal plasticity, *J. Appl. Phys., 36*, 3370–3386, 1965.

Gilman, J. J., Dislocation mobility in crystals, *J. Appl. Phys., 36*, 3195–3206, 1965.

Gilman, J. J., Progress in the microdynamical theory of plasticity, in *Proceedings of the 5th U.S. National Congress of Applied Mechanics*, pp. 385–403, American Society of Mechanical Engineers, New York, 1966.

Gilman, J. J., and W. G. Johnston, The origin and growth of glide bands in LiF crystals, in *Dislocations and Mechanical Properties of Crystals*, pp. 116–163, John Wiley, New York, 1957.

Gleiter, H., The migration of small angle boundaries, *Phil. Mag., 20*, 821–830, 1969.

Glen, J. W., The creep of polycrystalline ice, *Proc. Roy. Soc. London, Ser. A, 288*, 519–538, 1955.

Glen, J. W., The flow law of ice, *Publ. 47*, Ass. Int. Hydrol. Sci., pp. 171–183, 1958.

Glen, J. W., and M. F. Perutz, The growth and deformation of ice crystals, *J. Glaciol., 2*, 397–403, 1954.

Goetze, C., High-temperature rheology of Westerly granite, *J. Geophys. Res., 76*, 1223–1230, 1971.

Goguel, J., *Tectonics*, W. H. Freeman, San Francisco, Calif., 389 pp., 1962.

Gordon, R. B., Diffusion creep in the earth's mantle, *J. Geophys. Res., 70*, 2413, 1965.

Gordon, R. B., and L. A. Davis, Velocity and attenuation of seismic waves in imperfectly elastic rock, *J. Geophys. Res., 73*, 3917–3935, 1968.

Gow, A. J., The inner structure of the Ross Ice Shelf at Little America V, Antarctica, as revealed by deep core drilling, *Ass. Int. Hydrol. Sci. Publ. 61*, 272–284, 1963.

Gow, A. J., H. T. Ueda, and D. E. Garfield, Antarctic ice sheet: Preliminary results of first core hole to bedrock, *Science, 161*, 1011–1013, 1968.

Graton, L. C., Thoughts on the geyser mechanism (title only), *Trans. AGU, 30*, 471, 1949.

Green, H. W., Syntectonic and annealing recrystallization of fine-grained quartz aggregates, Ph.D. thesis, Univ. of Calif. at Los Angeles, 203 pp., 1968.

Green, H. W., Diffusional flow in polycrystalline materials, *J. Appl. Phys., 41*, 3899–3902, 1971.

Green, H. W., A CO_2-charged asthenosphere, *Nature Phys. Sci., 238*, 2–5, 1972.

Green, H. W., and S. V. Radcliffe, The nature of deformation lamellae in silicates, *Geol. Soc. Amer. Bull., 83*, 847–852, 1972a.

Green, H. W., and S. V. Radcliffe, Dislocation mechanisms in olivine and flow in the upper mantle, *Earth Planet. Sci. Lett., 15*, 239–247, 1972b.

Green, H. W., II, and S. V. Radcliffe, Deformation processes in the upper mantle, this volume.

Green, H. W., D. T. Griggs, and J. M. Christie, Syntectonic and annealing recrystallization of fine-grained quartz aggregates, in *Experimental and Natural Rock Deformation*, edited by P. Paulitsch, pp. 272–335, Springer-Verlag, Berlin, 1970.

Greenspan, H. P., *The Theory of Rotating Fluids*, p. 12, Cambridge Univ. Press, London, England, 1969.

Griggs, D. T., Deformation of rocks under high confining pressure, *J. Geol., 44*, 541–577, 1936.

Griggs, D. T., Creep of rocks, *J. Geol., 47*, 225–251, 1939.

Griggs, D. T., Experimental flow of rocks under conditions favoring recrystallization, *Geol. Soc. Amer. Bull., 51*, 1001–1022, 1940.

Griggs, D. T., Hydrolytic weakening of quartz and other silicates, *Geophys. J. Roy. Astron. Soc., 14*, 19–31, 1967.

Griggs, D. T., The sinking lithosphere and the focal mechanism of deep earthquakes, in *Nature of Solid Earth*, edited by E. C. Robertson, pp. 361–384, McGraw-Hill, New York, 1972.

Griggs, D. T., and D. W. Baker, The origin of deep-focus earthquakes, in *Properties of Matter Under Unusual Conditions*, edited by H. Mark and S. Fernbach, pp. 23–42, Interscience, New York, 1969.

Griggs, D. T., and J. D. Blacic, The strength of quartz in the ductile regime (abstract), *Eos Trans. AGU, 45,* 102, 1964.

Griggs, D. T., and J. D. Blacic, Quartz: Anomalous weakness of synthetic crystals, *Science, 147,* 292–295, 1965.

Griggs, D. T., and N. E. Coles, Creep of single crystals of ice, *Rep. 11,* 24 pp., U.S.A. Snow, Ice, and Permafrost Res. Estab., 1954.

Griggs, D. T., F. J. Turner, and H. C. Heard, Deformation of rocks at 500°C to 800°C, in *Rock Deformation, Mem. 79,* edited by D. T. Griggs and J. Handin, pp. 39–104, Geological Society of America, Boulder, Colo., 1960.

Groshong, R. H., Jr., Twinned calcite as a natural strain gage, *Geol. Soc. Amer. Abstr. Programs, 2,* 563, 1970.

Groves, G. W., and A. Kelly, Independent slip systems in crystals, *Phil. Mag., 8,* 877–887, 1963.

Guyot, P., and J. E. Dorn, A critical review of the Peierls mechanism, *Can. J. Phys., 45,* 983–1016, 1967.

Haasen, P., *Versetzungen und Plastizität von Germanium und insb. Festkörperprobleme,* vol. 3, pp. 167–208, edited by F. Sauter, Deutsche Physikalische Gesellschaft, 1964.

Haimson, B. C., Earthquake related stresses at Rangeley, Colorado, in *Proceedings of the 14th Symposium on Rock Mechanics,* Pennsylvania State University, State College, Pa., in press, 1972.

Haimson, B., and C. Fairhurst, In-situ stress determination at great depth by means of hydraulic fracturing, in *Rock Mechanics—Theory and Practice,* edited by W. Somerton, pp. 559–584, Amer. Inst. of Mining, Met., and Petrol. Eng., New York, 1970.

Hales, A. L., Gravitational sliding and continental drift, *Earth Planet. Sci. Lett., 6,* 31–34, 1969.

Handin, J., An application of high pressure in geophysics: Experimental rock deformation, *Trans. ASME, 75,* 315–324, 1953.

Handin, J., On the Coulomb-Mohr failure criterion, *J. Geophys. Res., 74,* 5343–5348, 1969.

Handin, J., and J. M. Logan, Experimental folding of rocks under confining pressure (abstract), *Geol. Soc. Amer. Abstr. Programs, 2,* 567, 1970.

Handin, J., R. V. Hager, Jr., M. Friedman, and J. N. Feather, Experimental deformation of sedimentary rocks under confining pressure: Pore pressure tests, *Amer. Ass. Petrol. Geol. Bull., 47,* 171–755, 1963.

Handin, J., M. Friedman, J. M. Logan, L. J. Pattison, and H. S. Swolfs, Experimental folding of rocks under confining pressure: Buckling of single-layer beams, this volume.

Hartman, P., and E. den Tex, Piezocrystalline fabrics of olivine in theory and nature (abstract), *Int. Geol. Congr. Rep. Sess. India 22nd,* 42–43, 1964.

Hashin, Z., and S. Shtrikman, A variational approach to the theory of elastic behavior of polycrystals, *J. Mech. Phys. Solids, 10,* 343–352, 1962.

Healy, J. H., W. W. Rubey, D. T. Griggs, and C. B. Raleigh, The Denver earthquakes, *Science, 161,* 1301–1310, 1968.

Heard, H. C., Effect of large changes in strain rate in the experimental deformation of Yule marble, *J. Geol., 71,* 162–195, 1963.

Heard, H. C., The flow law of polycrystalline halite (abstract), *Eos Trans. AGU, 51,* 424, 1970.

Heard, H. C., Steady-state flow in polycrystalline halite at pressure of 2 kilobars, this volume.

Heard, H. C., and N. L. Carter, Experimentally induced 'natural' intragranular flow in quartz and quartzite, *Amer. J. Sci., 266,* 1–42, 1968.

Heard, H. C., and C. B. Raleigh, Steady-state flow in marble at 500° to 800°C, *Geol. Soc. Amer. Bull., 83,* 935–956, 1972.

Heard, H. C., and W. W. Rubey, Tectonic implications of gypsum dehydration, *Geol. Soc. Amer. Bull., 77,* 741–760, 1966.

Heard, H. C., F. J. Turner, and L. E. Weiss, Studies of heterogeneous strain in experimentally deformed calcite, marble, and phyllite, *Univ. Calif. Publ. Geol. Sci., 46,* 81–101, 1965.

Helmstaedt, H., and O. L. Anderson, Petrofabrics of mafic and ultramafic inclusions from kimberlite pipes in southeastern Utah and northeastern Arizona (abstract), *Eos Trans. AGU, 50,* 345, 1969.

Hermance, J. F., A. Nur, and S. Bjornsson, Electrical properties of basalt: Relation of laboratory to in-situ measurements, *J. Geophys. Res., 77,* 1424–1429, 1972.

Herring, C., Diffusional viscosity of a polycrystalline solid, *J. Appl. Phys., 21,* 437–445, 1950.

Hess, H. H., History of ocean basins, in *Petrological Studies: A Volume in Honor of A. F. Buddington,* edited by A. E. J. Engel, H. L. James, and B. L. Leonard, pp. 599–620, Geological Society of America, Boulder, Colo., 1962.

Hess, H. H., Seismic anisotropy of the uppermost mantle under oceans, *Nature, 203,* 629–631, 1964.

Heuer, A. H., R. F. Firestone, J. D. Snow, H. W. Green, R. G. Howe, and J. M. Christie, An improved ion-thinning device, *Rev. Sci. Instrum., 42,* 1177–1184, 1971.

Hide, R., Free hydromagnetic oscillations of the earth's core and the theory of the geomagnetic secular variation, *Phil. Trans. Roy. Soc. London, Ser. A, 259,* 615–650, 1966a.

Hide, R., On the theory of the geomagnetic secular variation, in *Magnetism and the Cosmos,* edited by W. R. Hindmarsh, F. J. Lowes, P. H. Roberts, and S. K. Runcorn, pp. 141–147, Oliver and Boyd, Edinburgh, 1966b.

Higgins, G. H., and T. R. Butkovich, Effect of water content, yield, medium, and depth of burst on cavity radii, *Rep. UCRL-50203,* Lawrence Livermore Lab., Livermore, Calif., 1967.

Higgins, G. H., and G. C. Kennedy, The adiabatic gradient and the melting point gradient in the core of the earth, *J. Geophys. Res., 76,* 1870–1878, 1971.

Higgs, D. V., and J. Handin, Experimental deformation of dolomite single crystals, *Geol. Soc. Amer. Bull., 70,* 245–278, 1959.

Hill, R. W., The elastic behavior of a crystalline aggregate, *Proc. Phys. Soc. London, Sect. A, 65,* 349–354, 1952.

Hill, R., Elastic properties of reinforced solids: Some theoretical principles, *J. Mech. Phys. Solids, 11,* 257–372, 1963.

Hill, R. E. T., and A. L. Boettcher, Water in the earth's mantle: Melting curves of basalt-water and basalt-water-carbon dioxide, *Science, 167,* 980–981, 1970.

Hirano, R., Study on the aftershocks of the Kwanto earthquake, *J. Meteorol. Soc. Jap., ser. 2, 2,* 77, 1924.

Hirsch, P. B., A. Howie, R. B. Nicholson, D. W. Pashley, and M. J. Whelan, *Electron Microscopy of Thin Crystals,* Butterworth, London, 549 pp., 1965.

Hobbs, B. E., The structural environment of the northern part of the Broken Hill orebody, *J. Geol. Soc. Aust., 13,* 315–338, 1966a.

Hobbs, B. E., Microfabric of tectonites from the Wyangala Dam area, New South Wales, Australia, *Geol. Soc. Amer. Bull., 77,* 685–706, 1966b.

Hobbs, B. E., Recrystallization of single crystals of quartz, *Tectonophysics, 6,* 353–401, 1968.

Hobbs, B. E., The analysis of strain in folded layers, *Tectonophysics,* in press, 1972.

Hobbs, B. E., Deformation of non-Newtonian materials, this volume.

Hobbs, B. E., A. C. McLaren, and M. S. Paterson, Plasticity of single crystals of synthetic quartz, this volume.

Hoek, E., and Z. T. Bieniawski, Brittle fracture propagation in rock under compression, *Int. J. Fract. Mech., 1,* 137–155, 1965.

Hoffman, H. V., and F. M. Sauer, Free-field and surface motions, Operation Flintlock, Shot Piledriver, *Rep. POR-4000 (WT-4000),* Def. At. Support Agency, Washington, D.C., 1969.

Hörz, F., Statistical measurements of deformation structures and refractive indices in experimentally shock loaded quartz, in *Shock Metamorphism of Natural Materials,* edited by B. M. French and N. M. Short, pp. 243–253, Mono, Baltimore, 1968.

Hörz, F., Structural and mineralogical evaluation of an experimentally produced impact crater, *Contrib. Mineral. Petrol., 21,* 365–377, 1969.

Hörz, F., and T. Ahrens, Deformation of experimentally shocked biotite, *Amer. J. Sci., 267,* 1213–1229, 1969.

Houser, F. N., and F. G. Poole, Summary of physical and chemical nature of the granite rocks at the U15a site, Climax Stock, NTS, *Tech. Lett., Area 15-1,* U.S. Geol. Surv., Washington, D.C., 1961.

Hubbert, M. K., and W. W. Rubey, Role of fluid pressure in mechanics of overthrust faulting, *Geol. Soc. Amer. Bull., 70,* 115–166, 1959.

Hubbert, M. K., and D. G. Willis, Mechanics of hydraulic fracturing, *Trans. AIME, 210,* 153–166, 1957.

Institute of Radio Engineers, Standards on piezoelectric crystals, *Proc. Inst. Radio Eng., 49,* 1378, 1949.

International Tables for X-Ray Crystallography, vol. 1, edited by N. F. N. Henry and K. Lonsdale, p. 119, Kynoch Press, Birmingham, England, 1952.

Isacks, B., J. Oliver, and L. R. Sykes, Seismology and the new global tectonics, *J. Geophys. Res., 73,* 5855, 1968.

Ito, K., and G. C. Kennedy, The fine structure of the basalt-eclogite transition, *Mineral. Soc. Amer. Spec. Pap. 3,* 77–84, 1970.

Jackson, E. D., and T. L. Wright, Xenoliths in the Honolulu volcanic series, Hawaii, *J. Petrol., 11,* 405–430, 1970.

Jaeger, J. C., *Elasticity, Fracture and Flow,* 268 pp., Methuen, London, 1969.

Jaffe, H. W., P. Robinson, and C. Klein, Jr., Exsolution lamellae and optic orientation of clinoamphiboles, *Science, 160,* 776–778, 1968.

Jeffreys, H., Aftershocks and periodicity in earthquakes, *Gerlands Beitr. Geophys., 53,* 111–139, 1938.

Johnsen, A., Biegungen und Translationen, *Neues Jahrb. Mineral. Geol. Paleontol. Monatsh., Abt. 2,* 133–153, 1902.

Johnson, A. M., Development of folds within Carmel formation, Arches National Monument, Utah, *Tectonophysics, 8,* 31–76, 1969.

Johnson, M. H., and C. F. McKee, Relativistic hydrodynamics in one dimension, *Phys. Rev. D, 3,* 858–863, 1971.

Johnson, M. H., and E. Teller, General shock conditions, this volume.

Johnson, M. R. W., The structural history of the Moine thrust zone at Lochcarron, Wester Ross, *Trans. Roy. Soc. Edinburgh, 64,* 139–168, 1960.

Johnston, W. G., Yield points and delay times in single crystals, *J. Appl. Phys., 33,* 2716–2730, 1962.

Kamb, B., Theory of preferred crystal orientation developed by crystallization under stress, *J. Geol., 67,* 153–170, 1959a.

Kamb, B., Ice petrofabric observations from Blue glacier, Washington, in relation to theory and experiment, *J. Geophys. Res., 64,* 1891–1909, 1959b.

Kamb, B., The thermodynamic theory of non-hydrostatically stressed solids, *J. Geophys. Res., 66,* 259–271, 1961a.

Kamb, B., Author's reply to discussions of the paper 'The thermodynamic theory of non-hydrostatically stressed solids,' *J. Geophys. Res., 66,* 3985–3988, 1961b.

Kamb, B., Refraction corrections for universal stage measurements, 1, Uniaxial crystals, *Amer. Mineral., 47*, 227–245, 1962.

Kamb, B., Glacier geophysics, *Science, 146*, 353–365, 1964.

Kamb, B., Experimental recrystallization of ice under stress, this volume.

Kamb, B., and E. R. LaChapelle, Flow dynamics and structure in a fast-moving icefall (abstract), *Eos Trans. AGU, 49*, 312, 1968.

Kamb, B., and R. L. Shreve, Structure of ice at depth in a temperate glacier (abstract), *Eos Trans. AGU, 44*, 103, 1963.

Kasahara, J., I. Suzuki, M. Kumazawa, Y. Kobayashi, and K. Tida, Anisotropism of P-wave in dunite, *J. Seismol. Soc. Jap., 21*, 222–228, 1968.

Kats, A., Y. Haven, and J. M. Stevels, Hydroxyl groups in α-quartz, *Phys. Chem. Glasses, 3*, 69–75, 1962.

Keen, C. E., and D. L. Barrett, A measurement of seismic anisotropy in the northeast Pacific, *Can. J. Earth. Sci., 8*, 1056–1064, 1971.

Kehle, R. O., Determination of tectonic stresses through analysis of hydraulic well fracturing, *J. Geophys. Res., 69*, 259–273, 1964.

Keller, G. V., L. A. Anderson, and J. I. Pritchard, Geological survey investigations of the electrical properties of the crust and upper mantle, *Geophysics, 31*, 1078–1087, 1966.

Kelly, A., and G. W. Groves, *Crystallography and Crystal Defects*, 428 pp., Longman, London, 1970.

Kieslinger, A., Restspannung und Entspannung im Gestein, *Geol. Bauw., 24*, 95–112, 1958.

Kingery, W. D., and E. D. Montrone, Diffusional creep in polycrystalline sodium chloride, *J. Appl. Phys., 36*, 2412–2413, 1965.

Klassen-Neklyudova, M. V., *Mechanical Twinning of Crystals*, 213 pp., Consultants Bureau, New York, 1964.

Knopoff, L., Energy release in earthquakes, *Geophys. J. Roy. Astron. Soc., 1*, 44–52, 1958.

Knopoff, L., Thermal convection in the earth's mantle, in *The Earth's Mantle*, edited by T. F. Gaskell, pp. 171–196, Academic Press, New York, 1967.

Knopoff, L., Continental drift and convection, in *The Earth's Crust and Upper Mantle, Geophys. Monogr. Ser.*, vol. 13, edited by P. J. Hart, pp. 683–689, AGU, Washington, D.C., 1969.

Knopoff, L., Model for aftershock occurrence, this volume.

Kröner, E. P., Berechnung der elastischen Konstanten des Vielkristalls aus den Konstanten des Einkristalls, *Z. Physik, 151*, 504–518, 1958.

Kumazawa, M., The elastic constants of rock in terms of elastic constants of constituent mineral grains, petrofabric and interface structures, *J. Earth Sci. Nagoya Univ., 72*, 147–176, 1964.

Kumazawa, M., The elastic constants of single crystal orthopyroxene, *J. Geophys. Res., 74*, 5973–5980, 1969.

Kumazawa, M., and O. L. Anderson, Elastic moduli, pressure derivatives and temperature derivatives of single crystal olivine and single crystal forsterite, *J. Geophys. Res., 74*, 5961–5972, 1969.

Kumazawa, M., H. Helmstaedt and K. Masaki, Elastic properties of eclogite xenoliths from diatremes of East Colorado Plateau and their implication to the upper mantle structure, *J. Geophys. Res., 76*, 1231–1247, 1971.

Lambert, I. B., and P. J. Wyllie, Low-velocity zone of the earth's mantle: Incipient melting caused by water, *Science, 169*, 764–766, 1970.

Lang, A. R., and V. F. Miuscov, Dislocations and fault surfaces in synthetic quartz, *J. Appl. Phys., 38*, 2477–2483, 1967.

Langseth, M. G., Jr., X. LePichon, and M. Ewing, Crustal structure of the mid-ocean ridges, 5, Heat flow through the Atlantic Ocean floor and convection currents, *J. Geophys. Res., 73*, 5321, 1966.

Lappin, M. A., Structural and petrofabric studies of the dunites of Almklovadalen, Nordfjord, Norway, in *Ultramafic and Related Rocks*, edited by P. J. Wyllie, pp. 83–190, John Wiley, New York, 1967.

Lappin, M. A., The petrofabric orientation of olivine and seismic anisotropy of the mantle, *J. Geol., 79*, 730–740, 1971.

Laurance, N., Self-diffusion of the chloride ion in sodium chloride, *Phys. Rev., 120*, 57–62, 1960.

Laurent, J. F., and J. Bénard, Autodiffusion des ions dans les halogenures alcalins polycristallins, *J. Phys. Chem. Solids, 7*, 218–227, 1958.

LeComte, P., Creep in rock salt, *J. Geol., 73*, 469–484, 1965.

LePichon, X., Sea floor spreading and continental drift, *J. Geophys. Res., 73*, 3661–3697, 1968.

Longman, I. M., Formulas for computing the tidal accelerations due to the moon and the sun, *J. Geophys. Res., 64*, 2351–2355, 1959.

Low, J. R., and A. M. Turkalo, Slip band structure and dislocation-multiplication in silicon iron crystals, *Acta Met., 10*, 215–227, 1962.

MacDonald, G. J. F., Orientation of anisotropic minerals in a stress field, in *Rock Deformation, Mem. 79*, edited by D. T. Griggs and J. W. Handin, pp. 1–8, Geological Society of America, Boulder, Colo., 1960.

MacGregor, I. D., and J. L. Carter, The chemistry of clinopyroxenes and garnets of eclogite and peridotite xenoliths from the Roberts Victor Mine, South Africa, *Phys. Earth Planet. Interiors, 3*, 391–397, 1970.

Malkus, W. V. R., Precession of the earth as the cause of geomagnetism, *Science, 160*, 259–264, 1968.

Markovitz, H., Non-linear steady-flow behaviour, in *Rheology*, vol. 4, edited by F. R. Eirich, pp. 347–410, Academic, New York, 1967.

Marler, G. D., *The Story of Old Faithful Geyser*, The Yellowstone Library and Museum Association, 1953.

McConnell, R. K., Viscosity of the earth's mantle, in *The History of the Earth's Crust*, edited by R. A. Phinney, pp. 45–57, Princeton Univ. Press, Princeton, N.J., 1968.

McHenry, D., The effect of uplift pressure on the shearing strength of concrete, in *Proceedings of the 3rd Congres des Grands Barrages, Stockholm. C. R.*, vol. 1, question 8, R. 48, p. 1–24, 1948.

McKenzie, D. P., The viscosity of the mantle, *Geophys. J. Roy. Astron. Soc., 14*, 297–305, 1967.

McKenzie, D. P., The geophysical importance of high temperature creep, in *The History of the Earth's Crust*, edited by R. A. Phinney, p. 28, Princeton Univ. Press, N. J., 1968.

McKenzie, D. P., Relation between fluid injection plane solutions for earthquakes and the directions of the principal stresses, *Bull. Seismol. Soc. Amer., 59*, 591–601, 1969a.

McKenzie, D. P., Speculations on the consequences and causes of plate motions, *Geophys. J. Roy. Astron. Soc., 18*, I, 1969b.

McLaren, A. C., and B. E. Hobbs, Transmission electron microscope investigation of some naturally deformed quartzites, this volume.

McLaren, A. C., and P. P. Phakey, A transmission electron microscope study of amethyst and citrine, *Aust. J. Phys., 18*, 135–141, 1965a.

McLaren, A. C., and P. P. Phakey, Dislocations in quartz observed by transmission electron microscopy, *J. Appl. Phys., 36*, 3244–3246, 1965b.

McLaren, A. C., and P. P. Phakey, Diffraction contrast from Dauphiné twin boundaries in quartz, *Phys. Status Solidi, 31*, 723–737, 1969.

McLaren, A. C., and J. A. Retchford, Transmission electron microscope study of the dislocations in plastically deformed synthetic quartz, *Phys. Status Solidi, 33*, 657–668, 1969.

McLaren, A. C., J. A. Retchford, D. T. Griggs, and J. M. Christie, Transmission electron microscope study of Brazil twins and dislocations experimentally produced in natural quartz, *Phys. Status Solidi, 19*, 631–644, 1967.

McLaren, A. C., R. G. Turner, J. N. Boland, and B. E. Hobbs, Dislocation structure of the deformation lamellae in synthetic quartz; a study by electron and optical microscopy, *Contrib. Mineral. Petrol., 29*, 104–115, 1970.

McLaren, A. C., C. F. Osborne, and L. A. Saunders, X-ray topographic study of dislocations in synthetic quartz, *Phys. Status Solidi, A4*, 235–247, 1971.

McLellan, A. G., Non-hydrostatic thermodynamics of chemical systems, *Proc. Roy. Soc. London, Ser. A, 314*, 433–455, 1970.

McSkimin, H. J., Ultrasonic methods for measuring the mechanical properties of liquids and solids, in *Physical Acoustics*, vol. 1, part A, pp. 271–334, Academic, New York, 1964.

McSkimin, H. J., P. Andreatch, and R. N. Thurston, Elastic moduli of quartz versus hydrostatic pressure at 25°C and −195.8°C, *J. Appl. Phys., 36*, 1624–1632, 1965.

Michaud, L., Explosions nucléaires souterraines: Étude des rayons de cavité, *Rep. CEA R-3594*, Commissariat à Energie Atomique, Saclay, France, 1968.

Möckel, J. R., The structural petrology of the garnet peridotite of Alpe Arami (Ticino, Switzerland), *Leidse Geol. Meded., 42*, 61–130, 1969.

Moffatt, H. K., Turbulent dynamo action at low Reynolds number, *J. Fluid Mech., 41*, 435–452, 1970a.

Moffatt, H. K., Dynamo action associated with random inertial waves in a rotating conducting fluid, *J. Fluid Mech., 44*, 705–719, 1970b.

Moores, E. M., Petrology and structure of the Vourinos ophiolite complex of northern Greece, *Geol. Soc. Amer. Spec. Pap. 118*, 74 pp., 1969.

Moores, E. M., and F. J. Vine, The Troodos Massif, Cyprus, and other ophiolites as ocean crust: Evaluation and implications, *Phil. Trans. Roy. Soc. London, Ser. A, 268*, 443–446, 1971.

Morgan, W. J., Rises, trenches, great faults, and crustal blocks, *J. Geophys. Res., 73*, 1959, 1968.

Morris, G. B., R. W. Raitt, and G. G. Shor, Velocity anisotropy and delay-time maps of the mantle near Hawaii, *J. Geophys. Res., 74*, 4300–4316, 1969.

Morris, P. R., Averaging fourth-rank tensors with weight functions, *J. Appl. Phys., 40*, 447–448, 1969a.

Morris, P. R., Crystallite orientation analysis for rolled hexagonal materials, *Trans. Met. Soc. AIME, 245*, 1877–1881, 1969b.

Morris, P. R., Elastic constants of polycrystals, *Int. J. Eng. Sci., 8*, 49–61, 1970.

Morris, P. R., and A. J. Heckler, Crystallite orientation analysis for rolled cubic materials, *Advan. X-Ray Anal., 11*, 454–472, 1968.

Morris, P. R., and A. J. Heckler, Crystallite orientation analysis for rolled hexagonal materials, *Trans. Metal. Soc. AIME, 245*, 1877–1881, 1969.

Moses, P. L., Geothermal gradients known in greater detail, *World Oil, 152*, 79–82, 1961.

Muehlberger, W. R., Conjugate joint sets of small dihedral angle, *J. Geol., 69*, 211–219, 1961.

Muehlberger, W. R., and P. S. Clagaugh, Internal structure and petrofabrics of Gulf Coast salt domes, in *Diapirism and Diapirs, Mem. 8*, edited by J. D. Braunstein, and G. D. O'Brien, pp. 90–98, American Association of Petroleum Geologists, Tulsa, Okla., 1968.

Mügge, O., Ueber Translation und verwandte Erscheinungen in Kristallen, *Neues Jahrb. Mineral., Geol. Palaeontol. Monatsh., Abt. 1*, 71–158, 1898.

Müller W. F., and M. Défourneaux, Deformationsstrukturen in Quarz als Indikator für Stosswellen: Eine experimentelle Untersuchung an Quarz-Einkristallen, *Z. Geophys., 34*, 483–504, 1968.

Munson, R. C., Relationship of effect of waterflooding of the Rangely oil field on seismicity, in *Engineering Geology Case Histories*, no. 8, edited by Adams, pp. 39–49, Geological Society of America, Boulder, Colo., 1970.

Murray, G. E., Salt structures of Gulf of Mexico Basin—A review, in *Diapirism and Diapirs, Mem. 8,* edited by J. Braunstein and G. D. O'Brien, pp. 99–121, American Association of Petroleum Geologists, Tulsa, Okla., 1968.

Musgrave, A. W., and W. G. Hicks, Outlining of shale masses by geophysical methods, *Geophysics, 31,* 711–725, 1966.

Musgrave, M. J. P., *Crystal Acoustics,* p. 288, Holden-Day, San Francisco, Calif., 1970.

Nabarro, F. R. N., Deformation of crystals by the motion of single ions, in *Strength of Solids,* pp. 175, The Physical Society, London, 1948.

Nabarro, F. R. N., Steady state diffusional creep, *Phil. Mag., 16,* 231–237, 1967.

Nabarro, F. R. N., Z. S. Basinski, and D. B. Holt, The plasticity of pure single crystals, *Advan. Phys., 13,* 193–323, 1964.

Nakaya, U., Mechanical properties of single crystals of ice, *Res. Rep. 28,* 46 pp., U.S.A. Snow, Ice, and Permafrost Res. Estab., 1958.

Nettleton, L. L., Fluid mechanics of salt domes, *Bull. Amer. Ass. Petrol. Geol., 18,* 1175–1204, 1934.

Nichols, E. A., Geothermal gradients in mid-continent and Gulf Coast oil fields, *Trans. AIME, Petrol. Dev. Tech. Trans., 170,* 44–50, 1947.

Nicolas, A., J. L. Bouchez, F. Boudier, and J. C. Mercier, Textures, structures and fabrics due to solid state flow in some European lherzolites, *Tectonophysics,* in press, 1972.

Noll, W., On the continuity of the solid and fluid states, *J. Ration. Mech. Anal., 4,* 3–81, 1955.

Nur, A., and J. D. Byerlee, An exact effective pressure law for compression of rock with fluids (abstract), *Eos Trans. AGU, 52,* 342, 1971.

Nur, A., and G. Simmons, The effect of viscosity of a fluid phase on velocity in low porosity rocks, *Earth Planet. Sci. Lett., 7,* 99–108, 1969a.

Nur, A., and G. Simmons, The effect of saturation on velocity in low porosity rocks, *Earth Planet. Sci. Lett., 7,* 183–193, 1969b.

Nur, A., and G. Simmons, The origin of small cracks in igneous rocks, *Int. J. Rock Mech. Min. Sci., 7,* 307, 1970.

Nye, J. F., The flow law of ice from measurements in glacier tunnels, laboratory experiments and the Jungfraufirn borehole experiment, *Proc. Roy. Soc. London, Ser. A, 219,* 477–489, 1953.

Nye, J. F., *Physical Properties of Crystals,* 322 pp., Oxford University Press, Oxford, 1957.

Obert, L., Effects of stress relief and other changes in stress on the physical properties of rock, *Rep. Invest. 6053,* 8 pp., Bur. of Mines, U.S. Dep. of Interior, Washington, D.C., 1962.

O'Brien, G. D., Survey of diapirs and diapirism, in *Diapirism and Diapirs, Mem. 8,* edited by J. Braunstein, and G. D. O'Brien, pp. 1–9, Ameri-can Association of Petroleum Geologists, Tulsa, Okla., 1968.

Odé, H., Review of mechanical properties of salt relating to salt dome genesis, in *Saline Deposits, Spec. Pap. 88,* edited by R. B. Mattox, pp. 543–595, Geological Society of America, Boulder, Colo., 1968.

O'Hara, M. J., Mineral parageneses in ultrabasic rocks, in *Ultramafic and Related Rocks,* edited by P. J. Wyllie, p. 393, John Wiley, New York, 1967.

Omori, F., On the aftershocks of earthquakes, *J. Coll. Sci., Imp. Univ. Tokyo, 7,* 111–200, 1894.

Orange, A. S., Granitic rock: Properties in situ, *Science, 165,* 202–203, 1969.

Orowan, E., Mechanism of seismic faulting, in *Rock Deformation, Mem. 79,* edited by D. T. Griggs and J. Handin, pp. 323–345, Geological Society of America, Boulder, Colo., 1960.

Oxburgh, E. R., and D. L. Turcotte, Mid-ocean ridges and geotherm distribution during mantle convection, *J. Geophys. Res., 73,* 2643, 1968.

Paradis, A. R., and D. F. Hussey, *Graphical Display System,* Computer Center, Univ. of Calif., Berkeley, Calif., 1969.

Patel, J. R., and A. R. Chaudhuri, Macroscopic plastic properties of dislocation-free germanium and other semiconductor crystals, 1, Yield behaviour, *J. Appl. Phys., 34,* 2788–2799, 1963.

Paterson, M. S., X-ray line broadening in plastically deformed calcite, *Phil. Mag., 40,* 451, 1959.

Paterson, M. S., A high pressure, high temperature apparatus for rock deformation, *Int. J. Rock. Mech. Min. Sci., 7,* 517–526, 1970.

Paterson, M. S., and F. J. Turner, Experimental deformation of unstrained crystals of calcite in extension, in *Experimental and Natural Rock Deformation,* edited by P. Paulitsch, pp. 109–141, Springer-Verlag, Berlin, 1970.

Paterson, M. S., and L. E. Weiss, Experimental deformation and folding in phyllite, *Geol. Soc. Amer. Bull., 77,* 343–373, 1966.

Paterson, M. S., and L. E. Weiss, Folding and boudinage of quartz-rich layers in experimentally deformed phyllite, *Geol. Soc. Amer. Bull., 79,* 795–812, 1968.

Paulus, M., and F. Reverchon, Dispositif de bombardement ionique pour preparations micrographiques, *J. Phys. Radium, 22,* 103–107, 1961.

Perrett, W., Free field ground motion in granite, Operation Flintlock, Shot Piledriver, *Rep. POR-4001 (WT-4001),* Def. At. Support Agency, Washington, D.C., 1968.

Phakey, P., G. Dollinger, and J. Christie, Transmission electron microscopy of experimentally deformed olivine crystals, this volume.

Phillips, W. L., Jr., Effect of strain rate and temperature on the stress-strain characteristics of NaCl, LiF and MgO single crystals, *Trans. AIME, 224,* 434–436, 1962.

Pierce, C. B., Effect of hydrostatic pressure on ionic conductivity in doped single crystals of sodium chloride, potassium chloride, and rubidium chloride, *Phys. Rev., 123,* 744–754, 1961.

Pollack, H. N., Gravitational mechanism for seafloor spreading, *Science, 163,* 176–177, 1969.

Post, R. L., Jr., The flow laws of Mt. Burnette dunite at 750°–1150°C (abstract), *Eos Trans. AGU, 51,* 424, 1970.

Press, F., Chairman, Ad hoc panel on earthquake prediction, A proposal for a 10-year program of research, Office of Science and Technology, Washington, D.C., 1965.

Price, N. J., Mechanics of jointing in rocks, *Geol. Mag., 96,* 149–167, 1959.

Price, N. J., The initiation and development of asymmetrical buckle folds in non-metamorphosed competent sediments, *Tectonophysics, 4,* 173–201, 1967.

Rabb, D., Size distribution study of Piledriver particles, *Rep. UCRL-50765,* Lawrence Livermore Lab., Livermore, Calif., 1968.

Radcliffe, S. V., A. H. Heuer, R. M. Fisher, J. M. Christie, and D. T. Griggs, High voltage transmission electron microscopy study of rocks from Apollo 11, in *Proceedings of the Apollo 11 Lunar Science Conference,* vol. 1, edited by A. A. Levinson, pp. 721–746, Pergamon, New York, 1970.

Rädler, K.-H., Zur Elektrodynamik turbulent bewegter leitender Medien, *Z. Naturforsch. A, 23,* 1841–1851, 1968.

Raitt, R. W., Seismic refraction studies of the Mendicino fracture zone, *Rep. MPL-U-23/63,* Mar. Phys. Lab., Scripps Inst. Oceanogr., Univ. of Calif., San Diego, Calif., 1963.

Raitt, R. W., G. G. Shor, T. J. G. Francis, and G. B. Morris, Anisotropy of the Pacific upper mantle, *J. Geophys. Res., 74,* 3095–3109, 1969.

Raleigh, C. B., Fabrics of naturally and experimentally deformed olivine, Ph.D. thesis, Univ. of Calif. at Los Angeles, 1963.

Raleigh, C. B., Glide mechanisms in experimentally deformed minerals, *Science, 150,* 739–741, 1965a.

Raleigh, C. B., Structure and petrology of an Alpine peridotite on Cypress Island, Washington, U.S.A., *Contrib. Mineral. Petrol., 2,* 719, 1965b.

Raleigh, C. B., Plastic deformation of upper mantle silicate minerals, *Geophys. J. Roy. Astron. Soc., 14,* 45–56, 1967.

Raleigh, C. B., Mechanisms of plastic deformation in olivine, *J. Geophys. Res., 73,* 5391–5406, 1968.

Raleigh, C. B., Earthquake control at Rangely, Colorado (abstract), *Eos Trans. AGU, 52,* 344, 1971.

Raleigh, C. B., and S. H. Kirby, Creep in the upper mantle, *Mineral. Soc. Amer. Spec. Pap. 3,* 113–121, 1970.

Raleigh, C. B., and M. S. Paterson, Experimental deformation of serpentinite and its tectonic implications, *J. Geophys. Res., 70,* 3965–3985, 1965.

Raleigh, C. B., and J. L. Talbot, Mechanical twinning in naturally and experimentally deformed diopside, *Amer. J. Sci., 265,* 151–165, 1967.

Raleigh, C. B., S. H. Kirby, N. L. Carter, and H. G. Avé Lallemant, Slip and the clinoenstatite transformation as competing rate processes in enstatite, *J. Geophys. Res., 76,* 4011–4022, 1971.

Raleigh, C. B., J. H. Healy, and J. D. Bredehoeft, Faulting and crustal stress at Rangely, Colorado, this volume.

Ramberg, H., Contact strain and folding instability of a multi-layered body under compression, *Geol. Rundsch., 51,* 405–439, 1961a.

Ramberg, H., Relationship between concentric longitudinal strain and concentric shearing strain during folding of homogeneous sheets of rocks, *Amer. J. Sci., 259,* 382–390, 1961b.

Ramberg, H., Fluid dynamics of viscous buckling applicable to folding of layered rocks, *Amer. Ass. Petrol. Geol. Bull., 47,* 484–505, 1963a.

Ramberg, H., Evolution of drag folds, *Geol. Mag., 100,* 97–106, 1963b.

Ramberg, H., Selective buckling of composite layers with contrasted rheological properties, A theory for simultaneous formation of several orders of folds, *Tectonophysics, 1,* 307–341, 1964.

Ramberg, H., *Gravity Deformation and the Earth's Crust,* 214 pp., Academic, London, 1967.

Ramsay, J. G., *Folding and Fracturing of Rock,* 568 pp., McGraw-Hill, New York, 1967.

Ramsay, J. G., and R. H. Graham, Strain variation in shear belts, *Can. J. Earth Sci., 7,* 786–813, 1970.

Ree, F. H., T. Ree, and H. Eyring, Relaxation theory of creep of metal, *Proc. Amer. Soc. Civil Eng., J. Eng. Mech. Div., 86,* 41–59, 1960.

Reusch, E., Ueber eine besondere Gattung von Durchgängen im Steinsalz und Kalkspath, *Ann. Phys. Chemie Poggendorffs, 132,* 441–451, 1867.

Rigsby, G. P., Crystal orientation in glacier and in experimentally deformed ice, *J. Glaciol., 3,* 589–606, 1960.

Rikitaki, T., Oscillations of a system of disk dynamos, *Proc. Cambridge Phil. Soc., 54,* 89–105, 1958.

Rinehart, J. S., Geyser activity near Beowawe, Eureka County, Nevada, *J. Geophys. Res., 73,* 7703–7706, 1968.

Rinehart, J. S., Thermal and seismic indications of Old Faithful geyser's inner workings, *J. Geophys. Res., 74,* 566–573, 1969.

Rinehart, J. S., and A. Murphy, Observations on pre- and post-earthquake performance of Old Faithful geyser, *J. Geophys. Res., 74,* 574–575, 1969.

Ringwood, A. E., Composition and evolution of the upper mantle, in *The Earth's Crust and Upper Mantle, Geophys. Monogr. Ser.,* vol. 13, edited by P. J. Hart, pp. 1–17, AGU, Washington, D.C., 1969.

Rivlin, R. S., The hydrodynamics of non-Newtonian fluids, 1, *Proc. Roy. Soc. London, Ser. A, 193,* 260–281, 1948.

Roberts, G. O., Spatially periodic dynamos, *Phil. Trans. Roy. Soc., Ser. A, 266,* 535–558, 1970.

Robertson, P. B., M. R. Dence, and M. A. Vos, Deformation in rock-forming minerals from Canadian craters, in *Shock Metamorphism of Natural Materials,* edited by B. M. French and N. M. Short, pp. 433–452, Mono, Baltimore, 1968.

Robin, P.-Y., A note on effective pressure, submitted to *J. Geophys. Res.,* 1972.

Robinson, L. H., Jr., Effects of pore and confining pressures on the failure characteristics in sedimentary rock, Third Symposium on Rock Mechanics, *Colo. Sch. Mines Quart., 54,* pp. 177–199, 1959.

Roe, R. J., Description of crystallite orientation in polycrystalline materials, 3, General solution to pole-figure inversion, *J. Appl. Phys., 36,* 2024–2031, 1965.

Roedder, E., Liquid CO_2 inclusions in olivine-bearing nodules and phenocrysts from basalts, *Amer. Mineral. 50,* 1746–1782, 1965.

Romanes, J., Salt domes of North Germany, Symposium on salt domes, *J. Inst. Petrol. Technol., 17,* 242–258, 1931.

Rooney, R. P., R. Riecker, and M. Ross, Deformation twins in hornblende, *Science, 169,* 173–175, 1970.

Ross, M., J. J. Papike, and P. W. Weiblen, Exsolution in clinoamphiboles, *Science, 159,* 1099–1102, 1968.

Sander, B., Über Zusammenhänge zwischen Teilbewegung und Gefüge in Gesteinen, *Tschermaks Mineral. Petrogr. Mitt., 30,* 281–314, 1911.

Sander, B., *Gefügekunde der Gesteine,* 352 pp., Springer-Verlag, Berlin, 1930.

Sander, B., *Einführung in die Gefügekunde der geologischen Körper,* Teil 1, Springer-Verlag, Berlin, Vienna, 1948.

Scheidegger, A. E., The tectonic stress and tectonic motion direction in Europe and western Asia as calculated from earthquake fault-plane solutions, *Bull. Seismol. Soc. Amer., 54,* 1519–1528, 1964.

Schlocker, J., Petrology and mineralogy of Tatum Dome, Lamar County, Mississippi, *Tech. Lett., 28,* 120 pp., U.S. Geol. Surv., Washington, D.C., 1963.

Schmidt, W., Gefügestatistik, *Tschermaks Mineral. Petrogr. Mitt., 38,* 392–423, 1925.

Scholz, C. H., Static fatigue of quartz, *J. Geophys. Res., 77,* 2104–2114, 1972.

Schröter, W., H. Alexander, and P. Haasen, Die Inhomogeneität der Verformung von Germanium im Streckgrenzenbereich, *Phys. Status Solidi, 7,* 983–998, 1964.

Schwerdtner, W. M., Intragranular gliding in domal salts, *Tectonophysics, 5,* 353–381, 1968.

Scott, R. F., *Principles of Soil Mechanics,* 550 pp., Addison-Wesley, Reading, Mass., 1963.

Serata, S., and E. F. Gloyna, Development of design principle for disposal of reactor fuel waste into underground salt cavities, Reactor fuel waste disposal project, 173 pp., Univ. of Tex., Austin, 1959.

Shames, I. H., *Mechanics of Deformable Solids,* 532 pp., Prentice-Hall, Englewood Cliffs, N.J., 1964.

Sherby, O. D., and P. M. Burke, Mechanical behavior of crystalline solids at elevated temperature, *Progr. Mater. Sci., 13,* 324–390, 1967.

Sherby, O. D., and M. T. Simnad, Prediction of atomic mobility in metallic systems, *Trans. Amer. Soc. Met., 54,* 227–240, 1961.

Sherwin, J. A., and W. M. Chapple, Wavelengths of single layer folds: A comparison between theory and observation, *Amer. J. Sci., 266,* 167–179, 1968.

Shewmon, P. G., *Diffusion in Solids,* 272 pp., McGraw-Hill, New York, 1963.

Shor, G. G., Jr., and D. D. Pollard, Mohole site selection studies north of Maui, *J. Geophys. Res., 69,* 1627–1637, 1964.

Short, N. M., Effects of shock pressures from a nuclear explosion on mechanical and optical properties of granodiorite, *J. Geophys. Res., 71,* 1195–1215, 1966.

Short, N. M., Progressive shock metamorphism of quartzite ejecta from the Sedan nuclear explosion crater, *J. Geol., 78,* 705–732, 1970.

Short, N. M., and T. E. Bunch, A worldwide inventory of features characteristic of rocks associated with presumed meteorite impact craters, in *Shock Metamorphism of Natural Materials,* edited by B. M. French and N. M. Short, pp. 255–266, Mono, Baltimore, 1968.

Shumskii, P. A., The mechanism of ice straining and its recrystallization, *Assoc. Int. Hydrol. Sci. Publ. 47,* 244–248, 1958.

Siethoff, H., Der Beginn der plastischen Verformung von hoch phosphordotiertem Silizium, *Acta Met., 17,* 793–801, 1969.

Sigurgeirsson, T., Hitamaelingar i geysi, *Náttúrufraedingurinn, 19,* 27–33, 1949.

Simmons, G., and A. Nur, Granites: Relation of properties in situ to laboratory measurements, *Science, 162,* 789–791, 1968.

Skempton, A. W., Effective stress in soils, concrete, and rocks, in *Conference on Pore Pressure and Suction in Soils,* pp. 4–16, Butterworth, London, 1961.

Smith, H. L., Tunnel photography, Operation Flintlock, Shot Piledriver, *Rep. POR-4012 (WT-4012),* Def. At. Support Agency, Washington, D.C., 1969.

Snow, D. T., Rock fracture spacings, openings, and porosities, *J. Soil Mech. Found. Div., Proc. Amer. Soc. of Civil Eng., 94*(SM1), 73–91, 1968.

Sokolnikoff, I. S., *Mathematical Theory of Elasticity,* 476 pp., McGraw-Hill, New York, 1956.

Spang, J. H., and W. M. Chapple, Mechanical origin of folds in sedimentary rocks: A comparison between theory and observation (abstract), *Geol. Soc. Amer. Abstr. Programs, 2,* 754–755, 1970.

Stacey, F. D., *Physics of the Earth,* 324 pp., John Wiley, New York, 1969.

Starkey, J., The geometry of kink bands in crystals, *Contrib. Mineral. Petrol., 19,* 133–141, 1968.

Starr, A. T., Slip in a crystal and rupture in a solid due to shear, *Proc. Cambridge Phil. Soc., 24,* 489–500, 1928.

Stearns, D. W., Macrofracture patterns on Teton Anticline, northwestern Montana (abstract), *Eos Trans. AGU, 45,* 107, 1964.

Stearns, D. W., Certain aspects of fractures in naturally deformed rocks, in *Advanced Science Seminar in Rock Mechanics,* vol. 1, edited by R. E. Riecker, pp. 97–118, Air Force Cambridge Research Laboratory, Bedford, Mass., 1968.

Steenbeck, M., F. Krause, and K.-H. Rädler, Berechnung der mittleren Lorenz-Feldstärke $\overline{\nabla \wedge B}$ für ein elektrisch leitendes Medium in turbulenter, durch Coriolis Kräfte beeinflusster Bewegung, *Z. Naturforsch. A, 21,* 369–376, 1966.

Stein, D. F., and J. R. Low, Mobility of edge dislocations in silicon-iron crystals, *J. Appl. Phys., 31,* 362–369, 1960.

Steinemann, S., Experimentelle Untersuchungen zur Plastizität von Eis, *Beitr. Geol. Schweiz, Hydrol. Ser., 10,* 1–72, 1958.

Stepanov, A. V., and V. P. Bobrikov, Dependence of the optical elasticity limit determined on the basis of system (111): [011] upon temperature for rock salt crystals, *Sov. Phys. JETP, Engl. Transl., 1,* 177–181, 1955.

Stephens, D. R., E. M. Lilley, and H. Louis, Pressure-volume equation of state of consolidated and fractured rocks to 40 kb, *Int. J. Rock. Mech. Min. Sci., 7,* 257–269, 1970.

Sterrett, T. S., Drilling investigation of the lower part of the Piledriver cavity, *Rep. UCRL-50765,* Lawrence Livermore Lab., Livermore, Calif., 1969.

Stewart, A. J., Potassium-argon dates from the Arltunga Nappe Complex, Northern Territory, *J. Geol. Soc. Aust., 17,* 205–211, 1971.

Stöffler, D., Deformation und Umwandlung von Plagioklas durch Stosswellen in den Gesteinen des Nördlinger Ries, *Contrib. Mineral. Petrol., 16,* 51–83, 1967.

Stöffler, D., Progressive metamorphism and classification of shocked and brecciated crystalline rocks at impact craters, *J. Geophys. Res., 76,* 5541–5551, 1971.

Stokes, R. J., Mechanical properties of polycrystalline sodium chloride, *Proc. Brit. Ceram. Soc., 6,* 189–207, 1966.

Tammann, G., and W. Salge, Octahedral glide of NaCl at high temperatures: Decrease in the yield point with rising temperature, *Neues Jahrb. Mineral. Geol. Palaeontol., Abt. A, 57,* 117–130, 1927.

Taub, A. H., Relativistic Rankine-Hugoniot equations, *Phys. Rev., 74,* 328–334, 1948.

Teller, E., *Physics of High Energy Density: Relativistic Hydrodynamics in Supernovae,* Academic, New York, 1971.

Temple, P. G., Mechanics of large-scale gravity sliding in the Greek Peloponnesos, *Geol. Soc. Amer. Bull., 79,* 687–700, 1968.

Terzaghi, K., Die Berechnung der Durchlässigkeitsziffer des Tones aus dem Verlauf der hydrodynamischen Spannungserscheinungen, *Sitzungsber. Akad. Wiss. Wien, Math. Naturwiss. Kl., Abt. 2A, 132,* 105–125, 1923.

Thomas, C. R., Map of the Rangely oil and gas field, Rio Blanco and Moffat Counties, Colorado, no. 41, Oil and Gas Investigation, U.S. Geological Survey, Washington, D.C., 1945.

Thomas, L. A., and W. A. Wooster, Piezocrescence —The growth of Dauphiné twinning in quartz under stress, *Proc. Roy. Soc. London, Ser. A., 208,* 43–62, 1951.

Thompson, E. G., An experimental technique for the investigation of the flow of halite and sylvinite, Ph.D. dissertation, 101 pp., Univ. of Tex., Austin, 1965.

Tighe, N. J., Microstructure of fine-grain ceramics, in *Ultrafine-Grain Ceramics,* edited by J. J. Burke, N. L. Reed, and V. Weiss, pp. 109–121, Syracuse University Press, New York, 1970.

Toksöz, M. N., J. W. Minear, and B. R. Julian, Temperature field and geophysical effects of a downgoing slab, *J. Geophys. Res., 76,* 1113–1138, 1971.

Torrance, K. E., and D. L. Turcotte, Structure of convection cells in the mantle, *J. Geophys. Res., 76,* 1154–1161, 1971.

Trommsdorf, V., and H. R. Wenk, Terrestrial metamorphic clinoenstatite in kinks of bronzite crystals, *Contrib. Mineral. Petrol., 19,* 158–169, 1968.

Truesdell, C., The mechanical foundations of elasticity and fluid dynamics, *J. Ration. Mech. Anal., 1,* 125–291, 1952.

Truesdell, C., The natural time of a viscoelastic fluid: Its significance and measurement, *Phys. Fluids, 7,* 1134–1142, 1964.

Truesdell, C., Fluids of the second grade regarded as fluids of convected elasticity, *Phys. Fluids, 8,* 1936–1938, 1965.

Truesdell, C., *The Elements of Continuum Mechanics,* 279 pp., Springer-Verlag, Berlin, 1966.

Truesdell, C., *Rational Thermodynamics,* 208 pp., McGraw-Hill, New York, 1969.

Truesdell, C., and W. Noll, The non-linear field theories of mechanics, *Encyclopedia of Physics,* vol. 3, sec. 3, edited by S. Flügge, Springer-Verlag, Berlin, 602 pp., 1965.

Truesdell, C., and R. Toupin, The classical field theories, *Encyclopedia of Physics,* vol. 3, sect. 3, edited by S. Flügge, pp. 226–793, Springer-Verlag, Berlin, 1960.

Tullis, J., Quartz: Preferred orientation in rocks produced by Dauphiné twinning, *Science, 168,* 1342–1344, 1970.

Tullis, J., Preferred orientations of experimentally deformed quartzites, Ph.D. thesis, Univ. of Calif. at Los Angeles, 344 pp., 1971.

Tullis, J., J. M. Christie, and D. T. Griggs, Microstructures and preferred orientations of experimentally deformed quartzite, *Geol. Soc. Amer. Bull.*, *84*, 1973.

Tullis, J., and T. Tullis, Preferred orientation of quartz produced by mechanical Dauphiné twinning: Thermodynamics and axial experiments, this volume.

Tullis, T. E., Experimental development of preferred orientation of mica during recrystallization, Ph.D. thesis, Univ. of Calif. at Los Angeles, 262 pp., 1971.

Turner, F. J., Nature and dynamic interpretation of deformation lamellae in calcite of three marbles, *Amer. J. Sci.*, *251*, 276–298, 1953.

Turner, F. J., and C. S. Ch'ih, Deformation of Yule marble, 3, Observed fabric changes, *Geol. Soc. Amer. Bull.*, *62*, 887–905, 1951.

Turner, F. J., and L. E. Weiss, *Structural Analysis of Metamorphic Tectonites*, McGraw-Hill, New York, 545 pp., 1963.

Turner, F. J., and L. E. Weiss, Deformational kinks in brucite and gypsum, *Proc. Nat. Acad. Sci.*, *54*, 359–364, 1965.

Turner, F. J., D. T. Griggs, and H. C. Heard, Experimental deformation of calcite crystals, *Geol. Soc. Amer. Bull.*, *65*, 883–934, 1954.

Turner, F. J., D. T. Griggs, R. H. Clark, and R. Dixon, Deformation of Yule marble, 7, Development of oriented fabrics at 300°C–500°C, *Geol. Soc. Amer. Bull.*, *67*, 1259–1294, 1956.

Turner, F. J., H. C. Heard, and D. T. Griggs, Experimental deformation of enstatite and accompanying inversion to clinoenstatite, *Int. Geol. Congr. Rep. Sess. Norden 21st*, 399–408, 1960.

U.S. Naval Observatory, *The American Ephemeris and Nautical Almanac for the Year 1970*, pp. 31, 65, 145, 146, 466, U.S. Government Printing Office, Washington, D.C., 1968.

Vallon, M., Contribution à l'étude structurographique de la glace froide de haute latitude, *C. R. Acad. Sci. Paris*, *257*, 3988–3991, 1963.

Van Thiel, M., Shock wave data, *Rep. UCRL-50108*, vols. 1, 2, and 3, Lawrence Livermore Lab., Livermore, Calif., 1966.

Verma, R. K., Elasticity of some high-density crystals, *J. Geophys. Res.*, *65*, 757–766, 1960.

Vine, F. J., and D. H. Matthews, Magnetic anomalies over oceanic ridges, *Nature*, *199*, 194, 1963.

Vogel, D. E., Petrology of an eclogite- and pyrigarnite-bearing polymetamorphic rock complex at Cabo Ortegal, NW Spain, *Leidse Geol. Meded.*, *40*, 121–213, 1967.

Wackerle, J., Shock compression of quartz, *J. Appl. Phys.*, *33*, 922–937, 1962.

Walsh, J. B., The effect of cracks on the compressibility of rock, *J. Geophys. Res.*, *70*, 381–389, 1965*a*.

Walsh, J. B., The effect of cracks on the uniaxial elastic compression of rocks, *J. Geophys. Res.*, *70*, 399–411, 1965*b*.

Walsh, J. B., The effect of cracks in rocks in Poisson's ratio, *J. Geophys. Res.*, *70*, 5249–5257, 1965*c*.

Walsh, J. B., Seismic wave attenuation in rock due to friction, *J. Geophys. Res.*, *71*, 2591–2599, 1966.

Walsh, J. B., Attenuation in partially melted material, *J. Geophys. Res.*, *73*, 2209–2216, 1968.

Walsh, J. B., New analysis of attenuation in partially melted rock, *J. Geophys. Res.*, *74*, 4333–4337, 1969.

Walsh, J. B., and W. F. Brace, Elasticity of rock: A review of some recent theoretical studies, *Rock. Mech. Eng. Geol.*, *4*, 283–297, 1966.

Walsh, J. B., and E. R. Decker, Effect of pressure and saturating fluids on the thermal conductivity of compact rock, *J. Geophys. Res.*, *71*, 3053–3061, 1966.

Walsh, J. M., and R. H. Christian, Equation of state of metals from shock wave measurements, *Phys. Rev.*, *97*, 1544–1556, 1955.

Wang, C., Density and constitution of the mantle, *J. Geophys. Res.*, *75*, 3264–3284, 1970.

Wasiutyński, J., Studies in hydrodynamics and structure of stars and planets, *Astrophys. Norv.*, *4*, 57, 1946.

Watson, K. D., and D. M. Morton, Eclogite inclusions in kimberlite pipes at Garnet Ridge, northeastern Arizona, *Amer. Mineral.*, *54*, 267–285, 1969.

Wayman, C. M., *Introduction to Crystallography of Martensitic Transformations*, 193 pp., Macmillan, New York, 1964.

Weertman, J., Theory of steady-state creep based on dislocation climb, *J. Appl. Phys.*, *26*, 1213–1217, 1955.

Weertman, J., Steady-state creep through dislocation climb, *J. Appl. Phys.*, *28*, 362–364, 1957.

Weertman, J., Mechanism for continental drift, *J. Geophys. Res.*, *67*, 1133–1139, 1962.

Weertman, J., The thickness of continents, *J. Geophys. Res.*, *68*, 929–932, 1963.

Weertman, J., Sliding of nontemperate glaciers, *J. Geophys. Res.*, *72*, 521–523, 1967.

Weertman, J., Dislocation climb theory of steady-state creep, *Trans. ASME*, *61*, 681–694, 1968.

Weertman, J., The creep strength of the earth's mantle, *Rev. Geophys. Space Phys.*, *8*, 145–168, 1970.

Weiss, L. E., Flexural-slip folding of foliated model materials, *Pap. 68-52*, pp. 294–357, Geol. Surv. of Can., 1969.

Weiss, L. E., and F. J. Turner, Some observations of translation gliding and kinking in experimentally deformed calcite and dolomite, this volume.

Wenk, H.-R., and V. Trommsdorff, Koordinatentransformation, Mittelbare Orientierung, Nachbarwinkelstatistik. Gefügekundliche Rechenprogramme mit Beispielen, *Contrib. Mineral. Petrol.*, *11*, 559–585, 1965.

Wenk, H.-R., and W. R. Wilde, Orientation distribution diagrams for three Yule marble fabrics, this volume.

Wenk, H.-R., C. S. Venkitasubramanyan, D. W. Baker, and F. J. Turner, Preferred orientation in experimentally deformed limestone, *Contrib. Mineral. Petrol.*, in press, 1972.

White, D. E., Some principles of geyser activity, mainly from Steamboat Springs, Nevada, *Amer. J. Sci., 265,* 641–684, 1967.

White, D. E., Hydrology, activity, and heat flow of the Steamboat Springs thermal system, Washoe County, Nevada, *U.S. Geol. Surv. Prof. Pap. 458-C,* 1–109, 1968.

White, D. E., L. J. P. Muffler, and A. H. Truesdell, Vapor-dominated hydrothermal systems compared with hot-water systems, *Econ. Geol., 66,* 75–97, 1971.

Whittaker, E. T., and G. N. Watson, *A Course of Modern Analysis,* 4th ed., p. 36, MacMillan, New York, 1927.

Whitten, E. H. T., *Structural Geology of Folded Rocks,* Rand McNally, New York, 663 pp., 1966.

Williams, P. F., and W. D. Means, Experimental folding of foliated materials (abstract), *Eos Trans. AGU, 52,* 345, 1971.

Williams, W. S., Influence of temperature, strain rate, surface condition and composition on the plasticity of transition metal carbide crystals, *J. Appl. Phys., 35,* 1329–1338, 1964.

Wilson, C. J. L., The microfabric of a deformed quartzite sequence, Mount Isa, Queensland, Ph.D. thesis, 268 pp., Aust. Nat. Univ., Canberra, 1970.

Wolff, H., Richtungsabhängigkeit des Translationmechanismus von Steinsalzkristallen in höheren Temperaturen, *Z. Phys., 93,* 147–165, 1935.

Wones, D. R., and H. P. Eugster, Stability of biotite: Experiment, theory, and application, *Amer. Mineral., 50,* 1228–1272, 1965.

Wooster, W. A., N. Wooster, J. L. Rycroft, and L. A. Thomas, The control and elimination of electrical (Dauphiné) twinning in quartz, *J. Inst. Elec. Eng., 94,* 927–938, 1947.

Wyllie, P. J., Role of water in magma generation and initiation of diapiric uprise in the mantle, *J. Geophys. Res., 76,* 1328–1338, 1971.

Yoder, H. S., Jr., Role of water in metamorphism, *Geol. Soc. Amer. Spec. Pap. 62,* 505–524, 1955.

Yoon, D. N., and D. Lazurus, Pressure dependence of ionic conductivity in KCl, NaCl, KBr, and NaBr, *Phys. Rev. B, 5,* 4935–4945, 1972.

Young, C., Dislocations in the deformation of olivine, *Amer. J. Sci., 267,* 841–852, 1969.

Zinserling, K., and A. Shubnikov, Über die Plastizität des Quarzes, *Z. Krist., 85,* 454–461, 1933.

Zisman, W. A., Compressibility and anisotropy of rocks at and near the earth's surface, *Proc. Nat. Acad. Sci., 19,* 666–679, 1933a.

Zisman, W. A., Comparison of the statically and seismologically determined elastic constants of rocks, *Proc. Nat. Acad. Sci., 19,* 680–686, 1933b.